Daniel Ammann

Ion-Selective Microelectrodes

Principles, Design and Application

With 153 Figures and 42 Tables

Springer-Verlag
Berlin Heidelberg New York Tokyo

PD Dr. Daniel Ammann
Laboratorium für Organische Chemie
ETH-Zentrum, Universitätstraße 16
CH-8092 Zürich

QD
571
.A48
1986

ISBN 3-540-16222-4
Springer-Verlag Berlin Heidelberg New York Tokyo

ISBN 0-387-16222-4
Springer-Verlag New York Heidelberg Berlin Tokyo

Library of Congress Cataloging-in-Publication Data. Ammann, Daniel, 1947- Ion-selective
microelectrodes. Bibliography: p. Includes index. 1. Electrodes, Ion selective. I. Title.
QD571.A48 1986 541.3'724 85-30423
ISBN 0-387-16222-4 (U.S.)

This work is subject to copyright. All rights are reserved, whether the whole or part of the
material is concerned, specifically those of translation, reprinting, re-use of illustrations,
broadcasting, reproduction by photocopying machine or similar means, and storage in data
banks. Under § 54 of the German Copyright Law where copies are made for other than private use a fee is payable to "Verwertungsgesellschaft Wort", Munich.

© Springer-Verlag Berlin Heidelberg 1986
Printed in Germany

The use of registered names, trademarks, etc. in this publication does not imply, even in the
absence of a specific statement, that such names are exempt from the relevant protective
laws and regulations and therefore free for general use.

The publisher cannot assume any legal responsibility for given data, especially as far as directions for the use and the handling of chemicals are concerned. This information can be
obtained from the instructions on safe laboratory practice and from the manufacturers of
chemicals and laboratory equipment.

Typesetting, printing and bookbinding: Appl, Wemding
2152-3140-543210

For
Madeleine, Désirée, Fabienne, Noémie

Preface

The microelectrode technique is today the most widely used method in electrophysiology. Microelectrodes offer a unique approach to measurements of electrical parameters and ion activities of single cells. Several important breakthroughs in transport physiology have arisen from microelectrode studies. Undoubtedly, there is a progressively wide-spread use of conventional and ion-selective microelectrodes.

Due to their particular dimension and properties microelectrodes are exclusively applied to measurements on living matter. This must have many consequences to my thoughts on experiments with microelectrodes. In this book, my concern is focussing on the description of an intracellular method that should lead to reliable information on cellular parameters. The methodical basis for any meaningful application is treated extensively. However, technical perfection and accurate results are not the only concern when working on animals and human beings. Rather, my thoughts are governed by the intellectual and moral mastery of the experimental approach on living subjects.

A measurement with microelectrodes usually necessitates the sacrifice of an animal. This is an immense fact, and means that the knowledge gained by the experiment must justify the death of a living subject.

Through the availability of a technology as described here, ethics has to argue with scientific developments which have a causal reach into the future. It is not utopic to expect after the mastery of matter (nuclei of elements) and of heredity (recombinant DNA) a third scientific revolution: the intervention on human behaviour by the brain neurosciences. These extraordinary scientific outcomes include both the beautiful knowledge on life and the risk of going ecologically or anthropologically wrong.

I am aware to describe only one distinct approach to the reality of a living system out of many others. I agree with the striking example given by P. Feyerabend (Wissenschaft als Kunst, Suhrkamp, Frankfurt, 1984, p. 17 ff.) of the experiment of Brunelleschi, the narrow-minded interpretation of the experiment by Alberti and the subsequent beautiful criticism of Leonardo da Vinci. Brunelleschi's experiment was the following: he paints from the entrance of Santa Maria del Fiore at Firenze with great exactness and beauty the front of San Giovanni at the opposite side of the piazza. While painting with his back turned to the piazza he is looking into a polished mirror where the facade of San Giovanni and the sky is reflected. Afterwards, he makes a small hole in the middle of the realistic painting. The observer (the experimentator) standing in the door of Santa Maria del Fiore, exactly where the painting has been made, holds in one hand a mirror and is guiding with the other hand the painting to his eye. Now, he sees in the mirror the painted facade of

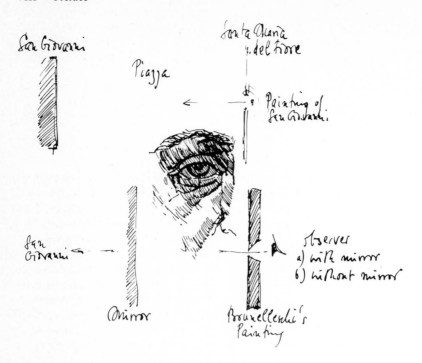

San Giovanni and the moving clouds at the sky. The mirror is removed, and, the view does not change although he views reality now. Undoubtedly, the painting-experiment of Brunelleschi has the character of a scientific approach since it tries to describe reality under very controlled experimental conditions. Based on this experiment Alberti erects a doctrine for painting. He reduces painting to a geometrical problem. Leonardo da Vinci, however, recognizes that such a theory is valid only under very restricted conditions. The perception of man is much more versatile than it can be described by one distinct approach.

Indeed, an intracellular experiment with microelectrodes is a similar fascinating event like Brunelleschi's painting. Its description of reality is however limited too. Thus, when I am performing microelectrode studies on animal cells I do not look at the living subject itself but at the electrical signals generated at the microelectrode which is connected to a complex electronic equipment. It is exclusively the resulting pen recording that allows me to interpret observations on the living matter. The real cell is hidden behind the impressive, however limited, information of the measured potentials. Accordingly, the result of the experiment is one specific scientific view of the cell and is only valuable for the conditions imposed. Although the results are describing life from the physiological point of view they overlook in many respects other aspects of the life of the cell, of the sacrificed animal, and of the environment it came from. Physiology and cell biology never offer a complete message of living structure. In other words, the cells and organisms have several aspects, the description given in this book is satisfying the current scientific approach only. Reality is a relative term, and a reduction to a single description appears to me a risk and a loss.

I am very grateful to Dr. Peter Anker, Dr. Daniel Erne, Dr. Urs Oesch, Dr. Ernö Pretsch and Prof. W. Simon for their comments and advices on the manuscript. Further I wish to thank Madeleine Ammann, Dr. Alan Berry, Catriona Tedford and Dr. Dorothée Wegmann for their most valuable help in correcting the manuscript. I also thank Brigitte Marti for the typing of the manuscript. Finally, it was important for me to visit the physiological laboratories of Dr. Armen Kurkdjian at Gif-sur-Yvette and of Prof. Florian Lang at Innsbruck. I am grateful for the discussions at these places.

Daniel Ammann
Zürich, November 1985

Table of Contents

1 Introduction

Severe demands and limitations are imposed on any intracellular analytical technique by the environment within the living cell. These difficulties arise both from the nature of the biological system and from the technical challenge of the intracellular approach. The cellular fluid consists of a highly complex sample of extremely small volume and is surrounded by a membrane which is crucially involved in the preservation of the state and function of the cell. In order to investigate this space in a single cell a highly selective and extremely miniaturized technique is required.

It is often rightly acknowledged that the introduction of conventional microelectrodes (reference microelectrodes) has opened new possibilities for experimental physiology, and fundamental parameters such as cell membrane potentials, membrane resistances, and net ion transports became accessible for the first time. Consequently, many basic properties of electrophysiological processes were elucidated. Nevertheless, all these investigations were seriously limited when attempts were made to attribute the observations to a specific type of ion. In this respect, the invention of ion-selective microelectrodes gave the opportunity to take a step forward in answering questions of cellular ion transport. In particular, the knowledge of intracellular ion activities is essential for studying transport mechanisms. These intracellular measurements, together with the simultaneous determinations of extracellular ion activities and of the resting potential can be used to identify, for each type of ion, electrochemical potential differences across the cell membrane. Hence, active and passive transport mechanisms can be distinguished. Indeed, potentiometry with ion-selective microelectrodes is the only method that gives simultaneous insight into both ion-activities and electrical parameters of the cell.

A historical perspective of the development of ion-selective microelectrodes clearly shows a trend towards liquid membrane microelectrodes. Their present priority is mainly due to the development of neutral carriers which serve as highly selective components in membrane solutions for microelectrodes. As a result, neutral carrier microelectrodes have almost completely replaced the solid state and glass membrane microelectrodes initially used.

Neutral carrier-based microelectrodes are now available for the direct potentiometric determination of the extracellular ion activities of H^+, Li^+ (therapeutic millimolar level), Na^+, K^+, and Ca^{2+} and the intracellular ion activities of H^+, Li^+ (therapeutic millimolar level), Na^+, K^+, Mg^{2+}, and Ca^{2+}. Classical ion-exchanger microelectrodes for K^+ and Cl^- are other, frequently used liquid membrane microelectrodes. The accessibility of all these ion activities at the cellular level is of outstanding value for the current research in electrophysiology. Adequately selective microelectrodes for extracellular Mg^{2+}, for various anions such as bicarbonate, phosphate, citrate, and malate as well as for larger organic cations and anions are yet to be realized. Promising developments will be presented in this book.

Various other techniques are available for the intracellular measurement of most of the ions that are accessible with microelectrodes. Of course, each method has its own advantages and limitations, but results obtained by different approaches usually complement each other. Some typical applications and features of ion-selective microelectrodes are the following:

- quantitative cytoplasmic measurements of ion activities or free ion concentrations (resting values)
- quantitative extracellular measurements of ion activities or free ion concentrations in the intercellular space
- detection of extremely local ion activities (detection volume of about 1 fl)
- simultaneous measurements of different types of ions in optimal proximity (multi-barrelled microelectrodes)
- measurements of ion activity transients (above the ms range)
- cell membrane surface recordings
- measurements in subcellular organelles (e. g. nucleus)
- depth-profile recording in tissues
- in vivo as well as in vitro measurements
- simultaneous measurements of electrical parameters of the cell (membrane potential and membrane conductivity).

The most striking and unique features of microelectrodes are the direct measurement of ion activities, the extremely local detection, and the simultaneous supply of electrical parameters. The extent of information and the confidence in the results can be considerably enhanced by the additional use of complementary methods such as optical techniques, nuclear magnetic resonance spectroscopy or electron probe X-ray microanalysis.

The microelectrode technique touches at least four widely discussed problems of today's science: the rapid growth of medical disciplines, the ethics of animal and human experimentation, the development of the neurosciences, and the uncertainty of intracellular in vivo measurements. Although the manuscript concentrates on analytical aspects, these subjects are included.

2 Classification of Ion-Selective Electrodes

Ion-selective electrodes can either be classified according to the membrane material used or the type of electrode body arrangement. Both classifications show that neutral carrier-based liquid membrane microelectrodes represent a very special type of electrode out of a vast variety of sensors. The unique features of carrier membranes as well as those of microelectrode bodies clearly illustrate the advantages of this type of electrode for intracellular studies in electrophysiology.

2.1 Types of Membranes

The following classification of membranes for ion-selective electrodes is commonly applied and satisfies the IUPAC recommendations [Guilbault et al. 76]:

Solid membranes
(fixed ion-exchange sites)
- homogeneous : glass membrane, crystal membrane
- heterogeneous: crystalline substance in inert matrix

Liquid membranes
(mobile ion-exchange sites)
- classical ion-exchanger
- charged carrier
- neutral carrier

Membranes in special electrodes
- gas-sensing electrode
- enzyme electrode
- microbial electrode
- tissue electrode

At present there are about 30 cations and anions which can be detected selectively by direct potentiometry with different types of ion-selective electrodes (Fig. 2.1 and [Cammann 79]). By means of special electrodes it is possible to directly measure gas molecules (e. g. NH_3 and CO_2) and organic species such as urea, penicillin and amino acids. In the last few years, the development of electrodes was mainly directed towards special electrodes (biosensors) as well as ion-selective neutral carrier-based liquid membrane electrodes. Although glass membrane electrodes are still widely

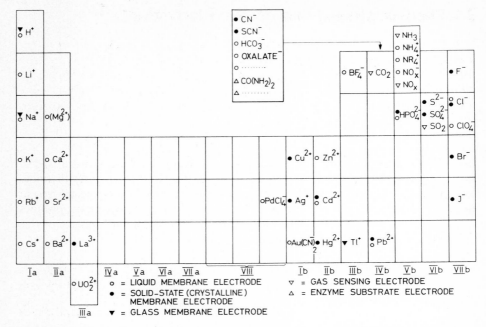

Fig. 2.1. Survey of ions and neutral species which can be measured with membrane electrodes (see also [Cammann 79]). For enzyme electrodes only urea is mentioned as a substrate. However, many other enzyme electrode systems are in development (for examples see [Schindler and Schindler 83])

used in biology and medicine for the determination of the activity of hydrogen and sodium ions [Koryta 80], neutral carrier-based liquid membrane electrodes are becoming increasingly important for the measurement of pH and of alkali and alkaline earth metal cations (see Fig. 2.1 and [Ammann et al. 83]). A trend towards neutral carrier membranes is especially pronounced in the case of microelectrodes. It has transpired that solid-state and glass membrane elctrodes can only be miniaturized with much technical skill.

Nonetheless, subtle but intricate methods have been developed for the construction of microelectrodes on the basis of the highly selective H^+ and Na^+ glass membranes [Thomas 78]. Many of the technical difficulties were overcome with the invention of liquid membrane microelectrodes [Orme 69; Walker 71]. The development of neutral carriers has opened up the possibility of preparing liquid membrane microelectrodes of different ion selectivities [Ammann et al. 81a]. The satisfactory selectivities of the carrier membranes, the availability of the membrane solutions, and the relatively easy construction of the microelectrodes have prompted the replacement of most of the other types of membranes, including the H^+ and Na^+ glasses.

2.2 Electrode Arrangements

The second classification of membrane electrodes is based on the size and shape of the electrode body (Fig. 2.2). Each electrode arrangement was developed for a specific application (e.g. flow-through electrodes for clinical analysis and on-line monitoring, catheter electrodes for continuous intravascular measurements and thin-film electrodes for intracortical studies). In most cases, attempts were made to miniaturize the electrode body as much as possible. Typical size-ranges for the various electrode arrangements are shown in Fig. 2.3. The lower limits correspond to the currently available smallest devices for each electrode type. The size is discussed in two respects. The active membrane area (upper frame of Fig. 2.3) is compared with the over-all dimensions of the sensor in the vicinity of the membrane (middle frame). For some electrodes the active area corresponds very closely to the size of

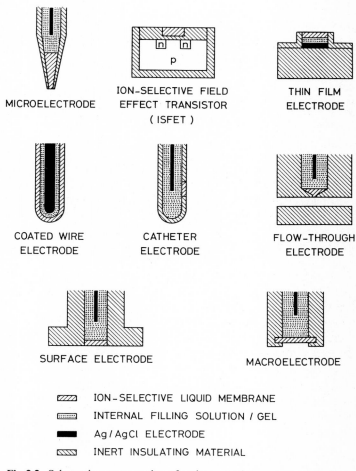

MICROELECTRODE

ION–SELECTIVE FIELD EFFECT TRANSISTOR (ISFET)

THIN FILM ELECTRODE

COATED WIRE ELECTRODE

CATHETER ELECTRODE

FLOW–THROUGH ELECTRODE

SURFACE ELECTRODE

MACROELECTRODE

ION–SELECTIVE LIQUID MEMBRANE
INTERNAL FILLING SOLUTION / GEL
Ag / AgCl ELECTRODE
INERT INSULATING MATERIAL

Fig. 2.2. Schematic representation of various membrane electrode bodies

SIZE OF ACTIVE MEMBRANE

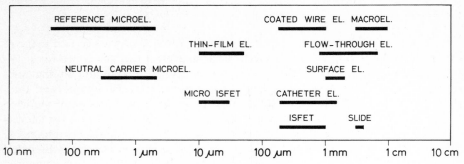

SIZE OF ELECTRODE BODY IN VICINITY OF THE MEMBRANE

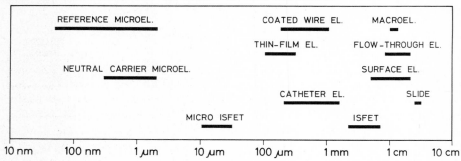

SIZE OF SAMPLE

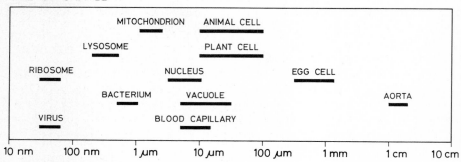

Fig. 2.3. Comparison of the size-ranges of the active membrane area and the electrode body about the membranes of various currently used electrodes with the size-ranges of some typical biological samples

Table 2.1. Smallest tip diameters of currently used microelectrodes

Type of microelectrode	Tip diameter	Reference
Reference microelectrodes	0.045 μm	[Brown and Flaming 77]
Ion-exchanger microelectrodes	0.1 μm	[Greger et al. 84a]
Neutral carrier microelectrodes	0.3 μm	[Tsien and Rink 80; Zeuthen 82]
Glass microelectrodes	0.5 μm	[Pucacco and Carter 78]
Solid-state microelectrodes	0.7 μm	[Armstrong et al. 77]
Gas-sensitive microelectrodes	1–2 μm	[Tsacopoulos and Lehmenkühler 77; Whalen 69]
Enzyme microelectrodes	2–10 μm	[Silver 75; Rehwald et al. 84]

the electrode body about the membrane (microelectrodes, micro-ISFETs, catheter electrodes, coated-wire electrodes), whereas in other types of electrodes the size is dictated by the electrode body itself and not by the active membrane surface (thin-film electrodes, ISFETs, surface electrodes, flow-through electrodes, slides, macro-electrodes). It is obvious that the size of the electrode body in the vicinity of the membrane is relevant to applications in a certain sample volume (e.g. biological samples, lower frame of Fig. 2.3). The comparison clearly shows that microelectrodes are by far the smallest electrodes available. None of the other electrode bodies can be used for measurements in the intracellular space. The smallest reference microelectrodes and, to a certain extent, the smallest neutral carrier microelectrodes seem to be useful for measurements in certain cell organelles. A further miniaturization of microelectrodes is possible. A reduction of the tip diameter of the smallest currently available reference microelectrodes (45 nm [Brown and Flaming 77]) is mainly a technical problem (pulling and filling of the micropipette). On the other hand, a further miniaturization of neutral carrier microelectrodes requires the development of membrane solutions of lower resistivity. At present, the limit of the tip size is about 0.3 μm (Table 2.1). As expected, liquid membranes based on classical charged ion-exchangers, which show a somewhat higher conductivity of the membrane phase, are useful in microelectrodes with tip diameters as small as 0.1 μm (Table 2.1). The smallest tip diameters reported for glass and solid-state membrane microelectrodes are about 0.5 μm. Both gas-sensitive and enzyme electrodes have been successfully minaturized to a realistic size for intracellular work (tip diameters are however > 1 μm (Table 2.1)).

2.3 Concluding Remarks

Neutral carrier microelectrodes represent a special type of ion-selective membrane electrodes. Both the type of membrane and the arrangement of the electrode body contribute to the particular suitability of these sensors for intracellular ion activity measurements. Typical tip diameters of microelectrodes are about 100 to 1000 times smaller than the size of most animal or plant cells. The smallest micropipettes described so far are 45 nm in diameter. They allow the measurement of membrane potentials of cell organelles. Neutral carrier microelectrodes are limited to tips of about 0.3 µm diameter. A further miniaturization would require membrane solutions of lower resistivity.

3 Natural and Synthetic Neutral Carriers for Membrane Electrodes

3.1 Carriers as a Class of Ionophores

The main component of a liquid membrane for electrodes is a complexing agent (ligand) that is responsible for the ion selectivity of the membrane. Many requirements are imposed on such molecules [Pretsch et al. 74; Morf et al. 79; Ammann et al. 83]. A deeper knowledge of the function of ion transport systems in biological membranes could provide useful guidelines for the molecular design of suitable components for liquid membranes.

In 1964 Moore and Pressman discovered that the antibiotic valinomycin exhibits a selectivity for alkali cations in rat liver mitochondria [Moore and Pressman 64]. This observation opened the field of ion-selective molecules to many scientific disciplines. Two years later it was shown by Štefanac and Simon [Štefanac and Simon 66] that valinomycin can be used in artificial membranes for analytical purposes.

In 1967 Pressman proposed that antibiotics which induce alkali-ion permeability in mitochondrial and other systems should be designated as ionophores (ion-bearers, ion-carrying agents) [Pressman et al. 67]. It was found that these ionophores can create specific pathways of ion permeability in biological and artificial membranes (Fig. 3.1). Depending on their function in a membrane, ionophores are subdivided into channel-forming molecules (channels) and ion-carriers (carriers). An elegant study of these two different types of ionophores was performed by reversibly solidifying and liquefying lipid bilayer membranes [Krasne et al. 71]. The carriers nonactin and valinomycin exhibited an abrupt loss of transport effectiveness when the membrane fluidity was lost (Fig. 3.2). This is in accordance with the free diffusion of a carrier in a membrane. In contrast, the channel-former gramicidin still mediates ionic conductance in frozen membranes (Fig. 3.2), which is in agreement with the assumption that the permeation of an ion through a channel remains unaltered if the mobility of the channel-forming molecule in the membrane is reduced [Krasne et al. 71]. The efficacy of gramicidin is only slightly reduced at decreased temperatures because of the smaller number of trans-membrane channels that are formed (Fig. 3.2).

The two classes of ionophores can be distinguished further by their ability to increase the permeability of artificial or native biological membranes:

channels: – relatively long-lasting stationary trans-membrane structures
          – membrane conductance can be observed to increase in discrete, quantum steps

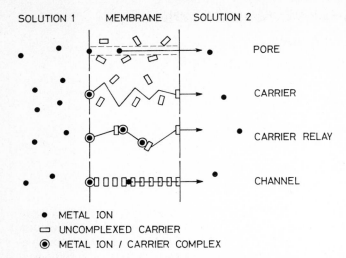

Fig. 3.1. Schematic representation of different transport processes in membranes

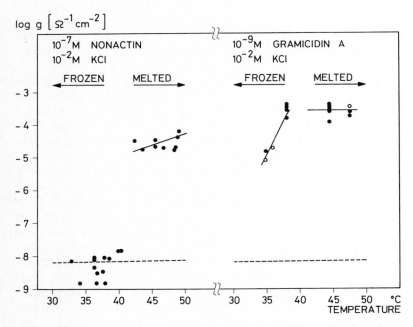

Fig. 3.2. The effect of temperature on nonactin- or gramicidin-mediated conductance g of bilayer membranes [Krasne et al. 71]. The dashed lines represent the average observed conductance for membranes in the absence of ionophores (for experimental details see [Krasne et al. 71])

- low or moderate selectivity (typically $1:10$ to $1:10^2$; in a few cases $1:10^3$)
- high transport rates (several ions pass through one opened channel)
- function restricted to thin bilayer membranes

carriers:
- diffusion across the membrane as a complex with individual ions
- membrane conductance is gradually influenced
- high selectivity (typically $1:10^3$ to $1:10^6$; in a few cases up to $1:10^{11}$)
- low transport rates (one ion is transported in the form of a carrier-complex; the free carrier has to diffuse back)
- function in both thin bilayer membranes (carrier mechanism (Fig. 3.1)) and thick artificial membranes (carrier relay mechanism (Fig. 3.1)).

So far, only carriers have become relevant as ion-selective components for membrane electrodes. Commonly used designs of liquid membrane electrodes require relatively thick membranes, typically about $2 \cdot 10^8$ pm, which is obviously still compatible with a carrier mechanism. In contrast, channel-forming ionophores are only active in bilayer membranes with a thickness of about $7 \cdot 10^3$ pm. To date, relatively few efforts have been made to prepare model compounds exhibiting channel properties. This is due to their limited applicability to membranes and the extreme synthetic difficulties encountered in their preparation (see [Tabushi et al. 82; Behr et al. 82]). In contrast, ion-carriers can be used in almost any type of liquid membrane, and accordingly, there have been many attempts to produce artificial carrier molecules.

The class of ion-carriers is usually further subdivided according to structural criteria (ability to dissociate into charged species and the type of selectivity (c.s.: cation-selective, a.s.: anion-selective)) (Fig. 3.3):

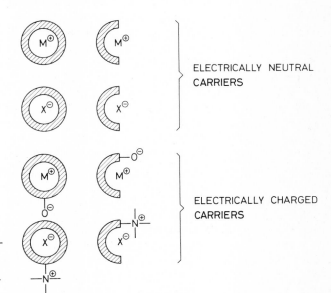

ELECTRICALLY NEUTRAL CARRIERS

ELECTRICALLY CHARGED CARRIERS

Fig. 3.3. Schematic representation of macrocyclic or non-macrocyclic, neutral or charged carriers with selectivity for cations or anions

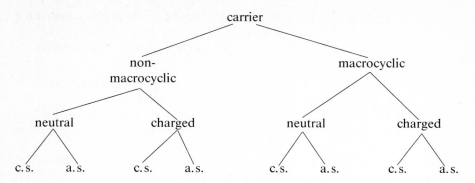

Until recently, the development of synthetic carriers has mainly been focused on cation-selective carriers. During the last few years strategies for carriers exhibiting selectivity for anions [Wuthier et al. 84; Schulthess et al. 84; Dietrich et al. 84] and uncharged species [Vögtle et al. 82] have also been developed. Most of the approved cation-carriers belong to one of the following classes:

- antibiotics
- crown compounds
- cryptands
- non-macrocyclic compounds (e. g. lipophilic oxa-amides (see 3.3)).

Of the large number of carrier antibiotics [Shamoo 75; Ovchinnikov et al. 74; Simon et al. 73] only the K^+-selective carrier valinomycin and the NH_4^+-selective macrotetrolides have found wide-spread use in potentiometry with ion-selective electrodes. Other antibiotics either exhibit selectivities that are too low for use in clinical or physiological applications, or their attractive complexing properties cannot be exploited in liquid membranes for ion-selective electrodes [Simon et al. 77 b; Covington and Kumar 76; Erne et al. 80 b]. After the introduction of crown ethers in 1967 [Pedersen 67], this first class of synthetic carriers expanded rapidly. However, for many years there were almost no reports on analytically relevant membrane electrodes based on crown compounds. Only during the past few years have new lipophilic crowns been successfully used as components in liquid membrane electrodes [Kimura et al. 79 a; Fung and Wong 80; Tamura et al. 82; Yamauchi et al. 82; Shono et al. 82; Kimura et al. 83]. The cryptands, introduced in 1969 by Lehn [Dietrich et al. 69], represent a class of compounds that are especially suited for a versatile architecture of ligand molecules. They have indeed found important applications in different areas (e. g. anion activation, anionic polymerization, phase-transfer catalysis, isotope separation, ion transport). However, due to the rather high stabilities and/or slow ion-exchange kinetics of the bi- and polycyclic cryptand complexes with cations or anions, no relevant contributions to the field of liquid membrane electrodes have been reported so far (see [Gajowski et al. 81]).

   In contrast, the non-macrocyclic molecules used in the microelectrodes described above were strictly designed with a view to their suitability as highly selective carriers for liquid membrane electrodes [Morf et al. 79]. Today, they represent the most successful and widely used class of neutral carriers for ion-selective electrodes [Ammann et al. 83]. Studies on the mechanisms of membranes containing these carriers have led to improved membrane technology for the preparation of analytically relevant neutral carrier electrodes [Ammann et al. 83].

3.2 Molecular Aspects of the Ion-Selectivity of Neutral Carriers

3.2.1 General Remarks

At present, it is not possible to design a neutral carrier based exclusively on theoretical and semi-empirical considerations. Experience has shown that the planning of tailored carrier molecules using available models fails. Keeping in mind the demands put on such molecules, the failure is not surprising: the ligands have to act as membrane-active carriers and simultaneously exhibit a selective recognition of a substrate (cation, anion, uncharged species). The difficulty in the direct design of a carrier mainly stems from the impossibility of predicting complex stabilities, ion-exchange rates, equilibrium conformations and stoichiometries of the carrier/cation complexes. Each of these parameters is strongly influenced by the conditions in the membrane phase.

The stepwise approach to a carrier structure appears to be more realistic. The procedure is based on both model calculations (Sect. 3.2.2) and the empirical knowledge gained from existing molecules (examples in Sect. 3.2.3). The merits of empirical considerations are often greater than those of model computations. Much of the practical experience is derived from trial and error studies with a promising structure. The procedure is clearly illustrated in the case of the optimization of a selective carrier: by slightly modifying the structure of the carrier molecule the necessary steps for improving the selectivity are recognized. Unfortunately, a successful procedure can only rarely be applied to the optimization of another type of carrier molecule.

3.2.2 Models for the Design of Neutral Carriers

Models that enlighten the relationship between structure and selectivity are important tools for predicting promising molecular structures. The models can also be used to confirm or discuss the structures of known carriers and their complexes (e.g. in comparison with spectroscopic or X-ray analysis data [Dobler 81]).

Models for the computation of the interaction between an ion and a carrier molecule can be classified in the following manner:

- calculations based on electrostatic models
- quantum chemical computations
    - semiempirical
    - ab initio

A further aid for the design of carriers is the visual examination of molecular models (ball-and-stick models, Dreiding models, Corey-Pauling-Koltun (CPK) models (space-filling models)). These mechanical models ar supplemented by computer-aided molecular modelling [Feldmann 76]. Computer programs are available for displaying, representing (colour images, surface display) and modelling structures.

3.2.2.1 Corey-Pauling-Koltun (CPK) Models

Crystallographic models (e.g. Dreiding models) represent the skeleton of a molecule. They yield information on the correct bond angles and bond lengths but do not indicate the actual space occupied by a molecule. The CPK space-filling model, on the other hand, shows few details of bond angles and distances in the backbone, but accurately reflects the surface contour of the molecule. Therefore, it represents the molecular topology as it is encountered by the ion to be complexed. This is the reason why the CPK model is popular for the design and illustration of the structures of carriers or carrier-complexes (see e.g. [Weber and Vögtle 81; Morf and Simon 71b]). The usefulness of the model is demonstrated by the example in Fig. 3.4. The CPK model of the K^+ complex of valinomycin gives a clear impression of the shape of the molecule. In spite of this, these models only offer images of the real structure which always has to be confirmed by more substantial facts (e.g. X-ray analysis).

3.2.2.2 Electrostatic Models

With the synthesis of the first nonmacrocyclic carriers, an electrostatic model describing the interaction of hypothetical ligands with cations was put forward [Morf and Simon 71a; Morf and Simon 71b]. The model proved to be successful in calculating the free energies of hydration [Morf and Simon 71a] and in predicting the molecular requirements for selective neutral carriers [Morf and Simon 71b]. The model is based on the fact that the change in the free energy associated with the transfer of a cation from an aqueous solution into the cavity formed by the carrier in the membrane is responsible for the ion-selective behaviour of the carrier membranes:

$$\Delta G_T^o = \Delta G_L^o - \Delta G_H^o, \tag{3.1}$$

where
ΔG_T^o: free energy for the transfer of a cation from the aqueous phase into the cavity formed by the carrier in the membrane [kJ mol^{-1}];
ΔG_L^o: free energy for the transfer of a cation from the gas phase into the carrier-complex within the membrane [kJ mol^{-1}];
ΔG_H^o: free energy of hydration, i.e. for the transfer of a cation from the gas phase into water [kJ mol^{-1}].

It is assumed that several electrostatic terms contribute to the free energy changes ΔG_L^o and ΔG_H^o. Useful model calculations of the free energy of transfer as a function

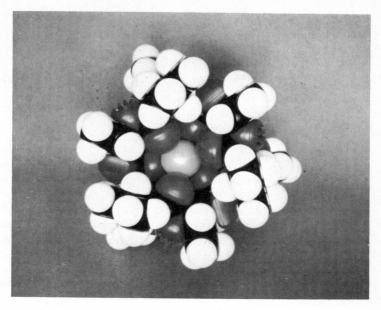

Fig. 3.4. CPK model of the K^+-complex of valinomycin

of the structure of hypothetical carriers have been made [Morf and Simon 71b; Simon et al. 73]. Valuable estimations of the influence of the coordination number, the properties of the ligand groups, the dimensions of the ligand, the steric interactions, and of the surrounding solvent have been made and applied to the design of synthetic carriers [Morf et al. 79]. Several examples of such model considerations are given in the discussion of selectivity-determining parameters (Sect. 3.2.3).

3.2.2.3 Quantum Chemical Model Calculations

The merit of classical electrostatic models is restricted because various parameters are not known. Obviously, more fundamental computations are necessary for a deeper insight into the features of the molecular design. Among the quantum chemical approaches the currently used semiempirical procedures often lead to non-realistic results for interaction energies and complex geometries [Schuster et al. 75; Bendl and Pretsch 82]. In contrast, ab initio calculations have been successfully used to assess the interaction of small molecules with various cations [Schuster et al. 75]. The interaction energies are defined by

$$E_I = E_C - E_L - E_M \qquad (3.2)$$

where
E_I : ion-ligand interaction energy [kJ mol^{-1}];
E_C : energy of the complex [kJ mol^{-1}];

E_L : energy of the carrier [kJ mol^{-1}];
E_M : energy of the cation [kJ mol^{-1}].

An ab initio calculation of E_I is based on the following parameters:

- coordinates of each atom (structure of the carrier and its complex)
- nuclear charge
- number of electrons of the system
- basis sets.

Ab initio programs rely on the SCF-LCAO-MO (Self Consistent Field, Linear Combination of Atomic Orbitals to Molecular Orbitals) approximation. A correct description of the interaction depends on the basis set used in the computation. Today, ab initio calculations of interaction energies cannot be applied to molecules of rather high molar masses. Since the complexes of typical carriers often exhibit a molar mass of more than 1000 g/mol, an approximating procedure has to be applied. For this purpose, additive schemes that reproduce the results of an ab initio calculation were designed [Clementi 76; Corongiu et al. 80; Welti et al. 82; Gresh et al. 79]. According to one model, the interaction energy E_I of an ion I and a carrier L is assumed to be the sum of the interaction energies of the metal ion M and the i-th atom of the carrier L:

$$E_I = \sum_i V_{iM}, \tag{3.3}$$

where
V_{iM}: pair potential of the i-th atom of the carrier L and the metal ion M.

The interactions by pairs can be described by a simple function:

$$V_{iM} = - A_{iM}/r_{iM}^6 + B_{iM}/r_{iM}^{12} + q_i\, q_M\, C_{iM}/r_{iM}, \tag{3.4}$$

where
A_{iM}, B_{iM}, C_{iM} : adjustable parameters;
r_{iM} : distance between the i-th atom of the carrier and the ion M;
q_i, q_M : net charges of the i-th atom and the metal ion M, respectively.

The constants A_{iM}, B_{iM} and C_{iM} are determined by a least square approach using the interaction energies obtained from ab initio calculations. It is also assumed that atoms of the same kind in a similar chemical environment can be described by the same set of parameters A_{iM}, B_{iM} and C_{iM}. Thus, the pair potentials (Eqs. (3.3) and (3.4)) are transferable, i. e. for a given class of atoms they can be applied to various molecules. If the pair potentials (obtained from ab initio SCF-LCAO-MO calculations on small systems) of all the involved classes are known, the potential energy surface of an ion surrounded by large carrier molecules can be calculated on the basis of this approximation.

This quantum chemical model, which was previously used for other studies, e. g. for the computation of the interaction between water and amino acids [Clementi et

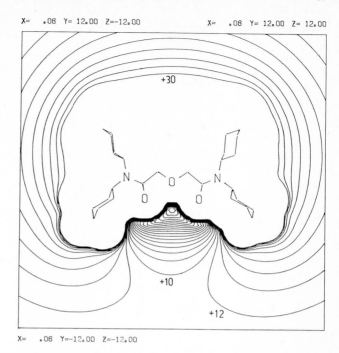

$X= \quad .08 \quad Y= 12.00 \quad Z=-12.00$ $\qquad$ $X= \quad .08 \quad Y= 12.00 \quad Z= 12.00$

$X= \quad .08 \quad Y=-12.00 \quad Z=-12.00$

Fig. 3.5. Isoenergy contour diagrams for the interaction of Na^+ with an oxa-diamide carrier. Distances (Na-O) of 224-230 pm were calculated for the position of the ion. The corresponding interaction energies are close to -105.8 kJ mol^{-1}. The contour-to-contour interval is 2 kcal mol^{-1} (8.36 kJ mol^{-1}) [Morf et al. 79]

al. 77] or DNA bases [Scordamaglia et al. 77], has now been successfully applied to computations of the interactions of metal ions with ion-carriers (Fig. 3.5) [Corongiu et al. 79; Corongiu et al. 80; Welti et al. 82]. At present, the use of such calculations for the design of new carriers is limited. The models may, however, help us to understand experimental observations or to select appropriate coordinating sites for an ion under study. A great deal of work is still necessary before ion selectivities can be reliably estimated in this way. It has to be kept in mind that interaction energies, not free energies, are calculated. In principal, the computation of free energies should be possible using molecular dynamics [Alder and Wainwright 59]. Furthermore, the conformational energies of the ligands should be taken into consideration. This can be achieved by using the techniques of molecular mechanics [Burkert and Allinger 82], but the precision is limited. The interaction of the complex with the surrounding solvent molecules can strongly influence the ion selectivities (see Sect. 4.1.2). Consequently, this term as well as the interaction of the complex-cation with the counterion should also be taken into account. Over all, quantum chemical model calculations are an aid in the design of carrier molecules while the computations allow us to make further predictions with respect to the structural properties of the ligands. However, the day of computer-designed carriers still seems to be far away.

3.2.3 Selectivity-Determining Parameters

3.2.3.1 General Remarks

Ideally, the shape of the cavity of a carrier and its substrate should be complementary. Different molecular parameters contribute to this feature of a carrier molecule. Evidently, the molecular properties are stored in the entire carrier-structure including its chemical environment. Even so, it was recognized that a theoretical treatment of the properties of structural subunits may be very successful in determining the influence of these structural parameters on the ion selectivity of the entire carrier molecule.

In practice, however, it is virtually impossible to change one parameter without affecting the others. For instance, the replacement of an ether oxygen atom by a thioether sulfur atom will not only influence the dipole moment of this binding site but at the same time change the polarizability, the size of the ligand atom, and the topology about the binding site. Thus, theoretical considerations are often difficult to realize in practice.

The following detailed list of constitutional parameters of a carrier molecule exhibiting specific molecular recognition indicates the complexity of carrier-design:

- parameters of the carrier topology:
  - constitution
  - conformation
  - flexibility
  - size
  - lipophilicity
  - chirality
- parameters of the binding sites of the carrier:
  - charge
  - dipole moment
  - polarizability
  - hydrogen bonds
  - size
  - number
  - arrangement.

In the following sections the impact of model calculations is demonstrated for selected parameters and, where possible, illustrated by the corresponding experimental studies.

3.2.3.2 Binding Sites

For the selective complexation of A-metal cations by a carrier molecule various properties of the binding sites are critical:

- polar coordinating groups should preferably contain oxygen as the ligand atom

- amine-nitrogen atoms show an affinity to Mg^{2+} [Williams 70] but they can induce H^+-selectivity because of protonation reactions [Schulthess et al. 81]
- increasing dipole moments increase the stability of a complex and simultaneously the preference for divalent over monovalent cations of the same size (if the orientation of the dipole remains constant) [Simon et al. 77b; Ammann et al. 81e].
- high polarizability of the ligand atom is advantageous for the preference of alkaline-earth cations.

Figure 3.6 illustrates that theoretical considerations can deal with one selected parameter (e.g. dipole moment), whereas in practice it is observed that a change in a binding site affects several properties of a molecule (e.g. dipole moment, polarizability, size). The replacement of the amide oxygen atom by a sulfur atom (Fig. 3.6, right) results in a drastic preference for the B-cation Cd^{2+} over the A-cations. It would be meaningless to interpret this effect in terms of, e.g., the dipole moment only.

3.2.3.3 Coordination Number and Cavity

A multidentate carrier molecule should be able to form a stable conformation that provides a cavity for the selective uptake of a cation. High stabilities can be expected if the optimal conformation for the cavity is inherent in the free ligand. Cryptands, for example, show similar structures in the free ligand and in the complex form [Lehn 73; Lehn 78] and are therefore strong complexing agents. As a result, the ion-exchange rates are often too slow to be compatible with the demands put on the exchange kinetics of a membrane carrier. Sufficiently fast exchange rates are easily achievable with open-chained electrically neutral molecules since their flexibility allows a stepwise replacement of the solvation shell by the ligand atoms of a carrier molecule. Such ligands do not usually exhibit a conformation that resembles the equilibrium conformation of the cavity in the complex. In spite of this, flexible acyclic carriers will form cavities using their polar binding sites. The coordination number and the cavity size determine the preference for a certain cation. Cations larger than the cavity will deform the optimal arrangement while smaller cations will not fill the vacancy. Unfortunately, the planning of tailored cavities is severely impeded by the uncertainty involved in predicting the equilibrium conformations and stoichiometries of the complexes (see illustrations in Fig. 3.7).

3.2.3.4 Arrangement of the Binding Sites

The size of the chelate rings formed by two ligand atoms and the central atom (cation) and the cyclic order (acyclic, bicyclic, polycyclic) influence the arrangement of the binding sites of a polydentate carrier. Preferably, two ligand atoms should form five-membered chelate rings with the non-solvated cation (chelate effect) [Schwarzenbach 52]. Similarly, the ring closure of a non-cyclic polydentate carrier to a macrocycle (macrocyclic effect [Cabbiness and Margerum 69]) or a macrobicycle (macrobicyclic effect [Lehn 73]) leads to a drastic increase in the stability of a complex

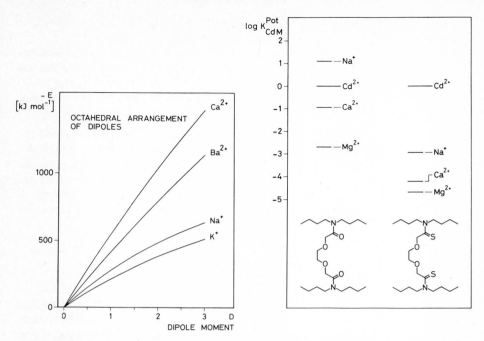

Fig. 3.6. Theoretically *(left)* and experimentally *(right)* observed influence of the properties of binding sites on the ion selectivity of a carrier molecule. *Left:* dependence of the free energy of the electrostatic interaction between a cation and a hypothetical hexadentate ligand (van der Waals radius of the binding sites: 140 pm) on the dipole moment of a binding site [Morf and Simon 78]. *Right:* effect of the replacement of amides by thioamides on the ion selectivity of carrier-based solvent polymeric membranes

(Fig. 3.8, upper). While five-membered chelate rings are generally advantageous, cyclic structures can destroy the carrier properties by forming complexes that are too stable (Sect. 3.2.4). Five-membered chelate rings can be built up from any structural units in which the two ligand atoms are separated by two atoms. However, units forming the same sized chelate ring can make different contributions to both the steric and electronic properties of a binding site. The carriers in Fig. 3.8 (lower) exhibit (from left to right) a change from a syn-periplanar to a syn-clinal arrangement of the ether oxygen atoms and an increase in the basicity of the ether oxygen atoms. It is not possible to separate the influence of these two effects on the ion selectivity. The example given in Fig. 3.8 (lower) can be explained as follows: carriers of this type with a high preference for Ca^{2+} over Na^+ and Ba^{2+} should conform to either one or both of the following specifications: small polar substituent constants and syn-clinal arrangement of the ether oxygen atoms. Conversely, large polar substituent constants and a syn-periplanar arrangement produce a preference for Na^+ and Ba^{2+} over Ca^{2+}. The deviations from this trend in the experimental data near both ends of the scale in Fig. 3.8 (lower) may be due to the bulkiness of these ligands (as measured by the average thickness of the ligand layer around the complexed cation (see Sect. 3.2.3.5)).

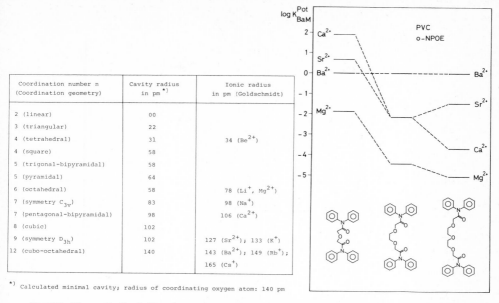

Coordination number n (Coordination geometry)	Cavity radius in pm [*]	Ionic radius in pm (Goldschmidt)
2 (linear)	00	
3 (triangular)	22	
4 (tetrahedral)	31	34 (Be^{2+})
4 (square)	58	
5 (trigonal-bipyramidal)	58	
5 (pyramidal)	64	
6 (octahedral)	58	78 (Li^+, Mg^{2+})
7 (symmetry C_{3v})	83	98 (Na^+)
7 (pentagonal-bipyramidal)	98	106 (Ca^{2+})
8 (cubic)	102	
9 (symmetry D_{3h})	102	127 (Sr^{2+}); 133 (K^+)
12 (cubo-octahedral)	140	143 (Ba^{2+}); 149 (Rb^+); 165 (Cs^+)

[*] Calculated minimal cavity; radius of coordinating oxygen atom: 140 pm

Fig. 3.7. Theoretically *(left)* and experimentally *(right)* observed influence of the coordination number on the ion selectivity of a carrier membrane electrode. *Left:* calculated minimal cavity radii [Morf and Simon 71b; Morf et al. 79] in comparison to ionic radii. *Right:* Influence of the number of ether oxygen atoms between N,N-diphenyl amide groups on the ion selectivity of oxa-amide carriers in membranes [Morf et al. 79]

3.2.3.5 Size of the Carrier

An important contribution to the stabilities of the complexes and the cation selectivities of a carrier is made by the electrostatic interactions between the charged complexes and the surrounding membrane solvent (Born equation (4.1)). A thin ligand layer about the central atom results in a preference for divalent over monovalent cations of the same size. The effect is particularly pronounced in solvents of high dielectric constants (Fig. 3.9). The theoretical result can be confirmed by potentiometric studies with membranes containing carriers of similar structure but surrounded by layers of ligand of variable thickness (Fig. 3.9, right).

3.2.4 Properties of Neutral Carriers for Analytically Relevant Membrane Electrodes

In order to function as a useful component in analytically relevant membrane electrodes, neutral carriers have to fulfil the following four requirements:
- induction of permselectivity
- induction of ion selectivity
- high lipophilicity
- fast ion-exchange rate

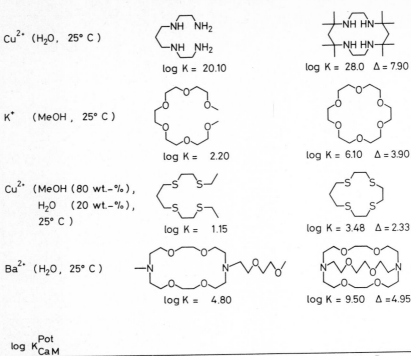

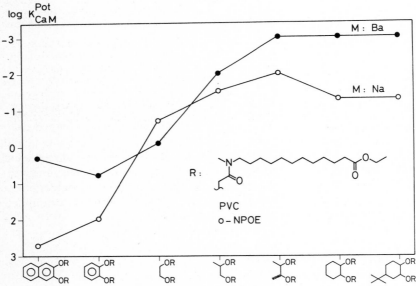

Fig. 3.8. Influence of the arrangement of the binding sites on the ion selectivity of carrier molecules. *Upper:* influence of the ring closure of acyclic polydentate ligands to macrocyclic and macrobicyclic molecules on the stability of the corresponding complexes [Morf et al. 79]. *Lower:* influence of the connective part between the ether oxygen atoms of dioxa-diamide carriers on the selectivity of the corresponding membrane electrodes [Ammann et al. 75a]

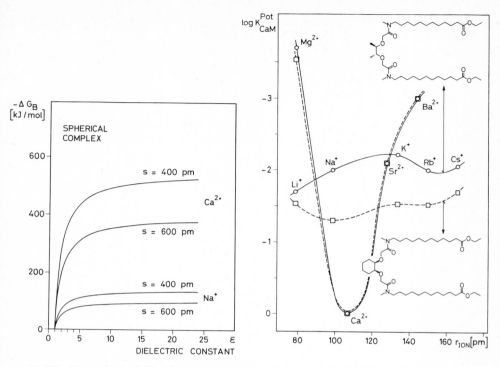

Fig. 3.9. Theoretically *(left)* and experimentally *(right)* observed influence of the average thickness s of a ligand layer about a metal cation on the ion selectivity of carrier membrane electrodes. *Left:* free energy of the electrostatic interactions between a cationic complex and the membrane solvent. The free energies ΔG_B (see Eq. (4.1)) are estimated for two metal ions of nearly the same size (Na$^+$, Ca^{2+}) but of different charge, for two values of the ligand layer thickness (s), and for a varying dielectric constant of the membrane solvent [Morf et al. 79]. *Right:* influence of the average thickness of a ligand layer on the ion selectivity of liquid membrane electrodes based on structurally similar carriers [Ammann et al. 75 a]

3.2.4.1 Permselectivity

The exclusion of sample counterions from a membrane phase is a definition of permselectivity. Ideal permselective behaviour of carrier membranes is shown by a Nernstian electrode response in potentiometric studies (see Chap. 5) and by a transference number of 1.0 for the primary ion in electrodialytic transport studies [Morf 81]. If such an experimental behaviour is observed it can be assumed that no sample anions are transported across the cation-selective carrier membrane. A mechanism for cationic permselectivity has been discussed extensively [Morf 81]. For thick liquid membranes the transport has to be interpreted on the basis of electroneutrality within the membrane phase. Electrodialytic transport experiments using labelled carriers and sample ions have yielded some insight into the underlying mechanisms [Thoma et al. 77]. The results demonstrate that the counterions of the cationic carrier-complexes in the membrane have to originate from the aqueous phase in contact with the membrane. It was suggested that anionic sites of the OH^--type are generated from water clusters which are present in the membrane phase [Thoma et al. 77]. The mechanism can be illustrated by the following carrier (L) induced ion-exchange reaction in the two-phase system membrane (m)/water (aq):

$$I^+(aq) + L(m) + H_2O(m) \rightleftharpoons IL^+(m) + OH^-(m) + H^+(aq) \tag{3.5}$$

The model claims that in order to maintain electroneutrality the presence of H_2O clusters containing OH^- within the membrane phase is necessary. So far, this theory

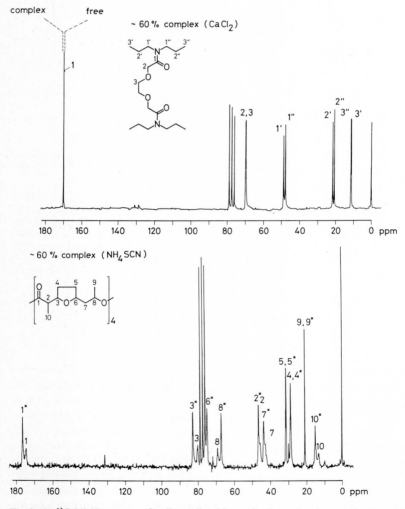

Fig. 3.10. ^{13}C NMR spectra of a dioxa-diamide carrier *(upper)* and nonactin *(lower)* in equilibrium with their Ca^{2+} and NH_4^+ complexes, respectively (solvent: $CDCl_3$; x: signals attributed to the complex) [Pretsch et al. 77]

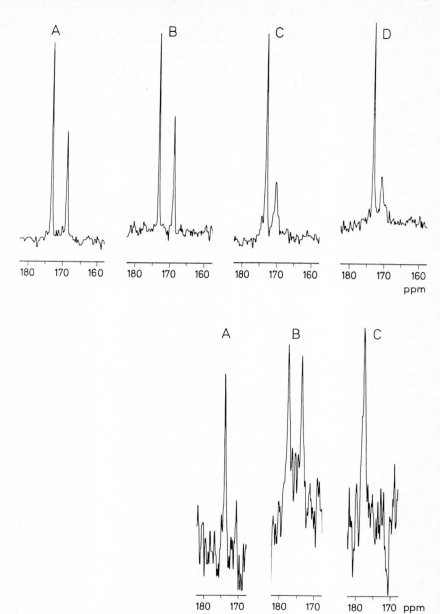

Fig. 3.11. ^{13}C NMR signals of the carbonyl carbon atoms of a dioxa-diamide carrier (N,N'-bis [(11-ethoxycarbonyl) undecyl] -N,N'-dimethyl-3,6-dioxaoctane diamide, *upper*) and nonactin *(lower)* in PVC membranes containing *o*-nitrophenyl-n-octyl ether as plasticizer [Büchi et al. 76a]. The membranes were equilibrated with the following solutions: *upper:* A: H$_2$O; B: 0.1 M CaCl$_2$; C: 0.03 M Ca(SCN)$_2$; D: 0.1 M Ca(SCN)$_2$; *lower:* A: H$_2$O; B: 0.1 M KSCN; C: 1 M KSCN

offers the most satisfying description of the permselectivity mechanism and the observed electromotive and transport behaviour of solvent polymeric membranes.

3.2.4.2 Ion- and Ligand-Exchange Kinetics

The required response times (milliseconds to several seconds) of liquid membrane electrodes can only be achieved if a series of rate-determining processes within the two-phase system sample/membrane are governed by the appropriate time constants. Complexation, decomplexation and various diffusion processes make important contributions that might limit the dynamic characteristics of neutral carrier electrodes. By an appropriate choice of both the carrier structure and the membrane composition (see Sect. 4.1.1) the kinetics of the individual steps can be optimized. For example, in order to keep the ion/carrier exchange reaction sufficiently fast, non-macrocyclic carriers are advantageous. Such molecules will undergo a fast, stepwise replacement of the hydration shell of the ion by the ligand atoms of the carrier. Using relaxation techniques, it was shown that a typical representative of the noncyclic dioxa-diamide carriers undergoes a fast complexation in homogeneous solution. The rate constant of complex formation of the carrier with Ca^{2+} in methanol ($k \geqslant 5 \cdot 10^8 \, l \, mol^{-1} \, s^{-1}$ [Erni and Geier, private communication; Ammann 75]) is close to that of the diffusion-controlled decomplexation reaction of the Ca^{2+} aquo-complex ($k_{H_2O} \approx 3 \cdot 10^8 \, s^{-1}$ [Diebler et al. 69]). This is in agreement with an analysis of the dynamics of the ligand substitution in metal complexes. With the exception of Be^{2+} and Mg^{2+} the alkali and alkaline-earth metal cations should show a characteristic rate which is almost independent of the nature of the substituting ligand. The rate constants of the ligand substitution are of the same order of magnitude as the rate constants for the substitution of the inner-sphere water molecules of the corresponding aquo-ions (typical value: $\sim 10^9 \, s^{-1}$ [Diebler et al. 69]). Indeed, most of the ion-selective antibiotics and a series of monocyclic crown compounds exhibit this kind of dynamic behaviour [Burgermeister and Winkler-Oswatitsch 77], whereas rather slow exchange kinetics have been found for some macrobicyclic systems [Lehn et al. 70].

Useful information about the exchange process of ligands and their complexes in homogeneous solutions as well as in membranes can be obtained by nuclear magnetic resonance studies. The ^{13}C NMR spectrum of a mixture of a typical dioxa-diamide carrier (Fig. 3.10, upper) and its Ca^{2+} complex in $CDCl_3$ shows the averaged signals for the ligand and the complex, indicating that the mean lifetime τ of these species is much smaller than 0.02 s ($\tau \ll 1/(v_{complex} - v_{ligand})$ [Pretsch et al. 77]). The macro-monocyclic macrotetrolide nonactin, however, exhibits a considerably slower exchange rate (above 0.2 s, separated sharp signals for complex and ligand, Fig. 3.10, lower [Pretsch et al. 77]). Carriers of the same type show similar behaviour in PVC membranes used for electrodes (Fig. 3.11) [Büchi et al. 76a]. The amide carbonyls of a dioxa-diamide carrier again show averaged signals for the ligand and the complex ($\tau \ll 0.02 \, s$) (Fig. 3.11, upper; note that the ester carbonyls of the carrier (signals at lower field) are not involved in coordination [Büchi et al. 76a]). As in $CDCl_3$, nonactin in membranes shows a slower ligand exchange than the acyclic carrier (Fig. 3.11, lower).

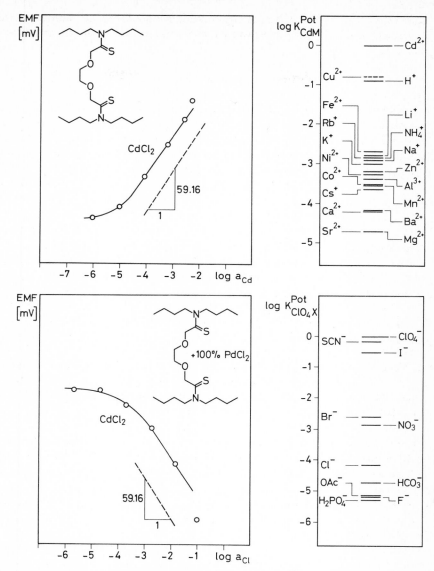

Fig. 3.12. EMF response to $CdCl_2$ solutions and selectivity factors (K_{CdM}^{Pot} for different cations M *(upper)*; $K_{ClO_4X}^{Pot}$ for different anions X *(lower)*) of membrane electrodes based on the carrier ETH 1062 with *(lower)* and without *(upper)* an equimolar amount of $PdCl_2$ [Hofstetter 82; Simon et al. 84b]

In addition, NMR sudies can provide information on the free energies of activation of the ligand-exchange reaction in a homogeneous solution [Hofstetter et al. 83]. The free energies of activation of the ligand-exchange reaction for the Zn^{2+} or Cd^{2+} complexes of a dioxa-dithioamide carrier (ETH 1062), as measured by NMR, were $\leqslant 45\,kJ\,mol^{-1}$ (acetonitrile). Using this carrier and a solution of $CdCl_2$ as the sample, cation permselectivity is observed in potentiometric measurements (Fig. 3.12, upper). The slope of the resulting electrode function is approximately 60 mV, suggesting that $CdCl^+$ is the permeating species [Hofstetter et al. 83]. It is known, however, that if the free energy of activation of the ligand-exchange reaction exceeds 65 kJ mol^{-1} (acetonitrile), the cationic carrier-complexes dominate the electromotive behaviour of the membrane and the positively charged sites act as anion-exchangers (examples of such complexes with free energies of activation of ligand-exchange reactions > 65 kJ mol^{-1} are the complexes of the carrier ETH 1062 with Pt^{2+} or Pd^{2+} [Hofstetter et al. 83]). In agreement with this, an electrode containing the $PdCl_2$ complex of ETH 1062 in the membrane phase did indeed response to the chloride ions in the $CdCl_2$ solutions (Fig. 3.12, lower).

3.2.4.3 Lipophilicity

A carrier molecule exhibiting an extremely low lipophilicity will not be useful as an ion-selective component. The rapid loss of a carrier from the membrane phase into the sample solution, resulting in a carrier-depleted membrane surface, prohibits potentiometric measurements. On the other hand, carriers of extremely high lipophilicity may lead to kinetic limitations of the carrier-induced transfer of ions between the aqueous and the membrane phase. As a result, the electromotive behaviour of the membrane electrode can be disturbed or even destroyed (Fig. 3.13) [Oesch et al. 79].

   A convenient but relatively rough estimate of the lipophilicity of a carrier molecule can be gained using the lipophilicity increments π_X proposed by Hansch and Leo for various structural fragments X of the carrier [Hansch and Leo 79]. The constant π_X of a structural unit X is defined as

$$\pi_X = \log P_X - \log P_H, \tag{3.6}$$

where P_X and P_H are the partition coefficients of the compound R-X and the reference compound R-H in the two-phase system 1-octanol/water. Thus, π_X represents the contribution of the fragment X to the total lipophilicity of the molecule. Values of π_X are additive, i.e. dividing the carrier molecule L into n parts, the overall lipophilicity P_{HANSCH} is given as

$$\log P_{HANSCH} = \sum_1^n \pi_X. \tag{3.7}$$

In an ideal case the P_{HANSCH} values correspond to the true partition coefficient K of the system 1-octanol/water. The Hansch parameters allow a convenient estimation of the relative lipophilicities of structurally similar molecules. They are less accurate for the evaluation of partition coefficients K, particularly if strong interactions (e.g. hydrogen bonding) or ionizable groups are involved [Hansch and Leo 79].

   A satisfying experimental estimation of K values may be obtained using thin layer chromatography (TLC) [Oesch et al. 85b]. A partition coefficient P_{TLC} of the two-phase system eluent (usually a H_2O/EtOH mixture)/stationary phase (usually

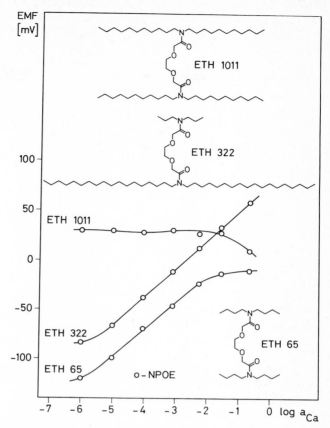

Fig. 3.13. Ca^{2+} electrode functions of PVC membranes with o-nitro-phenyl-n-octyl ether as a plasticizer and neutral carriers containing different lipophilic amide substituents. Although the carriers ETH 1011 (log $P_{HANSCH} = 17.5$) and ETH 322 (log $P_{HANSCH} = 16.5$) are of almost the same lipophilicity, ETH 1011 shows a kinetic limitation of the carrier-induced transfer of ions. The effect can be explained by assuming that the contact between the coordinating sites and the aqueous phase is reduced by the symmetrically-arranged long alkyl chains [Oesch et al. 79]

a reverse phase chromatography plate) can be determined for most carrier molecules.

By calibrating the chromatographic system with reference compounds of known K values, the P_{TLC} values of molecules of unknown lipophilicity can be evaluated. These P_{TLC} values are a good estimate of the lipophilicity as referred to the two-phase system 1-octanol/water [Oesch et al. 85b].

Figure 3.14 shows the correlation between the measured log P_{TLC} values and the estimated log P_{HANSCH} values (Eq. (3.7)). The fairly large deviations from the ideal correlation line indicate that both approaches are prone to deviations from the true K values. For molecules of rather high molar mass and complex structure (e.g. valinomycin) the chromatographic approach seems to be more reliable, since the choice of appropriate structural fragments π_X for such molecules is very difficult.

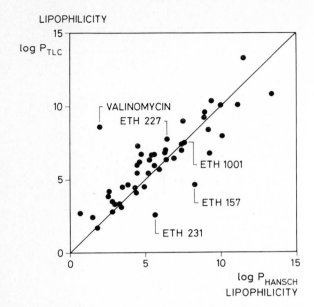

Fig. 3.14. Correlation of experimentally determined log P_{TLC} values with estimated log P_{HANSCH} values

3.3 Highly Cation-Selective Neutral Carriers for Liquid Membrane Electrodes

Using model calculations, CPK models, computer-aided molecular modelling, empirically gained knowledge on the relationship between structure and selectivity and adequate membrane technology it has been possible to design neutral carrier-based membranes with analytically relevant ion selectivities for H^+, Li^+, Na^+, K^+, Mg^{2+}, Ca^{2+}, Ba^{2+}, NH_4^+, Pb^{2+}, Cd^{2+} and UO_2^{2+}. The microelectrodes presented here contain antibiotics or non-macrocyclic oxa-amide carriers (Tab. 3.1). It should be noted that there are other attractive synthetic carriers such as lipophilic macrocyclic crown ethers. However, microelectrodes based on these crown ethers have yet to be described. In a recent review on neutral carriers the most important macrocyclic carriers are listed [Ammann et al. 83] (see also Chap. 9).

Table 3.1 Neutral carriers for cation-selective liquid membrane electrodes

Measuring ion, reference	Constitution, nomenclature		Identification code, source, chemical abstracts registry number
H+ [Schulthess et al. 81; Ammann et al. 81b]		tri-n-dode-cylamine	TDDA Fluka 91660 CAS-RN: 102-87-4
Li+ [Güggi et al. 75]		N,N'-diheptyl-N,N', 5,5-tetra-methyl-3,7-di-oxanonane diamide	ETH 149 Fluka 62557 CAS-RN: 58821-96-8
Li+ [Metzger et al. 84]		N, N-dicyclohexyl-N', N'-diisobutyl-cis-cyclohexane-1,2-dicarboxy-lic amide	ETH 1810
Na+ [Ammann et al. 76; Ammann and Anker 85]		N, N'-dibenzyl-N, N'-diphenyl-1,2-phenylene-di-oxy acetamide	ETH 157 Fluka 71733 CAS-RN: 61595-77-5
Na+ [Güggi et al. 76; Steiner et al. 79]		1,1,1-tris-[1'-(2'-oxa-4'-oxo-5'-aza-5'-methyl)-dodecanyl] pro-pane	ETH 227 Fluka 71732 CAS-RN: 61183-76-4
K+ [Pioda et al. 69; Oehme and Simon 76]		antibioticum valinomycin	valinomycin Fluka 94675 CAS-RN: 2001-95-8
Mg2+ [Erne et al. 80; Lanter et al. 80]		N, N'-diheptyl-N, N'-dimethyl-1,4-butane diamide	ETH 1117 Fluka 63082 CAS-RN: 75513-72-3
Ca2+ [Ammann et al. 75; Oehme et al. 76; Lanter et al. 82]		(-)-(R, R)-N, N'-bis-[11-(ethoxy-carbonyl)-undecyl] N, N', 4,5-tetrame-thyl-dioxaoctane diamide	ETH 1001 Fluka 21192 CAS-RN: 58801-34-6 and 58725-79-4

Table 3.1 (continued)

Measuring ion, reference	Constitution, nomenclature	Identification code, source, chemical abstracts registry number
Ba^{2+} [Güggi et al. 77; Läubli et al. 85]	N, N, N′, N′-tetraphenyl-3,6,9-trioxaundecane diamide	ETH 231 Fluka 11784 CAS-RN: 61595-78-6
	oxy-bis(2,1-phenyleneoxy-N, N-dicyclohexyl acetamide)	
NH_4^+ [Štefanac and Simon 66]	NONACTIN R=CH$_3$/MONACTIN R=C$_2$H$_5$	Macrotetrolides Nonactin: Fluka 74155 CAS-RN (Nonactin): 6833-84-7
Pb^{2+} [Lindner et al. 84]	N, N-dioctadecyl-N′, N′-dipropyl-3,6-dioxaoctane diamide	ETH 322
Cd^{2+} [Schneider et al. 80]	N, N, N′, N′-tetra-butyl-3,6-dioxaoctane dithioamide	ETH 1062 CAS-RN: 73487-00-0
UO_2^{2+} [Šenkyr et al. 79]	N, N′-diheptyl-N, N′,6,6-tetramethyl-4,8-dioxaundecane diamide	ETH 295 CAS-RN: 69844-41-3

3.4 Synthesis of Cation-Selective Neutral Carriers

Many reactions have proven useful for the synthesis of neutral carriers of the acyclic oxy-amide type and for their aza-, thia-, and macrocyclic derivatives (Fig. 3.15). The various steps involved allow a convenient preparation of the carriers shown in Table 3.1 and of many derivatives, i.e. other diols, diacids, amines, etc., which of course can be employed under the same or similar conditions. Details of the reactions including purification procedures and yields can be found in the literature: B [Ammann et al. 75a]; C [Šenkyr et al. 79]; D [Ammann et al. 75a]; E [Dietrich et al. 73]; F [Ammann et al. 75a]; G [Ammann et al. 75a]; H [Ammann et al. 75a; Pretsch et al. 80]; I [Erne et al. 82]; K [Ammann et al. 73]; L [Maj-Zurawska et al. 82]; P [Schneider et al. 80]; Q [Chmielowiec and Simon 78; Oesch et al. 79]; R [Erne et al. 80b]; S [Erne et al. 80a]; T [Erne et al. 80a].

3.5 Carriers for Anions

Currently, only one type of anion-selective microelectrode is in use [Walker 71; Baumgarten 81]. It contains a classical anion-exchanger membrane in which a lipophilic quaternary ammonium salt acts as the mobile exchanger site (e.g. Corning 477315 and 477913). Normally, these microelectrodes are used to selectively measure Cl^--ions, although they can also be used for other ions (for examples see [Wegmann et al. 85]). Regardless of the membrane composition, they all exhibit roughly the same selectivity sequence with a preference for lipophilic (R^-) and a rejection of hydrophilic anions [Koryta and Štulík 83]. The sequence is characterized by the Hofmeister series [Hofmeister 1888]:

$$R^- > ClO_4^- > SCN^- > I^- > NO_3^- > Cl^- > HCO_3^- > SO_4^{2-} > HPO_4^{2-} \qquad (3.8)$$

A theoretical treatment of liquid anion-exchanger membranes has shown that this kind of behaviour is typical of dissociated anion-exchangers, in which the complexation between the cationic sites and the counterions in the membrane phase is negligible. The selectivity is completely described by the distribution coefficients of the various anions between the sample solution and the membrane phase [Morf 81]. Indeed, selectivity sequences different from the Hofmeister series could not be achieved by using different anion-exchanger sites (e.g. ammonium, phosphonium, arsonium salts) or by varying the membrane compositions [Hartman et al. 78; Wegmann et al. 85]. From these studies, it cannot be expected that novel anion-selective microelectrodes will emerge from this class of membranes.

However, significant changes in the above selectivity pattern may become feasible by developing the following types of carrier molecules:

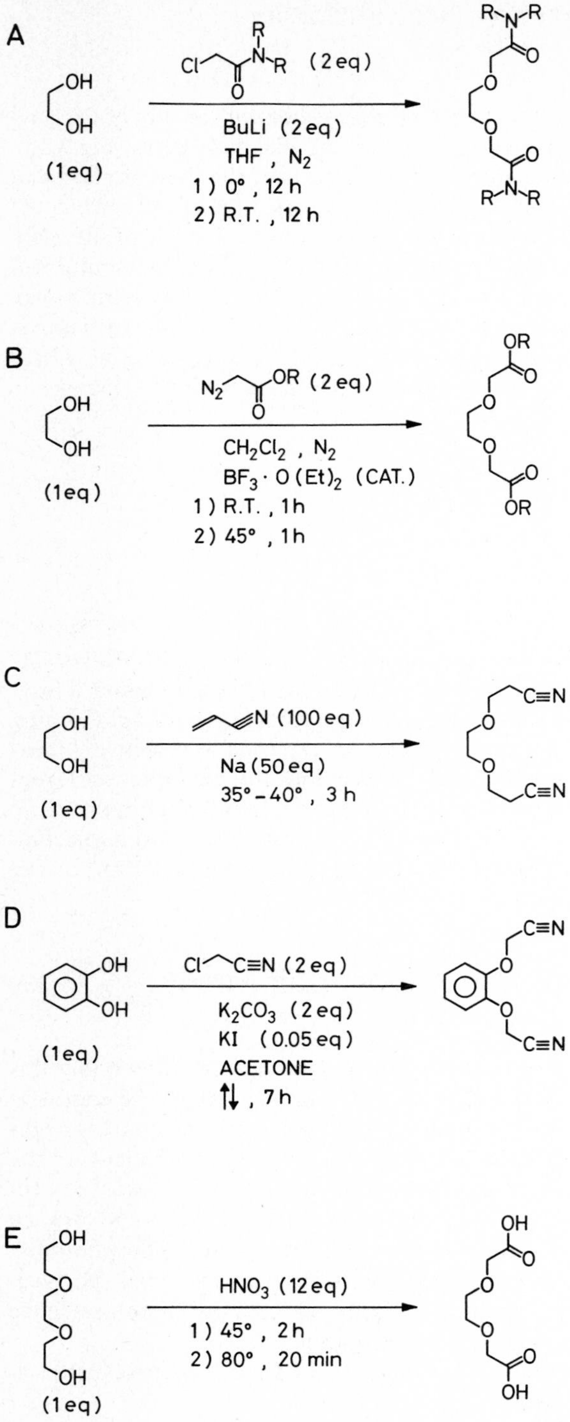
A
(2 eq)
BuLi (2 eq)
THF , N2
1) 0° , 12 h
2) R.T. , 12 h
(1eq)
B
(2 eq)
CH2Cl2 , N2
BF3 · O (Et)2 (CAT.)
1) R.T. , 1h
2) 45° , 1h
(1eq)
C
N (100 eq)
Na (50 eq)
35° – 40° , 3 h
(1eq)
D
Cl⌒C≡N (2 eq)
K2CO3 (2 eq)
KI (0.05 eq)
ACETONE
, 7h
(1eq)
E
HNO3 (12 eq)
1) 45° , 2h
2) 80° , 20 min
(1eq)

Fig. 3.15

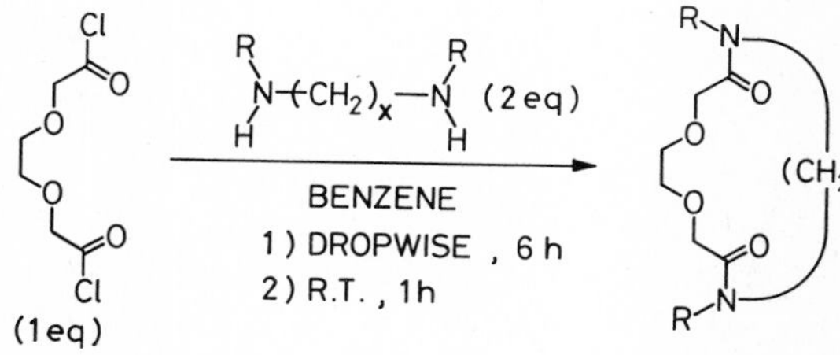

Fig. 3.15 (continued)

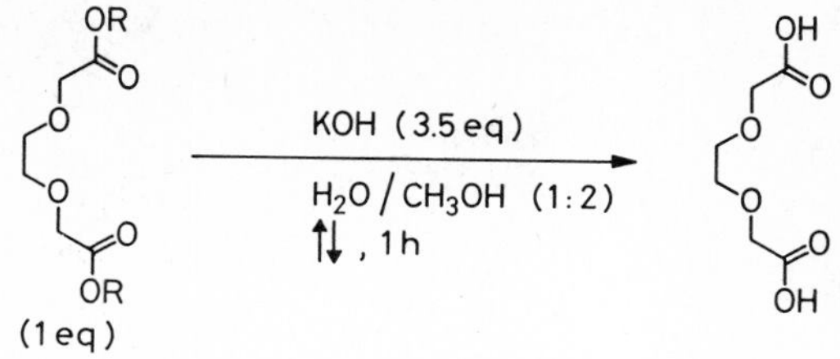

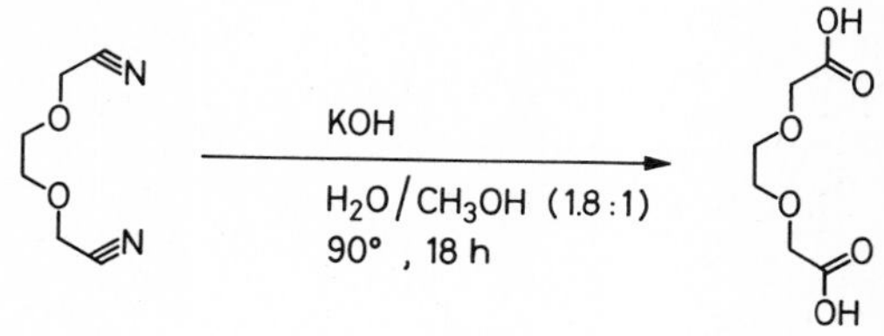

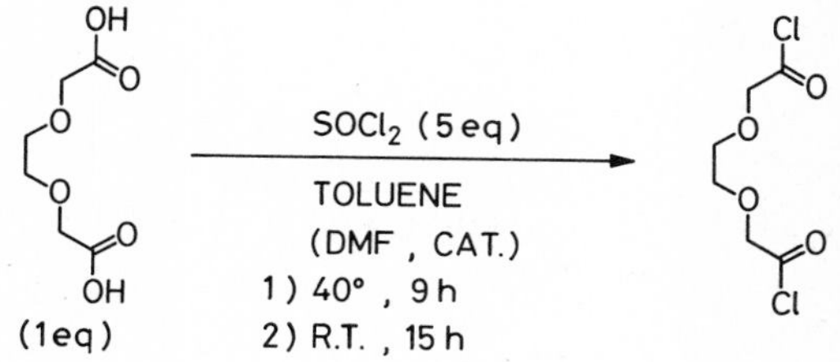

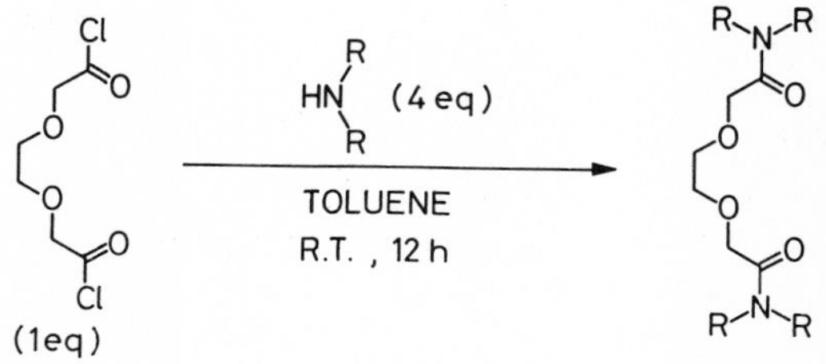

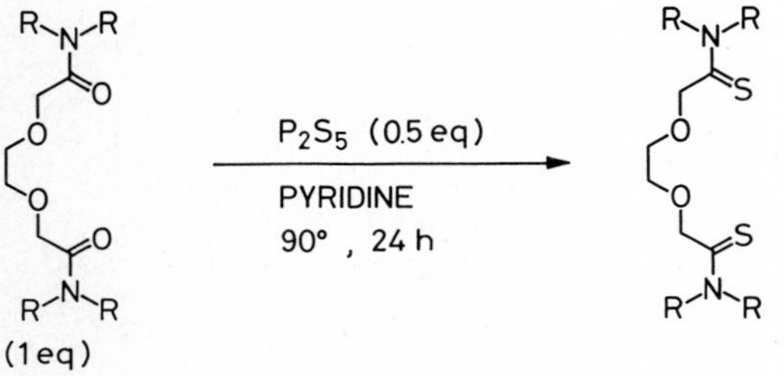

Fig. 3.15 (continued)

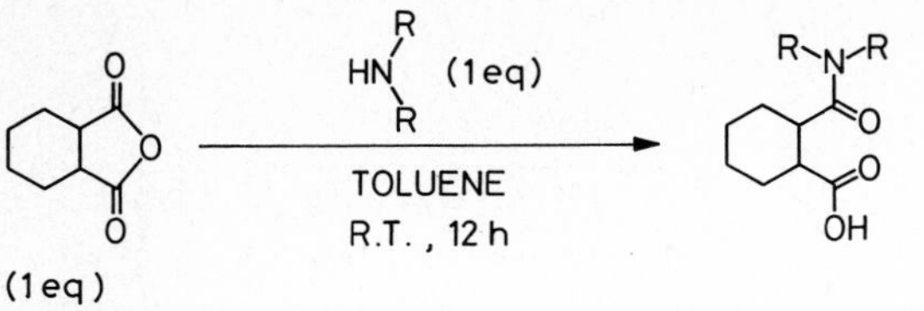

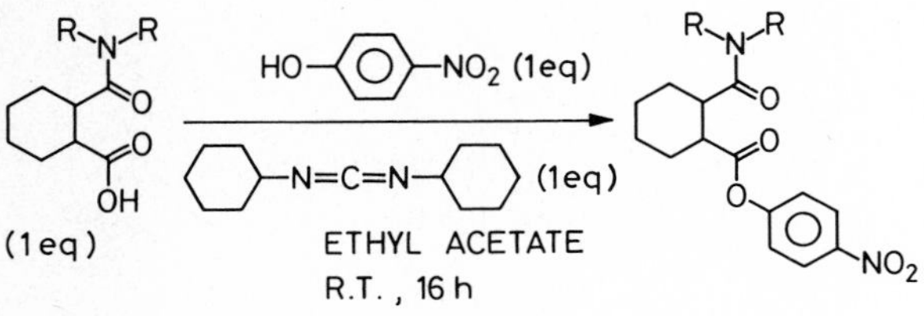

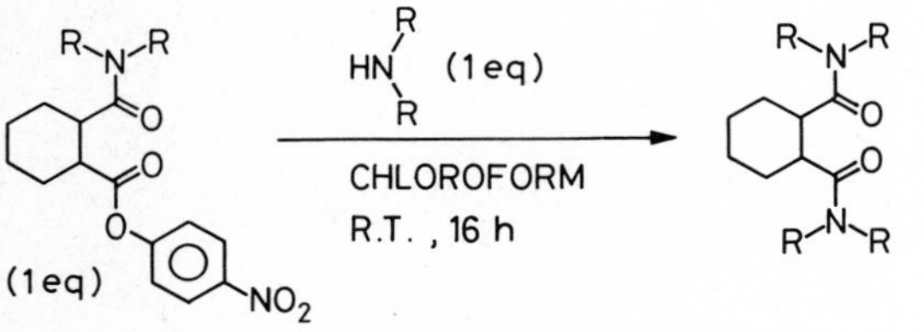

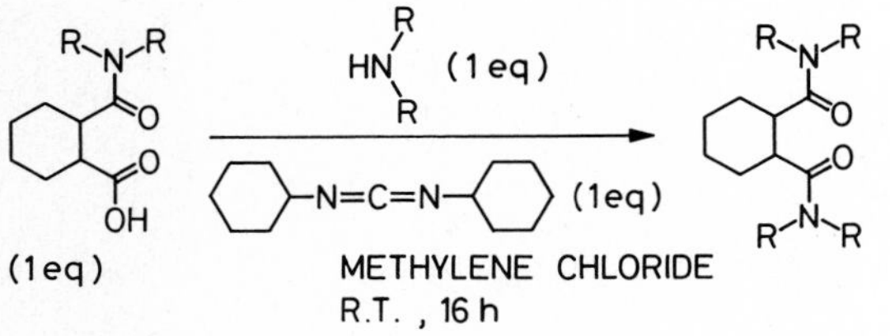

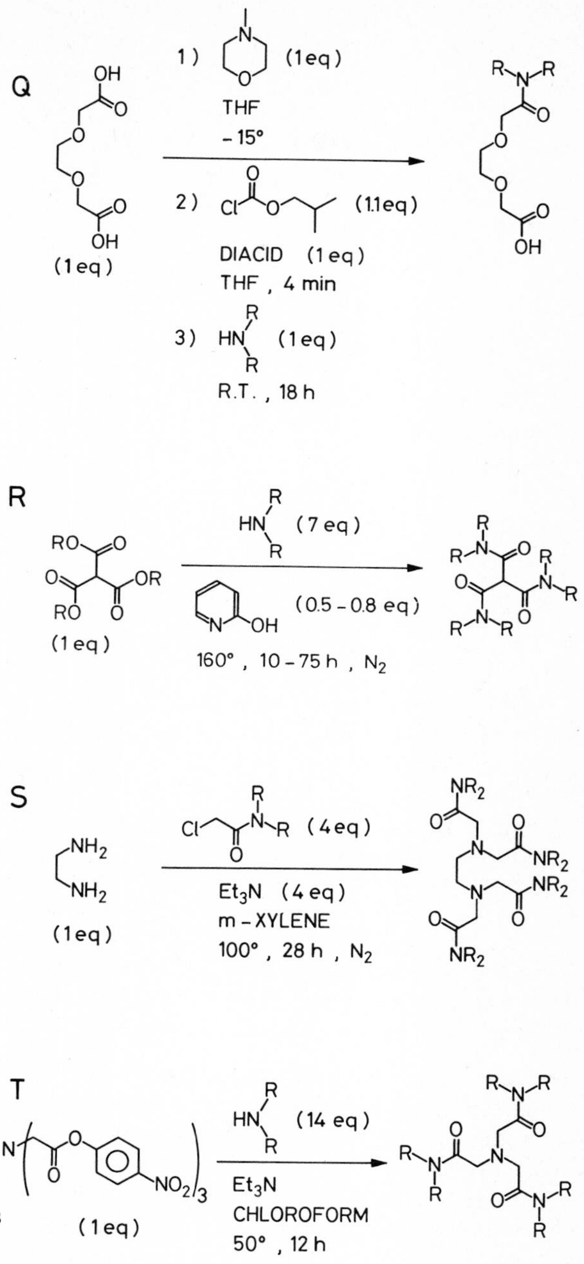

Fig. 3.15. Reactions for the synthesis of cation-selective neutral carriers

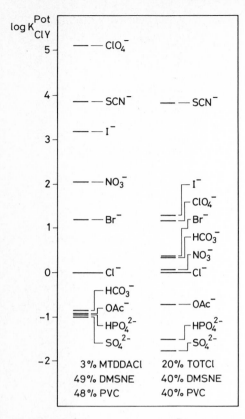

Fig. 3.16. Anion selectivities, log K_{ClY}^{Pot}, of a classical anion-exchanger membrane electrode based on methyl-tri-n-dodecylammonium chloride (MTDDACl) and of a neutral carrier membrane electrode based on tri-n-octyltin chloride (TOTCl) [Wuthier et al. 84]

- mobile, positively charged sites with a sufficiently strong and selective interaction with anions (associated anion-exchangers, charged carriers)
- electrically neutral anion-selective carriers.

Only recently have the first neutral anion-carriers [Wuthier et al. 84] and charged anion-carriers [Schulthess et al. 84; Schulthess et al. 85] been developed. The corresponding carrier membrane electrodes yield very large deviations from the Hofmeister series.

Tri-n-octyl tin chloride was found to behave as a selective carrier for anions [Wuthier et al. 84]. ^{13}C and ^{119}Sn NMR studies showed that a direct interaction between chloride ions and the organo-tin compound takes place. The interaction was confirmed by vapour pressure osmometry measurements. Evidence that tri-n-octyl tin chloride behaves as a neutral carrier was obtained from electrodialytic transport experiments [Wuthier et al. 84]. Solvent polymeric membrane electrodes containing this compound exhibit substantial selectivity changes compared to classical anion-exchanger membrane electrodes (Fig. 3.16). It follows, therefore, that organo-tin

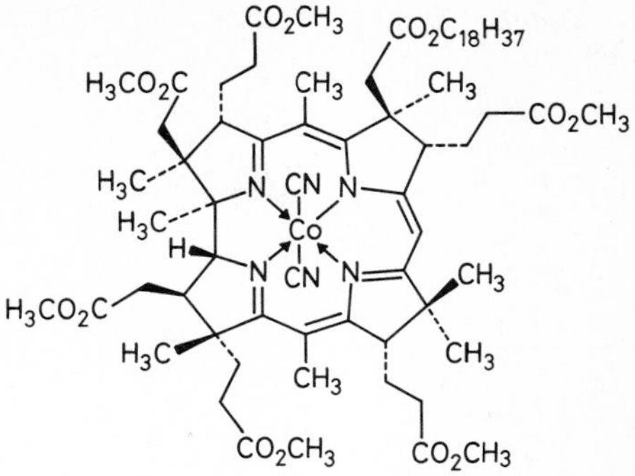

Fig. 3.17. Structures of a, b, d, e, f, g-hexamethyl-c-octadecyl Coα-Coβ-dicyano-cobyrinate (1) and the isomeric a, b, d, e, f, g-hexamethyl-c-octadecyl-Co-aquo-Co-cyano-cobyrinate perchlorates (2a and 2b) [Schulthess et al. 84]

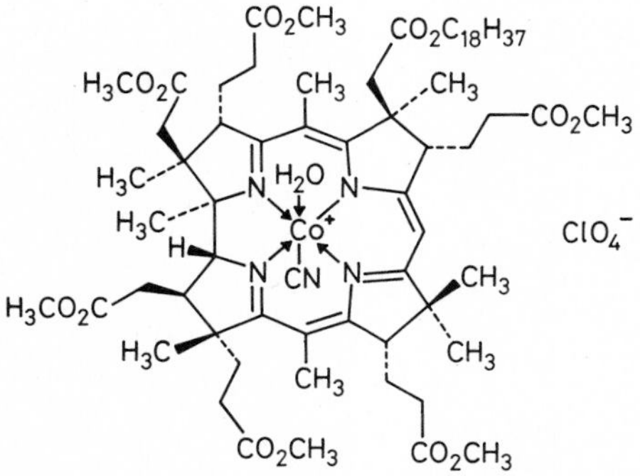

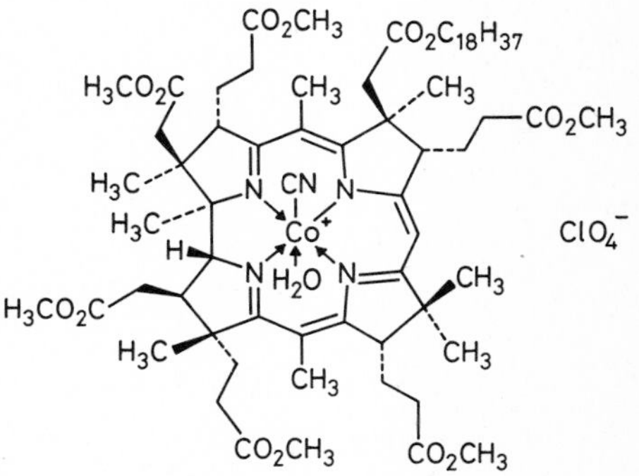

compounds acting as neutral carriers for anions should provide new opportunities for the development of analytically relevant anion selectivities. The ligand design will probably be focussed on structural modifications of tin and other organo-metallic molecules (e. g. germanium- and lead-organic compounds).

Lipophilic derivatives of vitamin B_{12} were found to be highly selective charged carriers for anions [Schulthess et al. 84]. Their ability to act as carriers is supported by the available information on the complexation behaviour of such compounds towards anionic ligands in aqueous solutions [Pratt 72]. For exploratory studies on the anion-selective properties of vitamin B_{12} derivatives the lipophilic Co(III)- cobyrinate octadecyl cobester and its ionic aqua-cyano perchlorate derivative were prepared (Fig. 3.17) [Schulthess et al. 84].

Incorporated into PVC membranes, these Co(III) corrins indeed exchange their axial ligands and behave as highly selective carriers for NO_2^- and SCN^- (Fig. 3.18). The NO_2^- electrode functions at a constant background of Cl^- ions (0.1 M or 1 M)

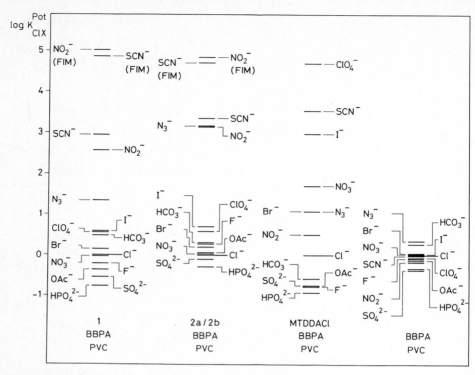

Fig. 3.18. Selectivity factors, log K_{ClX}^{Pot}, for solvent polymeric membranes with bis(1-butyl-pentyl) adipate (BBPA) as the membrane solvent. Membranes with the charged carriers 1 and 2a/2b (mixture of the isomers) Fig. 3.17) are compared with a carrier-free membrane (column 4) and a membrane containing a typical anion-exchanger (methyl-tri-n-dodecylammonium chloride (MTDDACl)) (column 3). Selectivity factors were obtained using the separate solution method in TRIS-buffered solutions (pH 7.45) of 0.1 M sodium salts or the fixed interference method (indicated by (FIM)) [Schulthess et al. 85]

but still shows detection limits below 10^{-5} M NO_2^- (Fig. 3.19) (corresponding to NO_2^-/Cl^- selectivities of more than 10'000; note that a classical anion-exchanger membrane electrode prefers NO_2^- to Cl^- only by a factor of about 3 (Fig. 3.18)). The future design of related charged carriers should be interesting and promising. Other corrin structures, porphyrins, and metal centres other than Co(III) will be the major guidelines.

3.6 Concluding Remarks

Ion-carriers and channel-formers belong to the same class of ionophores. So far, only carriers have been found to be useful ion-selective components in membrane electrodes. In order to be a suitable carrier, a molecule must fulfil a number of re-

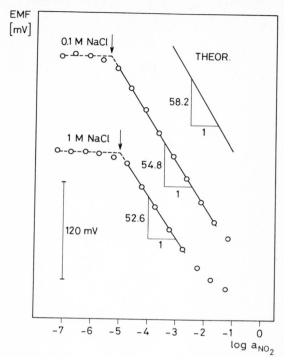

Fig. 3.19. EMF responses of a liquid membrane cell assembly containing the charged anion carriers 2a/2b (mixture of the isomers; Fig. 3.17) as a function of the activity of NO_2^- in the presence of a fixed amount of NaCl (0.1 M or 1 M). The arrows indicate the detection limit for NO_2^-. The selectivity factor log $K_{NO_2,Cl}^{Pot}$ is -4.8 [Schulthess et al. 85]

quirements and exhibit certain properties to be of use in membranes of analytically relevant electrodes.

Two major aids are available for the development of ion-selective carrier molecules. First, empirical knowledge is gained from trial and error studies and careful elucidation of the structure/selectivity relationship. Secondly, different model considerations can be applied: mechanical models, computer-aided molecular modelling, electrostatic model calculations and ab initio model computations. Molecular modelling and refined ab initio computations are increasingly influencing the design of carrier molecules. Nevertheless, the merits of empirical considerations still dominate.

The synthesis of cation-selective neutral carriers of the oxa-amide type has been studied in detail. The compounds are easily synthesized using standard reactions.

The design of anion-selective carriers is based, at least partially, on a different concept than the one used in the development of cation carriers. The first electrically neutral and electrically charged anion carriers have only recently been developed.

4 Liquid Membrane Electrodes Based on Neutral Carriers

For analytical applications neutral carriers are incorporated into artificial bulk liquid membranes for ion-selective electrodes. The membranes form an organic phase which separates the aqueous sample solution from the aqueous internal filling solution of the electrode. Ideally, a liquid membrane should be a homogeneous permselective barrier. Neutral carriers are highly suited for use in such artificial liquid membranes. They are capable of selectively extracting ions from an aqueous sample solution into the membrane phase and transporting these ions across the organic barrier by carrier translocation (Fig. 4.1).

Liquid membranes for ion-selective electrodes are composed of an ion-selective compound (neutral carrier), a membrane solvent (plasticizer), a membrane matrix (polymer) and, where useful, a membrane additive. For a discussion of the membrane materials it is advantageous to distinguish solvent polymeric membranes for macroelectrodes from membrane solutions for microelectrodes. Although solvent polymeric membranes are not directly used in microelectrodes, the properties of such membranes are discussed because many observations made on them are useful for the development of membrane solutions for microelectrodes.

Solvent polymeric membranes always contain a significant amount of a membrane matrix. Hence, the membrane solvents have to exhibit plasticizer properties for the respective polymeric matrix material. Only in a few cases does the membrane matrix behave like the membrane phase without appreciable concentrations of plasticizers (e.g. silicone rubber). In contrast, ion-selective solutions used as membrane phases in microelectrodes (membrane solutions) usually do not contain any polymer matrix. This obviously enables the use of a wider variety of membrane

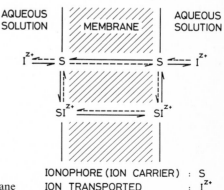

Fig. 4.1. Schematic representation of a cation-selective neutral carrier-based liquid membrane

solvents than in the case of solvent polymeric membranes. In certain membrane solutions for microelectrodes, polymers are nevertheless employed as membrane additives rather than as a membrane matrix. Typical solvent polymeric membranes can not be used in microelectrodes because of the need for membrane phases of low viscosity (microelectrode filling) and low resistance (electrical membrane resistance of the microelectrode). Thus, even for membranes based on the same carrier, characteristic differences in the composition exist between solvent polymeric membranes and membrane solutions. The corresponding macro- and microelectrodes often exhibit significantly different characteristics.

Properties of neutral carrier-based solvent polymeric membranes have been intensively investigated. During the past decade, many studies on molecular aspects (ion-ligand interaction in membranes [Büchi et al. 76a; Büchi et al. 76b]; ion-exchange kinetics [Ammann 75]; carrier-exchange kinetics [Hofstetter et al. 83; Simon et al. 84b]; bulk diffusion and interfacial diffusion [Oesch and Simon 79; Oesch and Simon 80]) as well as on technical aspects (membrane preparation [Jenny et al. 80a; Jenny et al. 80b; Anker et al. 81]; incorporation of membranes in electrode arrangements [Anker et al. 81; Ammann et al. 83]; construction of electrode bodies [Simon et al. 78; Jenny et al. 80a; Anker et al. 81; Osswald et al. 77]) of liquid membranes were carried out. The increase in knowledge gained from membrane technological studies has allowed the optimization of liquid membranes with respect to important properties such as selectivity, membrane resistance, lifetime and EMF stability in biological fluids. Indeed, the reliability of potentiometric ion activity measurements with neutral carrier membrane electrodes has been demonstrated for routine clinical and physiological work [Meier et al. 80; Meier et al. 82]. In more specialized applications, e.g. invasive approaches, other problems such as electrode sterilization, toxicity of membrane components, contamination of the sample, and risk of coagulation arise (Tab. 4.1). These are to some extent still unsolved. The current knowledge of the properties of sensor materials with respect to these problems is very limited. Consequently, for clinical applications, discrete ion determinations or continuous measurements using a drain from the bypass (ex vivo) should be performed whenever possible [Treasure et al. 85]. Although medical applications of biosensors are now being pushed with vehemence, problems related to in vivo measurements should be reliably solved first [Lowe 84].

Table 4.1. Risks of invasive applications of electrodes

Problem	Risk	Precaution, alternatives
Sterilization	– infection – immune response	– bench measurements – drain from bypass (ex vivo) – adequate sterilization
Contamination, toxicity	– poisoning – immune response – mutagenicity	– non-toxic, lipophilic membrane components – small exposed membrane area – large sample volume
Coagulation	– thrombus formation, infarcts	– use of anticoagulants (heparin) – small size of catheter – geometry of measuring device (e.g. no edges)

4.1 Parameters Determining Basic Properties of Solvent Polymeric Membranes

4.1.1 Membrane Composition

Because of the presence of a polymeric membrane matrix in solvent polymeric membranes, the choice of membrane compositions is restricted within certain limits. The quantity of matrix material in a membrane is not optional and the choice of membrane solvents is often determined by the matrix. Although various matrices have been studied, only a few are of practical importance. Especially useful are poly(vinyl chloride) and silicone rubber since they yield membranes of good mechanical stability and excellent electromotive properties. Furthermore, they usually guarantee a long-lasting chemical stability of the membrane components embedded in them (Fig. 4.2).

Often the means of altering the membrane composition are rather limited. Nevertheless, in using a selected set of membrane components it is still possible to influence such properties as selectivity, membrane resistance and lifetime of the resulting electrode by varying the relative amounts.

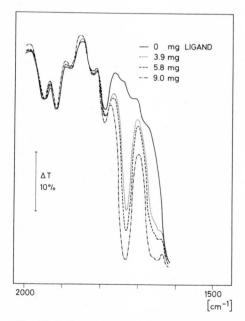 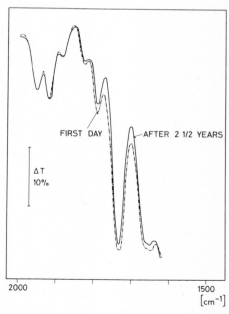

Fig. 4.2. Infrared spectra of PVC membranes containing a synthetic neutral carrier (N,N′-bis-[10-(ethoxycarbonyl)-decyl)]-N,N′-dimethyl-3,6-dioxaoctane diamide (constitution see Fig. 4.6) [Ammann et al. 72a]) and o-nitrophenyl-n-octyl ether as the membrane solvent. The ester carbonyl absorption band of the carrier molecule at 1730 cm^{-1} was used to monitor the carrier content. The spectra on the left demonstrate the dependence on the carrier concentration in the membrane, while the right-hand side compares a fresh membrane with the same membrane after 2.5 years of use and/or storage in an electrode body [Ammann 75]

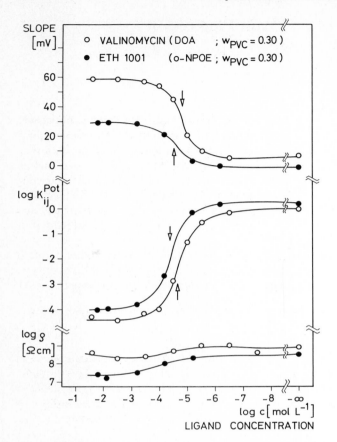

Fig. 4.3. Dependence of the electrode response (slope), selectivity factor (K_{ij}^{Pot}) and specific membrane resistance (ρ) on the carrier concentration for K^+- and Ca^{2+}-selective neutral carrier-based membranes [Oesch and Simon 80]. Slopes are given for measurements in KCl and $CaCl_2$ solutions. Selectivity factors correspond to $i = K^+$, $j = Li^+$ and $i = Ca^{2+}$, $j = Mg^{2+}$

Figure 4.3 shows an example of the influence of the neutral carrier concentration on the electromotive behaviour of the corresponding electrodes [Oesch and Simon 80]. Typical neutral carrier concentrations in solvent polymeric membranes are in the 10^{-2} to 10^{-1} M range. The results show that the observed limiting carrier concentrations of valinomycin and the Ca^{2+}-carrier ETH 1001 are about 1% of typical initial concentrations. At carrier concentrations of about 10^{-4} M a clear loss of electrode characteristics is observed (Fig. 4.3). The upper limit of the carrier concentration is usually determined by the restricted solubility of the carrier molecules in the membrane phase. In some cases, high carrier concentrations can cause inconvenient changes in the electrode response, for example, enhanced anion interference [Ammann 75].

Membrane compositions that yield optimal ion selectivities are usually not satisfactory in all respects for applications in clinical studies (solvent polymeric membranes) or physiological work (membrane solutions). Hence, membranes which

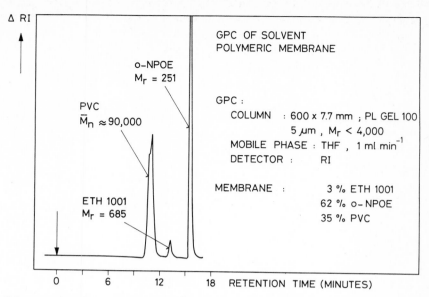

Fig. 4.4. Gel permeation chromatogram of a Ca^{2+}-selective neutral carrier solvent polymeric membrane [Oesch and Simon 85]

have been optimized with respect to a single property (e.g. selectivity) are only seldom the membranes of choice. Instead, useful membranes are the result of a well-balanced optimization of all the relevant requirements for an application in biological samples. The relevant properties which reflect the reliability and quality of sensors are selectivity, stability, response time and lifetime [Ammann et al. 83; Oesch et al. 85 c].

The performance of carrier membranes should not vary over time periods ranging from days (microelectrodes) to several months (clinical analyzers). This implies that the membrane composition should not be changed by a loss of membrane components into the sample solution or by an uptake of species from the sample into the membrane phase. A convenient analysis of the membrane composition can be made using high performance gel permeation chromatography. Membrane material dissolved in tetrahydrofuran is injected and separated into the various membrane components according to their molar mass (Fig. 4.4). This method allows the detection of a loss of membrane components or a contamination by sample species at a level of 1% of the total membrane weight [Oesch and Simon 85].

4.1.2 Membrane Solvent

The percentage of membrane solvent in poly(vinyl chloride) membranes and membrane solutions for microelectrodes is very high (60–90 wt.-%). Hence, the membrane solvent has to exhibit excellent plasticizer properties for use in solvent polymeric membranes. The high concentration of the membrane solvent substantially affects a number of parameters which influence the membrane electrode properties.

Thus, the choice of an appropriate membrane solvent is governed by several criteria:

- chemical stability
- chemical inertness with respect to the formation of carrier/ion complexes, i.e. no competitive coordinating sites
- chemical inertness with respect to hydrogen ion carrier properties, i.e. it should not contain any functional groups that can be protonated or deprotonated
- low vapour pressure
- adequate viscosity
- adequate dielectric constant
- solubilization properties for neutral carriers and membrane additives
- possibly: high lipophilicity
- possibly: plasticizer properties for a membrane matrix
- possibly: low toxicity
- possibly: biocompatibility.

The difficulties involved in satisfying each of these requirements are not the same.

Some typical aspects are discussed in the following:

The vapour pressure can be maintained sufficiently low by choosing lipophilic compounds of high molar mass.

Solvents of extremely high lipophilicity are not easy to develop since they often are too viscous and/or exhibit no plasticizer properties.

Long-term chemical stability is not a problem. Similarly the functional groups of membrane solvents can be chosen so that they do not disturb the complex formation between a carrier and an ion.

By avoiding functional groups that might be charged under measuring conditions (e.g. amines, carboxylic groups) the membrane solvents can be chosen so that H^+ or OH^- ions cause no interference and they do not act as disturbing ion-exchanger sites.

Little is known about the toxicity of membrane solvents. Remarkably high concentrations of plasticizers (which are also used in membrane electrodes or are structurally related to membrane solvents) have been found in the blood of patients undergoing hemodialysis treatment [Lewis et al. 78] or cardiac bypass surgery. For example, during a five hour dialysis session patients can receive up to 150 mg of the PVC-plasticizer bis (2-ethyl-hexyl) phthalate [Biggs and Robson 82]. This drastic loss of plasticizer from hemodialysis sets is mainly due to the insufficient lipophilicity of the currently used compounds and the strong extraction properties of protein-containing biological fluids [Oesch et al 85a]. Figure 4.5 shows that the solubility of membrane solvents in 7% bovine serum albumin solution or blood serum can be several orders of magnitude higher than in water. Improved plasticizers for medical devices [Biggs and Robson 82] as well as for membrane electrodes [Simon et al. 84a; Oesch et al. 85a] have been developed. The metabolism of phthalates [Draviam et al. 82], the mutagenicity of nitroaromatic compounds [Chiu et al. 78] and the carcinogenic effects of phthalates (see [Biggs and Robson 82]) have also been studied.

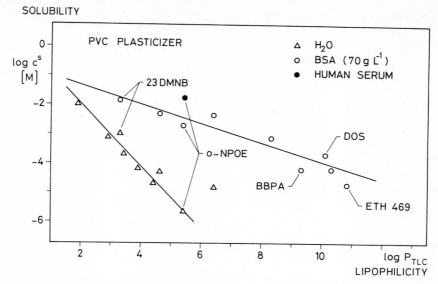

Fig. 4.5. Solubility of plasticizers of different lipophilicity (log P_{TLC}, see 3.2.4.3) in water, bovine serum albumin and human serum [Oesch et al. 85b]. 23 DMNB: 2,3-dimethyl-nitrobenzene, *o*-NPOE: *o*-nitrophenyl-n-octyl ether; BBPA: bis(1-butyl-pentyl) adipate; DOS: bis(2-ethyl-hexyl) sebacate; ETH 469: 1,10-bis(4′-(5″-nonyloxy-carbonyl)-butyryloxy)-decane [Oesch et al. 85a]

The carrier membrane selectivity depends a great deal on the dielectric constant of the membrane phase. The relative high content of the membrane solvent dictates to a large extent the dielectric constant of the membrane. Both membrane solvents of low dielectric constants ε (adipates, sebacates, phthalates ($\varepsilon \sim 4$)) and relatively high dielectric constants (nitroaromatics ($\varepsilon \sim 24$), carbonates ($\varepsilon \sim 65$)) are available. The drastic influence of the dielectric constant on the membrane selectivity stems from the contribution of the dielectric medium to the free energy of transfer. The relevant term is described by the Born equation [Morf 81]:

$$\Delta G_B = -\frac{(ze_o)^2}{2\,(r_{ion}+s)}(1-1/\varepsilon),\tag{4.1}$$

where ΔG_B describes the free energy of transfer of a neutral carrier complex (s: average thickness of the ligand layer about the ion) from the gas phase into a dielectric medium (membrane). Since the value of ΔG_B is proportional to the square of the charge of the ion to be complexed, an increased preference of the membrane for divalent over monovalent cations is expected with an increase in the polarity of the membrane solvent. Such a trend has been observed in potentiometric studies with neutral carrier-based liquid membranes (Fig. 4.6). Accordingly, the monovalent/divalent-selectivity of a carrier membrane can be changed by several orders of magnitude (Fig. 4.6, [Ammann et al. 75a]). Equation (4.1) contains another molecular parameter, namely the thickness s of the ligand layer; its influence depends on the ion-charges and the dielectric constant (for an illustration see Fig. 3.9).

PREFERENCE OF Na⁺ OVER Ca²⁺

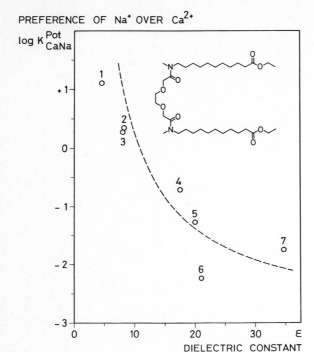

Fig. 4.6. Dependence of the monovalent/divalent (Na^+/Ca^{2+}) selectivity on the dielectric constant of the membrane solvent [Ammann et al. 72b]. 1: dibutyl sebacate; 2: tris(2-ethyl-hexyl) phosphate; 3: 1-decanol; 4: acetophenone; 5: 2-nitro-*p*-cymene; 6: *p*-nitro-ethylbenzene; 7: nitrobenzene. The selectivity factors were obtained from the EMF values measured in 0.1 M metal chloride solutions. The dashed curve was calculated on the basis of an equation derived from the Born term [Morf 81]

4.1.3 Membrane Matrix

Initially, liquid membranes were manufactured by soaking a porous material such as a glass frit (Štefanac and Simon 67] or a filter paper [Štefanac and Simon 67; Ammann et al. 72a] with an ion-selective solution. Drawbacks inherent to this approach (lifetime, EMF stability) were overcome by using polymeric membrane matrices plasticized with a membrane solvent [Bloch et al. 67].

The usefulness of a membrane matrix can be judged with respect to the following requirements:
- mechanical stability
- chemical stability
- chemical inertness, i. e. no competition with the complex formation of the neutral carrier (no potential coordinating sites, no charged sites)
- clean surface of the resulting membrane (no pores)
- low electrical resistance of the resulting membranes
- biocompatibility

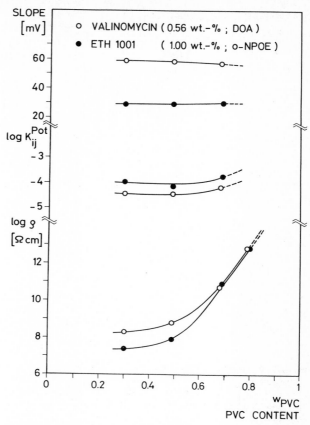

Fig. 4.7. Dependence of electrode response (slope), selectivity factor (K_{ij}^{Pot}) and specific membrane resistance (ρ) of neutral carrier-based membranes on their PVC content [Oesch and Simon 80]. Slopes are indicated for KCl and CaCl$_2$ solutions. Selectivity factors are for $i = K^+$, $j = Li^+$ and $i = Ca^{2+}$, $j = Mg^{2+}$

Among the different matrices used for neutral carrier-based membranes (such as silicone rubber [Pick et al. 73; Jenny et al. 80b; Anker et al. 83c], copolymers of poly(bisphenol-A carbonate) and poly(dimethyl)siloxane [Le Blanc and Grubb 76], poly(methyl methacrylate) [Fiedler and Růžička 73], polyurethane [Fiedler and Růžička 73], polystyrene [Fiedler and Růžička 73], and poly(vinylisobutyl ether) [Schäfer 76]), poly(vinyl chloride) [Bloch et al. 67; Craggs et al. 70; Moody et al. 70; Ammann et al. 83] is at present the most popular one. An early technique for membrane preparation is still in general use [Craggs et al. 70]. In brief, the membrane components (for this example: total weight 180 mg) are dissolved in about 2 ml of freshly distilled tetrahydrofuran. The solution is poured into a glass ring (23 mm i. d., height 1–2 cm) which rests on a glass plate. After covering with a paper tissue and letting the tetrahydrofuran evaporate overnight, the resulting solvent polymeric membrane, about 0.2 mm thick, is peeled away from the glass and appropriate discs for use in electrode bodies are stamped.

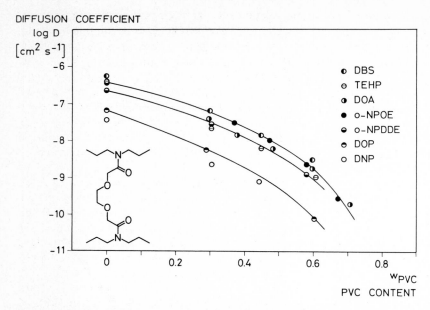

Fig. 4.8. Diffusion coefficient of a neutral dioxa-diamide carrier in PVC membranes using different plasticizers and PVC contents (for details see [Oesch and Simon 80])

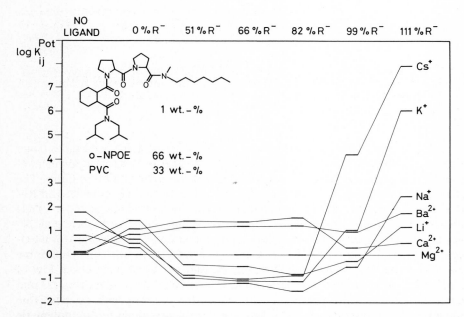

Fig. 4.9. Potentiometric selectivity factors of neutral carrier-based solvent polymeric membrane electrodes as a function of the concentration of potassium tetrakis(p-chloro-phenyl) borate (R^-) in the membrane phase (in mol-% relative to the carrier concentration)

Typically, solvent polymeric membranes contain about 30 wt.-% PVC ($w_{PVC} = 0.3$; w_{PVC}: weight of PVC relative to total weight of membrane) [Oesch and Simon 80; Ammann et al. 83]. In a few PVC-membranes a higher PVC-content can be advantageous (w_{PVC} about 0.5 [Wegmann et al. 85]). High w_{PVC}-values can result from the loss of plasticizer into the sample solution. There is a substantial increase in the specific membrane resistance with increasing poly(vinyl chloride) content (Fig. 4.7). In contrast, the selectivities and the electrode functions illustrated in Fig. 4.7 remain almost unchanged in the w_{PVC}-range considered. Thus, the limitations due to a high content of the matrix ($w_{PVC} > 0.7$) are not the result of a loss in carrier properties but are the result of unacceptably high electrical membrane resistances [Oesch and Simon 80].

An increase in w_{PVC} strongly influences the diffusion coefficients of the neutral carriers and other components in the membrane phase (Fig. 4.8). The diffusion coefficient is almost independent of the constitution of the diffusing species because conventional membrane components are of similar size. The relative shifts in the curves of Fig. 4.8 are due to the different viscosities of the plasticizers (DBS: $9 \cdot 10^2$ g cm^{-1}s^{-1}; DNP: $139 \cdot 10^2$ g cm^{-1} s^{-1} [Oesch and Simon 80]). For example, the dependence of the diffusion coefficients of carriers and plasticizers on the PVC content may become apparent by the rate of the loss of these components from the membrane phase (see discussion of the lifetime of membrane electrodes in 5.10).

4.1.4 Membrane Additives

The performance of most cation-selective neutral carrier membranes can be improved by membrane additives. The most commonly used membrane additives can be classified into:

- lipophilic anionic sites [Morf 81] (e. g. alkali tetraarylborates)
- highly lipophilic salts [Ammann et al. 85 a; Horvai et al. 85] (e. g. tetraalkylammonium tetraarylborates)
- polymers [Tsien and Rink 81; Lanter et al. 82] (e. g. poly(vinyl chloride) for membrane solutions of microelectrodes)
- dyes [Frömter et al. 81] (as markers in membrane solutions of microelectrodes).

Additives consisting of alkali salts with lipophilic anions have become of great importance for neutral carrier-based cation-selective membrane electrodes. The incorporation of such mobile cation-exchange sites into the membrane phase has proved to be beneficial in many respects. The additives reduce or eliminate the interference by lipophilic sample anions [Morf et al. 74a; Morf et al. 74b], give rise to significant selectivity changes (increased divalent/monovalent selectvity [Simon et al. 78a]), reduce the electrode response time [Lindner et al. 78], are capable of boosting cation sensitivity in the case of carriers with poor extraction properties [Erne et al. 80b] and considerably lower the electrical membrane resistance [Steiner et al. 79]. Sodium tetraphenylborate and its more lipophilic analogue potassium tetrakis(p-chloro-phenyl)borate are the most commonly used salts.

Figure 4.9 shows convincingly that the membrane additive KTpClPB exerts a drastic influence on the cation selectivity of the corresponding membrane electrodes (for other examples see [Ammann et al. 83; Metzger et al. 84]). In membranes without lipophilic borate sites (0%, Fig. 4.9) the neutral tetraamide carrier does not induce any selectivity compared with the ligand-free membrane (o-NPOE, Fig. 4.9). However, by adding KTpClPB up to about 100 mol-% (relative to the carrier concentration), the carrier membrane exhibits a pronounced preference for divalent over monovalent cations. Membranes containing more than 100 mol-% KTpClPB exhibit selectivity factors which roughly correspond to those of classical cation-exchanger membranes [Scholer and Simon 72]. This can be explained by the presence of an excess of TpClPB$^-$ cation-exchanger sites in the carrier membrane.

4.2 Ion-Selective Membrane Solutions for Microelectrodes

4.2.1 Composition of Membrane Solutions

The term membrane solution denotes an ion-selective organic membrane phase used in neutral carrier microelectrodes. The name was introduced to distinguish it from solvent polymeric membranes and to emphasize the absence of a membrane matrix. Many other terms have been used in the literature (ion-exchanger, ion-selective liquid, cocktail, etc.).

A major problem encountered in the development of neutral carrier-based membrane solutions arises from the necessity of keeping the electrical membrane resistance of the microelectrodes sufficiently low. The difficulties are compounded by the fact that the provisions for lowering the resistivity should not affect the ion selectivity of the membrane solution. Thus far, these problems have not yet been completely overcome (see Sect. 4.2.3). The following parameters must be taken into account in an effort to optimize the membrane composition:

- the membrane solvent
- the neutral carrier concentration
- the content and the kind of lipophilic salts.

The electrical membrane resistance $R_m[\Omega]$ of a microelectrode is related to the specific resistance (resistivity ρ [Ω cm]) of a membrane solution by equation (4.2) [Purves 81]:

$$R_m \approx \frac{2\,\rho}{\pi\,r\,\Theta},$$ (4.2)

where
r : radius of the tip at its narrowest point [cm];
Θ: angle of the taper of the microelectrode [rad].

According to Eq. (4.2) the specific resistance should not exceed 10^7 Ω cm, since for $r = 1$ μm and an assumed taper angle $\Theta \approx 8°$ [Purves 81] a membrane resistance of 10^{12} Ω is expected. Representative solvents of membrane solutions exhibit resistivi-

Table 4.2. Physico-chemical properties of currently used membrane solvents for neutral carrier microelectrodes (values from [Oesch 79; Oesch and Simon 80])

Membrane solvent	Neutral carrier microelectrode	log c(H$_2$O)^a [mol l^{-1}]	log P$_{Hansch}$	log P$_{TLC}$	log K^b	Viscosity [g cm^{-1}s^{-1}]	Dielectric constant
2,3-dimethyl-nitrobenzene	K$^+$	−3.0	2.9	3.3	–	358	–
o-nitrophenyl-n-octyl ether	H$^+$, Na$^+$, Ca^{2+}	−5.7	5.4	5.9	6.3	1380	24.2
tris(2-ethyl-hexyl) phosphate	Li$^+$	< −2.5	7.9	10.3	–	1250	10
dibutyl sebacate	K$^+$	−3.9	6.4	6.4	4.4	900	4.54
dioctyl phthalate	K$^+$	–	7.5	8.3	–	7400	–
propylene carbonate	Mg^{2+}	miscible	–	< −1.0	–	255	64.9

^{a} solubility in water ^{b} partition coefficient (water/membrane phase)

ties ρ in the range of 10^6 to 10^{10} Ωcm (dimethyl phthalate: $1.1 \cdot 10^{10}$ Ωcm; nitrobenzene: $2.5 \cdot 10^7 - 10^{10}$ Ωcm; propylene carbonate: $5 \cdot 10^5 - 10^6 \Omega$cm [Janz and Tomkins 72]). The values indicate that solvents with high dielectric constants are to be preferred. Indeed, they are even used in membranes with a selectivity for monovalent cations [Steiner et al. 79]; i.e. in order to attain sufficiently low membrane resistances a reduction in the monovalent/divalent selectivity is accepted.

The resistivities of the pure membrane solvents are lowered considerably by the addition of a relatively large amount of neutral carrier (5 wt.-% to 20 wt.-%) and by dissolving an alkali tetraarylborate in the membrane phase (about 1 wt.-%). In spite of these provisions, it is still highly desirable to develop neutral carrier membrane solutions of reduced specific resistance (see Sect. 4.2.3).

4.2.2 Solvents for Membrane Solutions

As already indicated in Sect. 4.1.2, a large number of demands are imposed on membrane solvents. Especially important for solvents in membrane solutions are the following properties:

- solubilization properties for high carrier concentrations
- solubilization properties for lipophilic salts
- moderate viscosity (microelectrode filling)
- adequate lipophilicity

So far, the polar solvent o-nitrophenyl-n-octyl ether has proved to be adequate for membrane solutions, and only in a few cases have other membrane solvents been used [Oehme and Simon 76; Lanter et al. 80]. Table 4.2 compares the physico-chemical properties of membrane solvents that have found application in neutral carrier microelectrodes.

4.2.3 Additives for Membrane Solutions

Most of the neutral carrier-based ion-selective membrane solutions for microelectrodes contain a lipophilic salt [Amman et al. 83]. The beneficial effects of these additives have been discussed above.

Unfortunately, the amount of alkali arylborates that can be added is strictly limited since high concentrations lead to drastic changes in the membrane selectivity
[Ammann et al. 83; Meier et al. 84]. Depending on the complex stoichiometry an increase above a given molar ratio of lipophilic salt to carrier yields a membrane with
the properties of a classical cation-exchanger [Ammann et al. 83; Meier et al. 84]. It has
been shown that this drawback needs not be taken into account if both the cations
and anions incorporated into the membrane are lipophilic [Pretsch et al. 85; Horvai
et al. 85]. Consequently, it can be expected that increasing amounts of these salts
will decrease the membrane resistances without changing the membrane selectivity.

The reduction of membrane resistances is very important to microelectrodes to
be used in physiological studies and would help to overcome the problems of
shielding and response times limited by electronic instrumentation. Several other
advantages would be gained from a general procedure for lowering the resistances
of neutral carrier-based microelectrodes. For example, the tip diameters could be
reduced and shunts that disturb sodium [Lewis and Wills 80] and calcium ion measurements [Tsien and Rink 81] could be eliminated. It would also be possible to use
non-polar membrane solvents for microelectrodes with a selectivity for monovalent
cations. At present, polar membrane solvents must be used to decrease the membrane resistance, although they also reduce the monovalent/divalent cation selec-

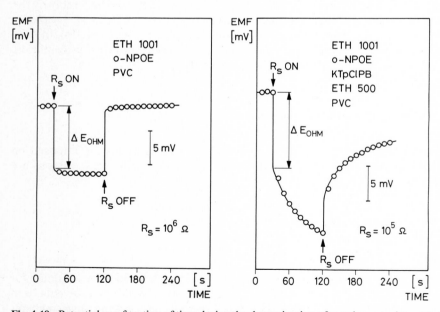

Fig. 4.10. Potential as a function of time during the determination of membrane resistances according to the voltage divider method (*left:* membrane 5, Table 4.3; *right:* membrane 12, Table 4.3) [Am-
mann et al. 85a]

Table 4.3. Composition of membranes (weight – %) and their properties in macroelectrodes (mol-% relative to ETH 1001 are given in parentheses)

Membrane components	Membrane																
	1	2	3	4	5	6	7	8	9	10	11	12	13	14	15	16	17
ETH 1001	0.55	0.55	0.55	0.54	1.0	1.0	1.0	1.0	1.0	1.0	0.9	0.92	0.94	0.94	0.93	0.95	0.94
o-NPOE	66.3	66.1	65.6	64.8	64.9	63.8	65.5	65.4	64.3	64.3	62.4	59.8	61.7	62.5	62.9	63.3	62.7
PVC	33.15	33.1	32.8	32.4	34.1	34.8	32.5	31.6	32.2	32.1	33.0	31.5	32.8	33.3	33.5	33.8	32.1
KTpClPB	–	0.25 (67)	0.25 (67)	0.26 (68)	–	0.4	–	–	–	0.1 (10)	0.45 (67)	0.44 (67)	0.46 (67)	0.46 (67)	0.47 (67)	0.45 (67)	0.46 (67)
ETH 500	–	–	0.8 (86)	2.0 (215)	–	–	1.0 (67)	1.9 (120)	2.4 (150)	2.4 (150)	3.25 (202)	7.34 (494)	–	–	–	–	–
ETH 2035	–	–	–	–	–	–	–	–	–	–	–	–	4.1 (199)	–	–	–	–
ETH 2039	–	–	–	–	–	–	–	–	–	–	–	–	–	2.8 (203)	–	–	–
ETH 2037	–	–	–	–	–	–	–	–	–	–	–	–	–	–	2.2 (203)	–	–
TBATPB	–	–	–	–	–	–	–	–	–	–	–	–	–	–	–	1.5 (200)	3.8 (500)
R [kΩ]	2000	132	50	32	700	63	82	50	50	30	35	24	28	28	26	31	24
Detection limit [log a units]	–7.1	–8.6	–8.9	–9.0	–7.6	–8.9					–8.8	–9.2	–8.5	–8.9	–7.9		
Slope (10^{-7}–10^{-3} M) [mV/decade]	24.0	28.3	27.9	27.9	26.4	28.5					27.6	28.5	26.9	28.3	25.8	~16	~15

tivity [Wuhrmann et al. 79; Steiner et al. 79]. Furthermore, lower membrane resistances would encourage the more widespread use of highly selective carrier microelectrodes instead of the ion-exchanger microelectrodes which exhibit modest selectivities but have low resistances (e.g., valinomycin-based membranes instead of the Corning 477317 ion-exchanger [Oehme and Simon 76]). Some neutral carriers have proven to be valuable to macroelectrode systems. However, they are not appropriate as membrane components in microelectrodes because the membrane resistances are extremely high [Ammann 81]. Finally, lower membrane resistances should permit a reduction in the amount of the neutral carrier needed in the membrane solution. In almost all of the solutions for membrane preparation, the carrier content is at least 10 wt.-% [Ammann et al. 81a]. Due to the severe demands on and the great interest in Ca^{2+}-selective microelectrodes, attempts have been made to reduce the electrical resistance of microelectrodes based on the Ca^{2+}-carrier ETH 1001 [Ammann et al. 85a]. These attempts concentrated on the influence of salts containing lipophilic cations and anions on the resistivity of membranes. The following salts were used: tetradodecylammonium tetrakis (p-chloro-phenyl) borate (ETH 500), tetraoctadecylammonium tetrakis (p-chloro-phenyl) borate (ETH 2035), methyltridodecylammonium tetrakis (p-chloro-phenyl) borate (ETH 2039) and tetraphenylphosphonium tetrakis (p-chloro-phenyl) borate (ETH 2037). The membrane resistances were evaluated using the voltage divider method [Purves 81]. The method relies on the measurement of the voltage attenuation due to a resistor parallel (shunt resistance R_S) to the electrode. The polarization of the membrane during the resistance measurement drastically influences the response (Fig. 4.10); therefore only the ohmic potential drop (ΔE_{ohm} in Fig. 4.10) was considered in the evaluation of the membrane resistances.

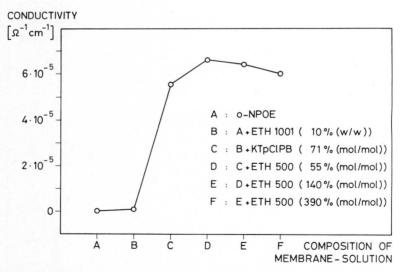

Fig. 4.11. Influence of lipophilic salts on the conductivity of membrane solutions for microelectrodes [Ammann et al. 85a]. Membrane solution B contains 10 wt.-% of the Ca^{2+}-selective neutral carrier ETH 1001. The additions of salts in the membrane solutions C-F are given in mol-% relative to the carrier concentration

Measured resistances of solvent polymeric membranes with different compositions clearly indicate that the possible reduction of the resistance by the addition of lipophilic salts into the membrane is limited (e.g. membranes 7–12, Table 4.3). The same conclusion can be drawn from conductometric measurements carried out on various membrane solutions (Fig. 4.11). This may be explained by the limited dissociation of these salts in o-NPOE. A further decrease in the resistance could not be achieved by using different lipophilic salts (membranes 13–16 compared to membrane 11, Table 4.3). An almost complete loss of the Ca^{2+} electrode performance occurs when the cation of the salt is rather hydrophilic. For example, membranes with a content of more than 100 mol-% KTpClPB or tetrabutylammonium tetraphenylborate (TBATPB) exhibit either cation-exchanger properties or deteriorated response (membranes 16, 17 compared to membrane 2). Furthermore, the membrane electrodes containing TBATPB were observed to drastically change the reference potential (E_o) and the slope as function of time. The measured potential was also heavily dependent on the stirring rate. This may indicate that the relatively poorly lipophilic TBA cations are replaced in the membrane by the more lipophilic Ca^{2+}-complexes of the neutral carrier.

The carrier concentration of the membrane has a significant effect on the membrane resistance only in the absence of lipophilic salts (membrane 1 compared to membrane 5). However, in the presence of lipophilic salts (e.g. ETH 500) the mem-

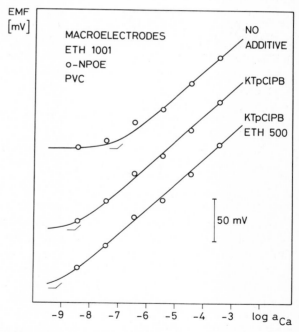

Fig. 4.12. Electrode functions of solvent polymeric membrane macroelectrodes in Ca^{2+}-buffered solutions with a constant background of 125 mM K^+. The detection limits depend on the addition of salts to the membrane phase and are: -7.1 log a_{Ca} (no additive, membrane 1 in Table 4.3), -8.6 log a_{Ca} (KTpClPB, membrane 2 in Table 4.3) and -9.2 log a_{Ca} (KTpClPB and ETH 500, membrane 12 in Table 4.3) [Ammann et al. 85a]

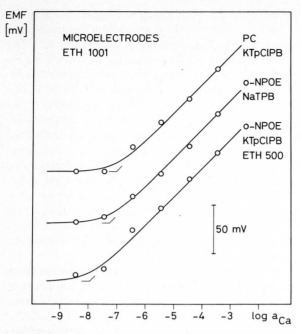

Fig. 4.13. Electrode functions of microelectrodes in Ca^{2+}-buffered solutions with a constant background of 125 mM K^+. The detection limits depend on the composition of the membrane solutions and are near -7.0 log a_{Ca} (propylenecarbonate/KTpClPB, Table 4.4), -7.2 log a_{Ca} (*o*-NPOE/NaTPB, Table 4.4) and -8.0 log a_{Ca} (*o*-NPOE/KTpClPB/ETH 500, Table 4.4) [Ammann et al. 85a]

brane resistance is almost independent of the carrier concentration (membrane 4 compared to membrane 11).

With respect to electrical resistance membrane 2 was so far the most attractive neutral carrier-based Ca^{2+} membrane. A further reduction of the membrane resistance (by a factor of 2–3) can only be achieved by using highly lipophilic salts as additives (membrane 4 compared to membrane 2 or membrane 12 compared to membrane 6). The lowest resistances finally obtained are only slightly higher than those of typical macro reference electrodes. A further advantage becomes obvious by considering electrode response functions (Fig. 4.12, Table 4.3) as the detection limit is shifted towards lower Ca^{2+} activities.

The same trend was observed for microelectrodes (Fig. 4.13, Table 4.4), although they exhibit higher detection limits than the corresponding macroelectrodes. Unfortunately, the membrane resistance of the membrane solution used so far [Oehme et al. 76] could not be decreased by adding large amounts of the lipophilic salt ETH 500 (Table 4.4). This additive only has a beneficial effect on the resistivity if the polarity of the membrane solvent is drastically increased at the same time (propylene carbonate: dielectric constant = 64.9). This, however, results in other electrode characteristics which are worse than in the case of membrane electrodes containing *o*-NPOE (Table 4.4).

Table 4.4. Properties of microelectrodes based on different membrane solutions

Composition of membrane solution [wt.-%]	Membrane resistance [GΩ]	Selectivity factors log K_{CaK}^{Pot}	Detection limit [log a_{Ca} units]
10 ETH 1001 1 NaTBP 89 o-NPOE [Oehme et al. 76]	32.3 ± 5.8 (n = 10)	− 5.5	− (7.6 ± 0.8) (n = 5)
7.4 ETH 1001 1.1 KTpClPB 24.8 ETH 500 66.7 o-NPOE	32.6 ± 5.9 (n = 5)	− 5.7	− (7.8 ± 0.3) (n = 3)
9 ETH 1001 1 KTpClPB 90 PC	5.8 ± 3.9 (n = 9)	− 4.7	− (6.8 ± 0.4) (n = 7)

The deviations of the measured resistances of individual microelectrodes containing the same membrane solution (Table 4.4) are most probably due to the poor reproducibility of the microelectrode geometry. In order to estimate the uncertainty in the microelectrode preparation technique, a series of reference microelectrodes were prepared and their membrane resistances were measured (3.5 M KCl reference electrolyte, saturated with AgCl). The resistances were in the area of 10 MΩ and the day-to-day reproducibility of the mean was about 15%. Scanning electron micrographs of several micropipettes proved that the tip diameters of the microelectrodes used in this study were 0.3–0.5 μm (Fig. 4.14).

In conclusion, the addition of large amounts of highly lipophilic salts to the membrane phase somewhat lowers the electrical membrane resistance of Ca^{2+} macroelectrodes but has no effect in the case of Ca^{2+} microelectrodes. The detection limits of both macro- and microelectrodes are slightly improved. It seems that the addition of these salts at least makes a drastic reduction in the content of the neutral carrier in membrane solutions for microelectrodes possible.

To date, the electrical membrane resistances of Ca^{2+} microelectrodes are unavoidably high and are responsible for the interfering electrical shunt pathways through the glass wall at the tip of very fine micropipettes (Sects. 7.7 and 9.6.3.3). This interference can be avoided by using relatively small amounts of a polymer (poly(vinyl chloride)) as a membrane additive [Tsien and Rink 81].

Finally, the use of dyes as membrane additives, should be mentioned. By ejecting small amounts of the coloured solution from the micropipette into a sample the position of the tip can be accurately determined [Frömter et al. 81] (see Sect. 7.4).

4.3 Concluding Remarks

Basically, three components make up a neutral carrier liquid membrane: the ion-selective neutral carrier, the membrane solvent, and the membrane matrix. Solvent polymeric membranes usually contain each component, whereas membrane solutions usually do not contain a membrane matrix.

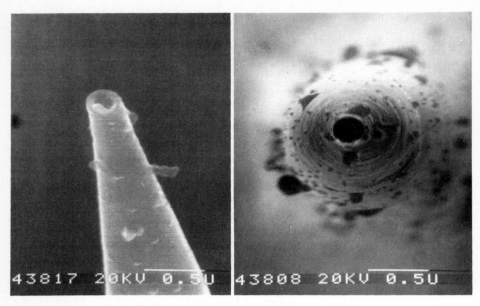

Fig. 4.14. Scanning electron micrographs of a typical single-barrelled microelectrode used for the evaluation of the reduction of membrane resistances. The micropipette is coated with a ca. 50 Å Pt/C layer. The estimated tip diameter is 0.28 µm (scale: 0.5 µm, magnification: ca. 50000) [Ammann et al. 85a]. The electron micrographs were obtained using a field emission scanning microscope (Hitachi S-700) by Dr. E. Wehrli, Department of Cell Biology, Swiss Federal Institute of Technology, Zurich, Switzerland

Current membrane technology, i. e. the use of fundamental knowledge of membrane properties for the development of membranes for particular applications, allows reliable potentiometric measurements to be made with neutral carrier-based electrodes in clinical chemistry and electrophysiology.

However, in considering invasive measurements in human beings or animals the knowledge of membrane properties such as toxicity, biocompatibility and the suitability for sterilization is still insufficient.

Membrane solvents substantially influence membrane properties such as selectivity, EMF stability and lifetime. The major demands on membrane solvents for solvent polymeric membranes and for membrane solutions are quite different. Optimal solvents for PVC membranes primarily have to exhibit excellent plasticizer properties and extremely high lipophilicities. On the other hand, the main requirements for solvents for membrane solutions are their ability to dissolve high concentrations of carriers and to favourably influence the electrical membrane resistances.

Most of the neutral carrier membranes contain a lipophilic salt as a membrane additive. Such salts can significantly improve their ion-selective behaviour. In membrane solutions, however, their central role is their beneficial influence on the membrane resistance.

The concentration of these commonly used salts (sodium tetraphenylborate, potassium tetrakis (*p*-chloro-phenyl) borate) in carrier membranes is strictly limited. Their advantageous effects can only be observed if the molar ratio between salt and neutral carrier lies within well-defined limits.

Additives consisting of highly lipophilic cations and anions have been investigated that do not affect the carrier properties, even when used in high concentrations. However, an optional reduction of the membrane resistance is not achieved by the addition of these salts.

5 Potentiometric Measurements of Ion Activities with Neutral Carrier-Based Electrodes

5.1 Cell Assemblies

Ion-selective electrodes permit the potentiometric measurement of the activity of a given ion in the presence of other ions. Two general arrangements of potentiometric liquid membrane electrode cells are shown in Fig. 5.1. The ion-selective electrode consists of a galvanic half-cell containing the liquid membrane, the internal filling solution and the internal reference electrode. The other half-cell is represented by an external reference electrode in contact with a reference electrolyte. Such a cell assembly is described in (Eq. 5.1):

Ag; AgCl, reference electrolyte| } reference
(salt bridge)| } half-cell
sample‖

membrane‖   ) membrane   (5.1)
internal filling solution, AgCl; Ag   } electrode
) half-cell

In conventional macroelectrode cell assemblies (Fig. 5.1, upper) the contact between the reference electrode and the sample solution is usually maintained by a salt bridge solution. The silver/silver chloride half-cell in macro- and microelectrodes is often replaced by other reference elements such as calomel half-cells (Hg; Hg_2Cl_2, KCl (satd.)). In contrast, microelectrode cell assemblies always include Ag/AgCl reference half-cells without salt bridges (Fig. 5.1, lower). These half-cells are used because the miniaturized microelectrode bodies preclude the incorporation of salt bridges. A reference electrolyte in direct contact with a Ag/AgCl element needs a certain Cl^- concentration. This may be disadvantageous because of cell contamination (see Sect. 7.6).

A cell potential (electromotive force (EMF), potential difference at zero current) is established when the membrane electrode and the external reference electrode are both in contact with the sample solution. The EMF is the sum of a series of local potential differences generated at the solid-solid, solid-liquid and liquid-liquid interfaces of the cell (see scheme in (5.1)). Ideally, only the potential difference between the liquid membrane and the sample solution should depend on the ion activities in the sample. All other contributions to the total measured EMF should be constant.

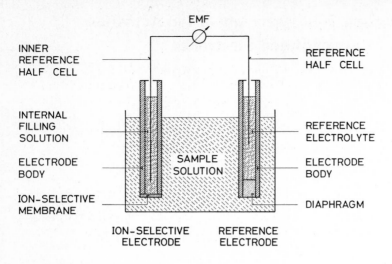

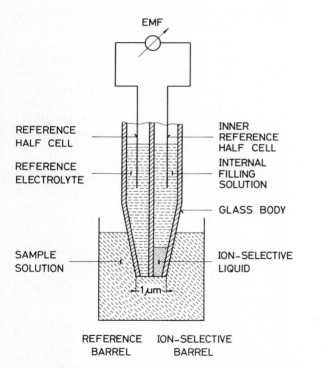

Fig. 5.1. Schematic cross-section of a macroelectrode *(upper)* and a double-barrelled microelectrode *(lower)* liquid membrane cell assembly

5.2 The Membrane Potential

For ideal membrane electrode cell assemblies, the measured EMF expresses a membrane potential E_M that is generated across the ion-selective membrane. The electrical potential difference E_M can be divided into three fundamental contributions, i.e. a potential difference within the bulk of the membrane (the diffusion potential E_D) and two interfacial potential differences (the boundary potentials E_B' and E_B'') at the sample solution and internal filling solution sides, respectively [Morf and Simon 78; Morf 81]:

$$E_M = E_D + E_B' + E_B''. \tag{5.2}$$

The diffusion potential E_D is generated by the diffusion of ions within the membrane phase. At zero-current conditions E_D is described by Eq. (5.3),

$$E_D = \frac{RT}{F} \int_0^d \sum_i \frac{t_i}{z_i} \, d\ln a_i, \tag{5.3}$$

where t_i denotes the transference number of the ion I (for the meaning of other symbols see Eq. (5.6)). A general explicit description of the diffusion potential is given by the Henderson formalism [Henderson 07; Morf and Simon 78; Morf 81] (see Eq. (5.35) in Sect. 5.7.). The boundary potentials E_B' and E_B'' are related to the exchange processes of the ion I at the phase boundaries between the membrane and the external solutions. If a thermodynamic equilibrium is assumed to exist at each interface the boundary potentials are described by the equilibrium distribution of the ion I:

$$E_B' = \frac{RT}{z_iF} \ln \frac{k_i a_i'}{c_i(0)}, \tag{5.4}$$

$$E_B'' = \frac{RT}{z_iF} \ln \frac{k_i a_i''}{c_i(d)}. \tag{5.5}$$

The ion activities a_i' and a_i'' refer to the sample solution and internal filling solution, respectively; $c_i(0)$ and $c_i(d)$ are the boundary values of the respective membrane concentrations and k_i denotes the distribution constant [Wuhrmann et al. 73; Morf and Simon 78; Morf 81]. In order to obtain a useful description of the total membrane potential E_M for practical purposes, expressions for E_D and E_B have to be developed in which the concentrations and potentials referring to the membrane phase are replaced by the corresponding values of the the external aqueous solutions. Such theoretical descriptions are available for permselective carrier membranes. They lead to results for E_M that can easily the converted into an equation of the Nicolsky type (see Eq. (5.7) in Sect. 5.3) [Morf 81].

5.3 The Nernst and the Nicolsky-Eisenman Equations

By selectively transferring the ion I to be measured from the sample solution to the membrane phase, a potential difference is generated between the internal filling solution and the sample solution. Ideally this potential difference should be a linear function of the logarithm of the activity of the ion. If all other potential differences are assumed to be constant, the EMF of the cell assembly is described by the Nernst equation:

$$EMF = E_o + s \log a_i, \tag{5.6}$$

where
EMF : electromotive force (cell potential) [mV];
E_o : reference potential [mV];
s   : Nernstian slope,

$$s = 2.303 \frac{R \, T}{z_i F} = 59.16/z_i; \quad ([mV]; 25\,°C)$$

R : gas constant ($8.314 \, JK^{-1}mol^{-1}$)
T : absolute temperature [K]
F : Faraday equivalent ($9.6487 \cdot 10^4 \, C \, mol^{-1}$)
z_i : charge number of the ion I.

In practice, however, such ideal electrode behaviour is not observed. In particular, deviations from the Nernst equation (5.6) become significant at low activities. It is therefore necessary to consider additional contributions to the measured EMF due to the presence of interfering ions J in the sample solution. A successful semi-empirical approach to treating real membrane electrode systems is given by the extended Nicolsky-Eisenman equation [Nicolsky 37a; Nicolsky 37b; Eisenman 67]:

$$EMF = E_o + s \log \left[a_i + \sum_i K_{ij}^{Pot}(a_j)^{z_i/z_j} \right], \tag{5.7}$$

with $E_o = E_i^o + E_R + E_D,$

where
E_i^o : constant potential difference including the boundary potential difference between the internal filling solution and the membrane.
E_R : constant potential difference consisting of the potential difference between the metallic lead to the membrane electrode and the internal filling solution of this electrode, and of the potential difference between the reference electrolyte and the silver lead within the reference electrode. E_R is independent of changes in the sample composition.
E_D : liquid-junction potential difference generated between reference electrolyte (or salt bridge and sample solution (see Sect. 5.7).
K_{ij}^{Pot} : potentiometric selectivity factor (see Sect. 5.5).

The sum $E_I^0 + E_R$ encompasses, for a given set of conditions, all the contributions that are independent of the sample composition. E_D is variable and sample dependent (see Sect. 5.7). It is evident that the Nicolsky-Eisenman formalism reduces to the Nernst equation for a perfectly selective electrode (all of the K_{ij}^{Pot} being zero) or for samples containing no other ions with the same sign of charge as the primary ion I. According to Eq. (5.6), a change in the EMF of 1 mV corresponds to a change in the activity of the ion I^{z_i} of $z_i \cdot 4\%$.

5.4 Electrode Function and Detection Limit

The electrode function describes the dependence of the EMF of an ion-selective membrane electrode cell assembly on the logarithm of the ion activity in the sample solution (Fig. 5.2). According to the Nernst equation (5.6), the electrode function exhibits a range of linear response at relatively high activities (typically 10^{-5} to 10^{-1} M). Deviations from linearity at high cation activities may be due to an interference caused by sample anions (see Sect. 5.11). At rather low activities of the ion to be detected (typically $< 10^{-5}$ M) the response curve flattens out until there is no longer a dependence of the EMF on changes in the ion activity (Fig. 5.2). The observed detection limit is governed by the presence of other ions in the sample. For example, impurities in the reagents (salts, water) used, the leaching of ions from the membrane and reference electrodes or intentionally introduced background electrolytes (calibration solutions) may be the source of such contaminations.

Typical detection limits of neutral carrier-based membrane electrodes are about 10^{-6} M. However, these values often do not reflect the true detection limits of the electrode because the quality of the salts and water used can strongly influence these limits in the absence of metal buffers [Simon et al. 78]. With metal-buffered calibration solutions it is possible to eliminate various effects which lead to a contamination of extremely dilute sample solutions. The detection limit is then exclu-

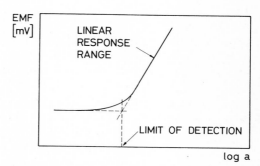

Fig. 5.2. Schematic diagram of the electrode function (calibration curve) of an ion-selective electrode. The detection limit is defined by the intercept of the two asymptotes of the Nicolsky response curve [Guilbault et al. 76]. For an ideal electrode function with slope s the detection limit corresponds to the activity at which the experimental response curve deviates by $s \cdot \log 2$ mV from the extrapolated Nernstian response

sively governed by the interfering ions. At present, the lowest detection limits of neutral carrier electrodes measured in metal or pH-buffered solutions are $5 \cdot 10^{-9}$ M for Ca^{2+} [Simon et al. 78] and $< 10^{-10}$ M for H^+ [Schulthess et al. 81].

Whereas a certain signal-to-noise ratio describes the detection limits of techniques with linear relationships between the signal and the amount of sample, the signal-to-noise ratio is meaningless for the detection limits of potentiometric measurements in which the EMF signal is proportional to the logarithm of the activity. The noise describes the relative precision of the EMF measurement rather than the detection limit. Therefore, a comparison of the detection limits of electrodes with those of other techniques is difficult (see Chap. 9).

5.5 Selectivity Factors

The weighing factors K_{ij}^{Pot} introduced in the Nicolsky-Eisenman equation (5.7) allow a specification of the potentiometrically observable ion selectivities of a membrane electrode. The selectivity factors are a measure of the preference by the sensor for the interfering ion J relative to the ion I to be detected. Obviously, for ideally selective membranes all of the K_{ij}^{Pot} values must be zero. A selectivity factor < 1 indicates a preference for the measuring ion I relative to the interfering ion J.

The highest ion selectivities of neutral carrier liquid membrane electrodes with preference for a certain A-cation are of the order of 10^6. H^+-selective neutral carrier membranes exhibit selectivities as high as 10^{11}.

The K_{ij}^{Pot} values should not be considered to be constant parameters that characterize membrane selectivity at all measuring conditions. Instead, one should allow for certain variations in the selectivity factors of a given membrane electrode. The K_{ij}^{Pot} values are dependent on both the method used and on the conditions (concentrations, metal buffers, etc.) of the measurement. Several discussions of selectivity factors have already been published [Moody and Thomas 71; Simon et al. 78; Amman et al. 79].

Selectivity factors are determined experimentally using the following techniques:

a) the separate solution method (SSM): the EMF values obtained for the measuring ion and the interfering ion, both determined in pure single electrolyte solutions, are compared using the Nicolsky-Eisenman equation:

$$K_{ij}^{Pot} = 10^{\frac{(E_j - E_i) z_i F}{2.303 \, RT}} \cdot \frac{a_i}{(a_j)^{z_i/z_j}} \cdot \tag{5.8}$$

The most striking advantage of this method is its simplicity. Unfortunately, the selectivity data are often not ideally representative for mixed sample solutions.

b) the fixed interference method (FIM): selectivity factors are obtained by graphically evaluating the electrode function of the measuring ion in solutions of a fixed concentration of an interfering ion. The value of a_i used to calculate K_{ij}^{Pot} is obtained from the intersection of the extrapolated parts of the linear portions of the response curves corresponding to E_i and E_j. At the intersection point, E_i is equal to E_j and Eq. (5.8) is reduced to

$$K_{ij}^{Pot} = \frac{a_i}{(a_j)^{z_i/z_j}} \cdot \tag{5.9}$$

This method is recommended by the IUPAC [Guilbault et al. 76; IUPAC 79] since it describes the experimental conditions more adequately (e.g. the determination of K_{CaNa}^{Pot} values at a fixed background of 10 mM Na^+ in view of intracellular Ca^{2+} measurements). It is, however, much more time consuming than the SSM technique.

c) the fixed primary ion method (FPM): the concentration of the interfering ion J is varied at a constant concentration of the primary ion I. Thus, the FPM is the reverse of the FIM. It is used less frequently for the characterization of carrier membranes, except when the dependence of an EMF response on the pH value of a sample solution is to be determined [Ammann et al. 74; Ammann et al. 75b; Zhukov et al. 81].

In the absence of metal buffers, both the selectivity factors and detection limits often reflect the quality of the reagents used, and therefore need not be representative of the inherent electrode selectivity. This situation arises in particular when highly selective membrane systems have to be characterized. For example, solvent polymeric membranes based on the Ca^{2+} carrier ETH 1001 yield selectivity factors log K_{CaNa}^{Pot} of -3.4 (SSM, 0.1 M unbuffered metal chloride solutions), -5.0 (FIM, 1 M NaCl background, Ca^{2+}-unbuffered solutions), and -6.1 (FIM, 0.1 M NaCl background, Ca^{2+}-buffered solutions) [Simon et al. 78]. It is evident that such a highly selective electrode is most appropriately characterized by the FIM measurements in Ca^{2+}-buffered solutions. It must therefore be stressed that the experimental conditions of the selectivity determinations should be clearly indicated, otherwise, a comparison of the selectivity factors of different sensors may be misleading.

In studying the selectivity behaviour of ideal neutral carrier-based liquid membranes it is helpful to devide the equilibrium (cf. Eq. (5.4)),

$$I^{z+}(aq) + nL(m) \xrightleftharpoons{K_{I,n}^{extr}} IL_n^{z+}(m), \tag{5.10}$$

into the following steps [Morf 81]:

a) transfer of the free carrier (L) from the membrane (m) to the boundary layer of the aqueous solution (aq):

$$L(m) \xrightleftharpoons{1/k_L} L(aq), \tag{5.11}$$

where k_L is the distribution coefficient of the carrier L between the membrane phase and the aqueous phase;

b) formation of cationic complexes in the aqueous phase:

$$I^{z+}(aq) + nL(aq) \overset{\beta_{IL,n}^w}{\rightleftharpoons} IL_n^{z+}(aq),\tag{5.12}$$

where $\beta_{IL,n}^w$ is the stability constant of the complex IL_n^{z+} in the aqueous phase;

c) transfer of the complexes into the membrane:

$$IL_n^{z+}(aq) \overset{k_{IL,n}}{\rightleftharpoons} IL_n^{z+}(m),\tag{5.13}$$

where $k_{IL,n}$ is the distribution coefficient of the complex IL_n^{z+}.

For the overall extraction coefficient $K_{I,n}^{extr}$ it follows that

$$K_{I,n}^{extr} = \beta_{IL,n}^w \cdot k_{IL,n} \cdot \left(\frac{c_L}{k_L}\right)^n,\tag{5.14}$$

where c_L is the concentration of the free carrier within the membrane.

When the measuring ions I^{z+} and the interfering ions J^{z+} are of the same charge, the selectivity factor K_{ij}^{Pot} is described by the ion-exchange equilibrium:

$$IL_n^{z+}(m) + J^{z+}(aq) \overset{K_{ij}^{Pot}}{\rightleftharpoons} I^{z+}(aq) + JL_n^{z+}(m).\tag{5.15}$$

Assuming different stoichiometries for the complexes, the selectivity factor is then given by [Morf and Simon 78; Morf et al. 79; Morf 81]:

$$K_{ij}^{Pot} = \frac{\sum\limits_n K_{J,n}^{extr}}{\sum\limits_n K_{I,n}^{extr}} = \frac{\sum\limits_n \beta_{JL,n} \cdot k_{JL,n} \left(\dfrac{c_L}{k_L}\right)^n}{\sum\limits_n \beta_{IL,n} \cdot k_{IL,n} \left(\dfrac{c_L}{k_L}\right)^n}\tag{5.16}$$

If the distribution coefficients, $k_{IL,n}$ and $k_{JL,n}$, are assumed to be equal [Morf and Simon 78], Eq. (5.16) simplifies to

$$K_{ij}^{Pot} = \frac{\beta_{JL}^w}{\beta_{IL}^w},\tag{5.17}$$

for neutral carriers which predominantly form 1:1 complexes. Equation (5.17) is valid for most natural neutral carriers (e.g. valinomycin). It was shown that for membrane electrodes based on these carriers the selectivity among ions of the same charge is only slightly influenced by the extraction properties of the membrane solvent. Thus, in an ideal case, the ion selectivity of neutral carrier membranes is completely governed by the complex formation properties of the carrier, i.e. the selectivity factor can be expressed in terms of the ratio of the complex formation constants in water. Such a simple correlation between selectivity factors and stability constants can not be expected for the synthetic non-macrocyclic carriers because cation-carrier complexes of different stoichiometries are involved and/or the central ion is not completely covered by the ligand shell and is therefore partly solvated [Morf et al. 73; Wuhrmann et al. 73]. Indeed, the K_{ij}^{Pot} values as described by the extraction equilibrium in Eq. (5.15) are found to deviate from the ratios of the corresponding complex stabilities measured in the homogeneous phase for membranes containing synthetic acyclic neutral carriers [Kirsch and Simon 76].

Selectivity factors are influenced by many other parameters. The dependence of the membrane selectivity on the dielectric constant of the membrane phase (Sect. 4.1.2) and on the concentration of lipophilic anionic sites (Sect. 4.1.4) has already been discussed. If the mean stoichiometry $\bar{n}$ of the measuring ion and the interfering ion is different, the selectivity factor becomes dependent on the free-carrier concentration c_L within the membrane phase [Morf 81]:

$$\frac{d\ln K_{ij}^{Pot}}{d\ln c_L} \approx \frac{z_i}{z_j}\bar{n}_j - \bar{n}_i. \tag{5.18}$$

5.6 Activity Coefficients. The Debye-Hückel Formalism

As indicated by the Nernst equation (5.6) and the Nicolsky-Eisenman equation (5.7), the EMF of an ion-selective electrode cell assembly is directly dependent on the ion activity, and only indirectly dependent on the ion concentration. For high ionic strengths I (Eq. (5.21)), as encountered in biological samples, concentrations cannot be replaced by activities, i. e.

$$a_i = \gamma_i \cdot c_i \text{ with } \gamma_i \neq 1, \tag{5.19}$$

where γ_i is the activity coefficient.

Two kinds of activities have to be distinguished. Mean activities a_{ix} (and mean activity coefficients $\gamma_\pm$) have been unambiguously defined and are measurable. For a given cation, the value of $\gamma_\pm$ depends on the kind of anion present and vice versa. On the other hand, single-ion activities a_i and a_x (and single-ion activity coefficients γ_+ and γ_-) cannot be thermodynamically defined and cannot be experimentally measured. The electroneutrality principle only permits a thermodynamic description of the mean activities of neutral combinations of ions. For 1:1 electrolytes the mean and single activities are assumed to be related in the following manner [Kortüm 72]:

$$a_{ix} = \sqrt{a_i \cdot a_x} \tag{5.20}$$

The EMF of cell assemblies with a liquid-junction (see Fig. 5.1) are usually described in terms of mean activity coefficients. If the sample is a 1:1 electrolyte of monovalent ions, the Nernst equation (5.6) can be rewritten as

$$EMF = \frac{RT}{F} \ln c_i \cdot \gamma_\pm + \text{const.} \tag{5.21}$$

Thus, mean activities can be obtained experimentally but the potential difference at the boundary of an ion-selective membrane electrode is thermodynamically related to single-ion activities [Kortüm 72]. The fact that ion-selective membrane electrode cell assemblies nevertheless assess mean ion activities can be verified in a discussion of the following simple cell:

K$^+$-selec-tive membrane electrode (I)	KCl (a_K')	liquid-junction	KCl (a_K'')	K$^+$-selective membrane electrode (II)	(5.22)

The potential differences at each K$^+$-selective half-cell are given by

$$EMF_K^I = \frac{RT}{F} \ln a_K' + E_K^{I\,\circ}, \tag{5.23}$$

$$EMF_K^{II} = \frac{RT}{F} \ln a_K'' + E_K^{II\,\circ}, \tag{5.24}$$

where a_K' and a_K'' are the single-ion activities of the two KCl solutions. The liquid-junction potential is generally described by Eq. (5.3). For an ideal equitransferent electrolyte such as KCl with $t_K = t_{Cl} = 0.5$, the integral becomes

$$E_J = \frac{RT}{F} \int_{a'}^{a''} (0.5 \, d \ln a_K - 0.5 \, d \ln a_{Cl}), \tag{5.25}$$

or

$$E_J = \frac{RT}{2F} \ln \frac{a_K'' \cdot a_{Cl}'}{a_K' \cdot a_{Cl}''}. \tag{5.26}$$

The total EMF of the cell in (5.22) is then given by

$$EMF = EMF_K^I - EMF_K^{II} + E_J$$

$$= \frac{RT}{2F} \ln \frac{a_K' \cdot a_{Cl}'}{a_K'' \cdot a_{Cl}''} + \text{const.} \tag{5.27}$$

According to equation (5.20) the mean ion activity of KCl is:

$$a_{KCl} = \sqrt{a_K \cdot a_{Cl}}, \tag{5.28}$$

and therefore,

$$EMF = \frac{RT}{F} \ln \frac{a'_{KCl}}{a''_{KCl}} + \text{const.} \tag{5.29}$$

Hence, the EMF of cell assemblies with liquid-junctions depends on the logarithm of the mean ion activities.

In practice, the situation, compared to that of the above example, is much more complicated (cell assemblies of the type given in (5.1)). The sample contains many different electrolytes, and often, the cation to be measured is not the same as the cation of the reference electrolyte. A theoretical treatment shows that a complex expression for the relevant activity coefficient of the measuring cation is obtained when the cations of the sample and the reference electrolyte are both involved [W. E. Morf, private communication]. However, if physiological samples with relatively high Cl^- concentrations and KCl reference electrolytes are considered, the single-ion activity coefficients of the metal chloride salts can be used as a good approximation. Several procedures have been proposed to evaluate γ_+ and γ_- [Kortüm 72; Morf 81]. The single-ion activities used to plot the electrode functions and to compute the liquid-junction potentials were obtained according to the Debye-Hückel formalism. The mean and single-ion activity coefficients are related in the following manner (Fig. 5.3):

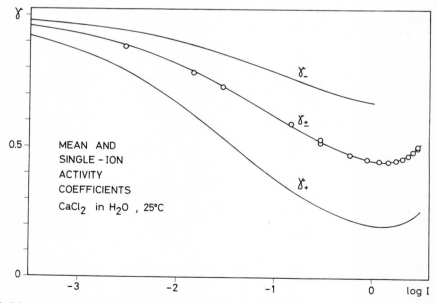

Fig. 5.3. Mean and single-ion activity coefficients for calcium chloride as functions of the logarithm of the ionic strength I [Meier et al. 80]. The circles denote measured mean activity coefficients $\gamma_\pm$. The curves were plotted using Eqs. (5.32) ($\gamma_\pm$, for parameters see Table 5.1), (5.30) (γ_+) and (5.31) (γ_-)

$$\log \gamma_+ = \frac{z_+}{z_-} \cdot \log \gamma_\pm, \tag{5.30}$$

$$\log \gamma_- = \frac{z_-}{z_+} \log \gamma_\pm. \tag{5.31}$$

To derive single-ion activities for a solution of known concentrations, the mean activities and the liquid-junction potentials have to be known. For many salts [Meier 82] the Debye-Hückel formalism yields a good approximation of the experimentally determined mean activity coefficients of an ion and its counterion [Hamer and Wu 72; Robinson and Stokes 68]:

$$\log \gamma_\pm = \frac{-A \, |z_+ \, z_-| \, \sqrt{I}}{1 + Ba \, \sqrt{I}} + C \, I, \tag{5.32}$$

with

$$I = \frac{1}{2} \sum_i c_i z_i^2, \tag{5.33}$$

where

A   : 0.5108 (H_2O; 25 °C) [$mol^{-1/2} \, l^{1/2}$];
B   : 0.328 (H_2O; 25 °C) [$mol^{-1/2} \, l^{1/2} \, Å^{-1}$];
z_+, z_- : charge numbers of the cation and anion of the electrolyte under consideration;
I   : ionic strength of the solution [$mol \, l^{-1}$];
a, C   : electrolyte-specific constants used to fit the theoretical relationship to measured values of $\gamma_\pm$ (see Table 5.1.) (a: [Å]; C: [$mol^{-1} \, l$];
c_i : concentration of any ion in the sample solution [$mol \, l^{-1}$];
z_i : charge of any ion in the sample solution.

Table 5.1. Debye-Hückel parameters a and C for aqueous electrolytes (equation (5.32), 25 °C) [Meier 82]

Electrolyte	a [Å]	C [$mol^{-1} l$]
HCl	4.30	0.120
LiCl	3.90	0.113
LiOAc	4.15	0.055
NaCl	4.00	0.040
KCl	3.65	0.015
NH_4Cl	3.50	0.016
$MgCl_2$	5.20	0.060
$CaCl_2$	5.00	0.040
$BaCl_2$	4.40	0.040

5.6 Activity Coefficients. The Debye-Hückel Formalism

77

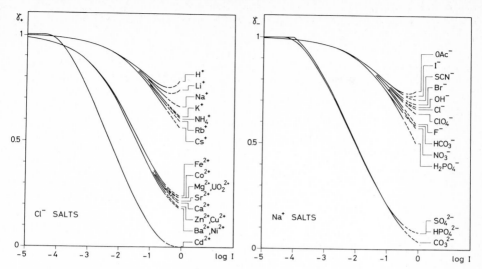

Fig. 5.4. Single-ion activity coefficients as a function of the ionic strength for cations (Cl⁻ salts, *left*) and anions (Na⁺ salts, *right*). The curves were plotted using equations (5.32), (5.31) and (5.30) [Meier et al. 80]

For a single electrolyte and ionic strengths $I \leqslant 1$ mol l⁻¹ Eq. (5.32) yields $\gamma_\pm$ values which are in good agreement (deviation 1%) with experimental mean activity coefficients [Meier 82].

Electrode functions of membrane electrodes are probably best characterized by plotting the measured EMF against the logarithm of the primary ion activities. Here, single-ion activities arising from pure electrolytes were considered throughout. Consequently, the concentrations were converted to single-ion activities using Eqs. (5.32), (5.30) and (5.31). The measured EMF values were corrected for the liquid-junction potentials and should therefore depend on single-ion activities rather than on mean activities (see e.g. (5.23)). For extra- as well as intracellular conditions the prevalent counterions were chloride and sodium (Fig. 5.4).

Undoubtedly, single-ion activity coefficients determined from mixed electrolytic solutions containing, for example, other physiologically relevant electrolytes and proteins should more adequately describe the conditions in biological fluids. Unfortunately, the evaluation of individual ion activities in electrolytic mixtures poses considerable difficulties [Butler 68; Mohan and Bates 75]. The applicability of ion-selective electrodes to the determination of mean activity coefficients has been critically discussed elsewhere [Bates et al. 83]. Potentiometry has also been suggested as a method for the evaluation of single-ion activity coefficients in biological samples [Uemasu and Umezawa 83]. Furthermore, ion activity standards have been proposed for aqueous extracellular electrolytic mixtures [Mohan and Bates 75] and alkali metal chlorides in bovine serum albumin solutions [Reboiras et al. 78]. Successful activity scales have also been introduced for pH measurements in the physiological H⁺ activity range [Sankar and Bates 78; Vega and Bates 78]. The discussion of activity coefficients is strongly associated with the calibration problems of electrode cell assemblies (see Sect. 7.3).

5.7 Liquid-Junction Potentials. The Henderson Formalism

Liquid-junction potentials are always generated when electrolytic solutions of different ionic compositions are in contact with one another. Typical membrane electrode cell assemblies (scheme in (5.1)) exhibit such a potential at the junction of the reference electrolyte with the sample solution. The potentiometric cells described here are therefore galvanic cells with diffusion potentials.

The total flux of ions at the liquid-junction of the reference electrode is:

$$J_i = \underbrace{u_i\, c_i\, RT \frac{d\ln a_i}{dx}}_{\text{diffusion}} - \underbrace{z_i\, u_i\, c_i\, F \frac{d\Phi}{dx}}_{\text{migration}} + \underbrace{c_i\, v_x.}_{\text{convection}} \qquad (5.34)$$

Under zero-current conditions and for electroneutral systems, the liquid-junction potential corresponds to a diffusion potential which is exclusively determined by the diffusion term of (5.34) and is generally described by Eq. (5.3).

Although it is often assumed that the liquid-junction potential E_J generated at the interface (diaphragm, free-flowing junction, etc.) between the reference electrolyte (or salt bridge) and the sample solution is independent of the sample composition, it may in fact vary considerably. Therefore, an evaluation of E_J on the basis of theoretical descriptions is of great practical interest. Equations for the calculation of E_J are available from the theories of Planck and Henderson [Morf 77]. The Henderson approximation is more frequently applied because of its simplicity [Henderson 07; Morf 77; Meier et al. 80; Morf 81]:

$$E_J = \frac{\sum\limits_i z_i\, u_i\, (a_i - a_i')}{\sum\limits_i z_i^2\, u_i\, (a_i - a_i')} \frac{RT}{F} \ln \frac{\sum\limits_i z_i^2\, u_i\, a_i'}{\sum\limits_i z_i^2\, u_i\, a_i}, \qquad (5.35)$$

where

z_i : charge number of the ion I;
u_i : absolute mobility of the ion I [$cm^2\, s^{-1} J^{-1}\, mol$] (see Table 5.2);
a_i : single-ion activity of the ion I in the sample solution [$mol\, l^{-1}$];
a_i' : single-ion activity of the ion I in the reference electrolyte (or salt bridge) of the reference electrode [$mol\, l^{-1}$].

Obviously, the function E_J is superimposed on the EMF of the membrane electrode. It is therefore advantageous to minimize the contributions of the liquid-junction potential to the measured EMF. This can be achieved by either choosing a reference electrolyte which closely resembles the sample solution (i.e. $a_i' \approx a_i$) or by using highly concentrated equitransferent reference electrolytes ($u_m \approx u_x$). In the former case, the value of the liquid-junction potential is approximately zero. However, E_J is very sensitive to changes in the composition and concentrations of the sample solution [Morf 81] (e.g. reference electrolyte and sample solution: 0.01 M KCl: $E_J = 0.00\, mV$; reference electrolyte: 0.01 M KCl, sample solution: 0.01 M NaCl: $E_J = 4.39\, mV$). The second approach is of greater practical relevance. The use of eq-

Table 5.2. Absolute mobilities of ions in aqueous solutions at 25 °C [Meier et al. 80; Morf 81; Handbook 75]

Ion I^{z+}	Absolute mobility[a] $u_i \cdot 10^9$ [cm^2 s^{-1} J^{-1} mol]
H^+	37.59
Li^+	4.24
Na^+	5.47
K^+	8.00
NH_4^+	8.00
Mg^{2+}	2.90
Ca^{2+}	3.22
Ba^{2+}	3.49
OH^-	20.6
Cl^-	8.11
NO_3^-	7.58
OAc^-	4.38

a
$$u_i = \frac{\Lambda_i}{|z_i| \, F^2} = \frac{D_i}{RT}, \qquad (5.36)$$

Λ_i : equivalent ionic conductivity at infinite dilution [Handbook 75] [Ω^{-1} cm^2 equiv.$^{-1}$]

D_i : diffusion coefficient [cm^2 s^{-1}].

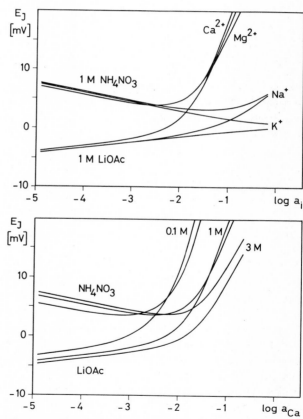

Fig. 5.5. The liquid-junction potential as a function of the single-ion activity of sample cations for 1 M NH$_4$NO$_3$ and 1 M LiOAc reference electrolytes *(upper)* and as a function of the Ca^{2+} single-ion activity for different concentrations of these reference electrolytes *(lower)* [Meier et al. 80]. The calculations were carried out using Eq. (5.35), single-ion activities and the mobility data of Table 5.2

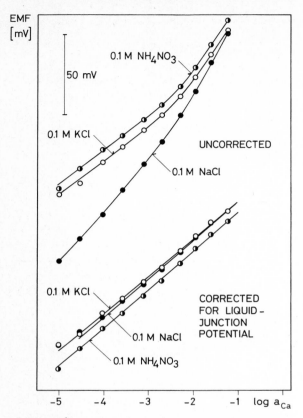

Fig. 5.6. Ca^{2+} electrode functions for a membrane electrode cell assembly with reference electrolytes (salt bridges) purposely chosen to be non-ideal: 0.1 M NH_4NO_3, KCl and NaCl. The measured (uncorrected) EMF are plotted at the top. After correcting for changes in the liquid-junction potentials E_J using Eq. (5.35) the points were replotted (bottom). The slopes of the corrected responses are practically indistinguishable according to the 95% confidence limit [Meier et al. 80]. For the sake of clarity, the points for uncorrected and corrected EMF values were separated

uitransferent solutions may lead to E_J values which are not equal to zero. However, the electrolyte guarantees the highest possible constancy for the liquid-junction potential. The most commonly used equitransferent solutions are saturated (4.2 M at 25 °C) or 3 M KCl solutions. Other suitable electrolytes are LiOAc, NH_4NO_3, NH_4Cl and KNO_3 (see mobility data in Table 5.2). Figure 5.5 illustrates the influence of the type of reference electrolyte and its concentration on the liquid-junction potential [Meier et al. 80].

For certain special applications such as ion-monitoring in extracorporeal bypasses, an isotonic reference solution has to be used for physiological and safety reasons. In such cases, changes in the liquid-junction potential often cannot be neglected in the evaluation of clinically relevant data, i.e. corrections according to the Henderson formalism become necessary. For example, if the concentration of Na^+ in the sample solution is varied within its physiological extracellular range (135 to

Table 5.3 Representative stabilities, reproducibilities and drifts of EMF obtained with neutral carrier-based electrodes

Electrode system	Stability of EMF, std. dev. [mV]	Reproducibility of EMF, std. dev. [mV]	Drift of EMF [mV/h]	References
Flow-through system, Na^+ solvent polymeric membrane	± 0.12[a]	± 0.05[b]	0.03	[Ammann et al. 85b]
Flow-through system, K^+ solvent polymeric membrane	± 0.03[a]	± 0.10[b]	0.01	[Ammann et al. 85b]
Flow-through system, Ca^{2+} solvent polymeric membrane	± 0.03[a]	± 0.09[b]	0.01	[Ammann et al. 85b]
Flow-through system, H^+ solvent polymeric membrane	± 0.08[a]	–	-0.03	[Ammann et al. 85b]
Ca^{2+} microelectrode	± 0.30[c]	–	–	[Oehme et al. 76]
H^+ microelectrode	–	± 0.23[d]	0.6	[Ammann et al. 81b]

[a] 5 h (n = 100); aqueous electrolytic solutions
[b] EMF differences (n = 5) between aqueous electrolytic solution and serum
[c] 12 h 10^{-1} M $CaCl_2$ solution
[d] EMF differences (n = 10) between two buffered aqueous electrolytic solutions

150 mM), E_J only changes by 0.07 mV for a 3 M KCl reference electrolyte solution whereas it changes by 0.5 mV for an isotonic reference solution (150 mM NaCl) [Ammann et al. 85b]. Figure 5.6 effectively shows the usefulness of mathematical corrections.

5.8 Stability and Reproducibility of EMF

The stability and reproducibility of EMF measurements made with currently available membrane electrode cell assemblies clearly depend on the type of electrode used (Table 5.3). Carefully optimized flow-through electrode systems for clinical use yield standard deviations which are less than 0.1 mV, i.e. the relative standard deviation in the determination of activities can be expected to be better than $z_i \cdot$ 0.4%. On the other hand, due to technical (geometry at the tip, silanization) and electrical reasons (extremely high resistances of the membranes), microelectrodes exhibit poorer characteristics (Table 5.3).

The quality of a measurement also depends on the kind of experiment being performed. For instance, the reproducibility of so-called precision measurements was found to be about 10 times better than that of routine measurements (Fig. 5.7) and approaches the limits imposed by the electronic measuring equipment (e.g. $\pm$

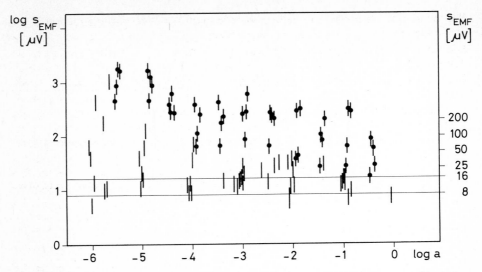

Fig. 5.7. Errors associated with EMF measurements [Meier et al. 80]. Vertical bars with circles: standard deviations of routine measurements performed with Na^+- and Ca^{2+}-selective electrodes using different bridge electrolytes. Error bars without circles: standard deviations of precision measurements conducted on Li^+-, Na^+-, K^+-, Ca^{2+}-, and Cl^--selective electrodes; special precautions were taken to ensure absence of drift, temperature fluctuations, and contamination by the bridge electrolyte. The electronic noise under the same conditions but with the ion-selective electrodes being replaced by a thermostated resistor of 100 MΩ is. $12 \pm 4 \mu V$ (for further details see [Meier et al. 80])

$12 \mu V$ [Meier et al. 80]). Furthermore, the reproducibility in solutions of high ion activities (around 0.1 M) is better than that obtained in micromolar solutions (by a factor of about 10) (Fig. 5.7).

5.9 Response Time

The response time of electrochemical cells containing ion-selective membrane electrodes is considered to be one of the most critical and limiting factors in the clinical and physiological applications of potentiometric sensors. Important issues for the study of response times are:

- design of the experimental setup [Lindner et al. 76; Koryta and Štulík 83];
- definition of practical response times [Lindner et al. 78; Uemasu and Umezawa 82; Pungor and Umezawa 83; Lindner et al. 85];
- evaluation of the rate-limiting step [Buck 81; Buck 76];
- theoretical description of the response curves [Lindner et al. 78; Morf 81];
- effect of the neutral carrier membrane composition on the response time [Morf et al. 75; Lindner et al. 78].

Response times inherent to ion-selective electrodes are only measurable if the overall response time of the potentiometric system is governed by the properties of the membrane electrode, i.e. if the time constant of the response function of the electrode is much larger than the time constants of the electrochemical cell and the electronic EMF-measuring device. Thus, the overall response time does not necessarily only reflect the properties of the membrane electrode. It can also express features of the measuring technique or properties of the electronic equipment.

Indeed, the overall response time is affected by a series of factors [Wang and Copland 73; Morf 81; Koryta and Štulík 83], for example the time constant of the measuring instrument [Ujec et al. 81], the impedance of the equivalent electrical circuit of the membrane (for microelectrodes [Oehme et al. 76]; for ISFETs [Oesch et al. 81]), the rate of the ion-transfer reaction across the membrane/sample interface [Fleet et al. 74; Lindner et al. 76], the diffusion of the ion to be measured through the stagnant layer in the sample [Markovic and Osburn 73; Morf et al. 75], the diffusion within the ion-selective membrane [Morf et al. 75; Morf and Simon 78] and the establishment of a liquid-junction potential at the reference electrode [Morf et al. 75].

Furthermore, a drift in the EMF can arise from the loss of membrane components into the sample or from poisoning of the membrane phase or surface by species from the sample solution.

Extensive studies were carried out to determine the dynamic behaviour of membrane electrodes [Fleet et al. 74; Morf et al. 75; Lindner et al. 76; Buck 75; Buck 81; Lindner et al. 78; Morf and Simon 78; Morf 81]. For neutral carrier-based membranes two types of response behaviour have to be distinguished. Neutral carrier membranes modified with ion-exchanger sites (e.g. sodium tetraphenylborate) have to be treated differently from unmodified ones. The response times of the first group can generally be represented by an exponential function [Lindner et al. 78] which actually describes the response of the cell assembly (electrode parameters are not included):

$$EMF_t = EMF_\infty + s \log \left[1 - (1 - \frac{a_i^o}{a_i})e^{-t/\tau'} \right], \tag{5.37}$$

$$\text{with } \tau' = \frac{\delta^2}{2D'}, \tag{5.38}$$

where

EMF_t : cell potential at the time t [mV];

EMF_∞ : equilibrium potential at the time $t = \infty$ [mV];

a_i^o, a_i : activities of the primary ion in the bulk of the sample solution at $t < 0$ and $t \geq 0$ [mol l^{-1}], respectively;

D' : mean diffusion coefficient in the adhering aqueous layer [cm^2 s^{-1}];

δ : thickness of the adhering aqueous layer [cm].

Since the composition of the modified carrier membranes remains approximately constant, the diffusion processes of ions passing through the membrane become negligible in the absence of interfering ions [Morf 81]. As a result, the dynamic response characteristic is governed by the transport processes in the aqueous diffusion layer, i.e. the response time depends on the shape and condition of the membrane surface as well as the composition of the sample. Furthermore, it is markedly influenced by the direction of the change in activity in the sample solution [Morf and Simon 78]. According to Eqs. (5.37) and (5.38) the response time of neutral carrier membranes containing lipophilic salts can be reduced by minimizing the aqueous diffusion layer (fast stirring, flowing sample, microelectrodes) and by using samples of higher activities [Morf and Lindner 75].

For unmodified neutral carrier membranes the diffusion of ions into the membrane may occur. A steady state within the membrane is usually attained rather slowly compared to the diffusion outside the membrane [Morf and Simon 78]. As a result, the rate-limiting process is determined to a large extent by the dynamic behaviour of processes within the membrane itself. The reponse curve can be approximated by a square-root type of function [Lindner et al. 78]:

$$\mathrm{EMF}_t = \mathrm{EMF}_\infty + s \log \left[1 - \left(1 - \frac{a_i^0}{a_i} \right) \frac{1}{\sqrt{t/\tau}} \right], \tag{5.39}$$

$$\text{with } \tau = \frac{D\, k^2 \delta^2}{D'^2}, \tag{5.40}$$

where
D : mean diffusion coefficient in the membrane $[\mathrm{cm^2\, s^{-1}}]$;
k : partition coefficient between the aqueous and the membrane phase.

Consequently, the following points have to be kept in mind when considering a reduction of the response time [Morf and Simon 78]:

– reduction of δ: stirring of the sample solution (Fig. 5.8), flow-through electrode systems, minimization of the membrane surface (microelectrodes: tip diameter and adhering film $< 10^{-4}$ cm)
– reduction of D: since the diffusion coefficient of the carrier-complex salt ILX is approximately proportional to the mobility of the sample anion X in the membrane ($D \sim ((z+1)/z)\, u_x RT$ (5.41)) [Morf and Simon 78], the permeability of the membrane with respect to sample anions should be low. This can be achieved by using highly viscous membrane phases (a viscous membrane solvent and/or a high percentage of the membrane matrix)
– reduction of k: the extraction of salt into the membrane is kept low if nonpolar membrane phases are used (Fig. 5.9). In addition, the neutral carrier should be a weak complexing agent and should be incorporated into the membrane at low concentrations. The sample solution should not contain any lipophilic anions since they enhance the salt extraction into the membrane phase [Lindner et al. 78].

As in the case of modified carrier membranes an activity change towards higher activities is advantageous (Fig. 5.10).

In conclusion, most of the analytically relevant neutral carrier-based electrodes exhibit a dynamic behaviour that is dictated by diffusion processes [Morf and Simon 78].

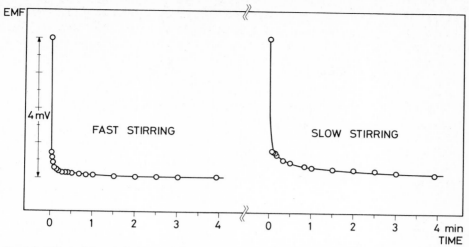

Fig. 5.8. Effect of the stirring rate on the EMF as a function of time for valinomycin-based solvent polymeric membrane electrodes after a stepwise change in the KCl activity [Morf et al. 75]. Points are experimental. Curves are calculated using Eq. (5.39)

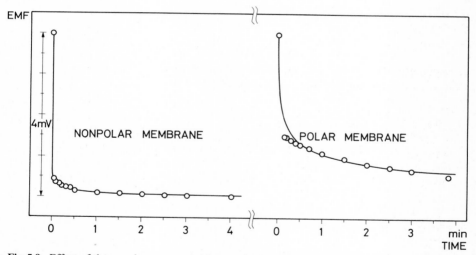

Fig. 5.9. Effect of the membrane composition on the EMF as a function of time for valinomycin-based solvent polymeric membrane electrodes after a stepwise change in the KCl activity [Morf et al. 75]. Points are experimental. Curves are calculated using Eq. (5.39)

The response times of microelectrodes with high internal resistances (due to the relatively low concentrations of charge-carrying species and the small membrane areas ($\sim 10^{-8}$ cm^2)) never reflect dynamic mass transfers towards and/or within the membrane. Instead they are manifestations of the RC time constant of the measuring circuit or of the ion-selective microelectrode itself:

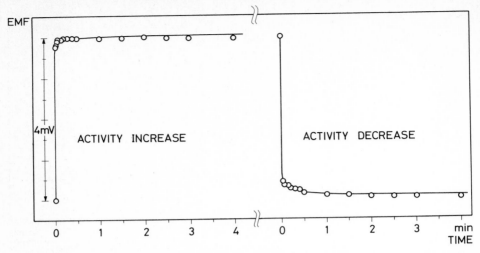

Fig. 5.10. Effect of the change in sample activity on the EMF as a function of time for valinomycin-based solvent polymeric membrane electrodes after a stepwise change in the KCl activity [Morf et al. 75]. Points are experimental. Curves are calculated using Eq. (5.39)

$$EMF_t = EMF_\infty - (EMF_\infty - EMF_0)e^{-t/RC},\tag{5.42}$$

where
EMF$_0$: EMF prior to the change in activity;
R   : internal resistance of the electrochemical cell;
C   : total capacitance of the amplifier input.

Although the IUPAC Commission on Analytical Nomenclature has recommended that response times should be stated in terms of t_{90} values [IUPAC 79] (time required for the EMF to undergo 90% of its total change, Fig. 5.11) many different definitions (t_α ($\alpha = 50, 90, 95, 99$), t*, (t($\triangle$t, $\triangle$E)) are still in use. The large number of conventions has brought about a measure of confusion into the discussion of response times. The initially proposed convention to use t* values [Guilbault et al. 76] (t*: time required for the EMF to approach within 1mV of its final value) was discarded since the t* values yield different results for mono- and divalent ion-selective electrodes and also for activity steps of different size [Lindner et al. 78; Uemasu and Umezawa 82]. In practice, it is often not easy to reliably determine the final value of the EMF after an activity step. It is therefore difficult to correctly specify the t_α values [Uemasu and Umezawa 82]. Nevertheless, they are an useful description of the time needed to approach the equilibrium state. Only recently has a new definition of the response time that is independent of the knowledge of the final value of the EMF been proposed [Uemasu and Umezawa 82] (see also [Lindner et al. 76]). This new method, using differential quotients t ($\triangle$t, $\triangle$E) (Fig. 5.12), offers several advantages [Uemasu and Umezawa 82; Lindner et al. 85]: the equilibrium potential need not be known; simple determinations are even possible for electrodes with very slow responses; parameters such as time constants can be determined; and finally the differential quotient t ($\triangle$t, $\triangle$E) is a more useful parameter in practice.

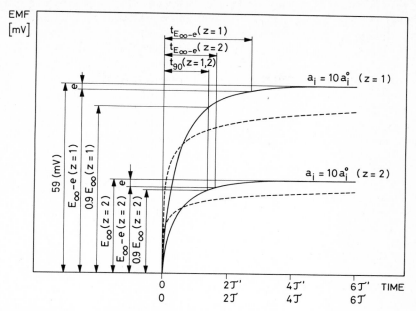

Fig. 5.11. Theoretical response times for mono- und divalent ion-selective electrodes calculated using Eqs. (5.39) *(dashed line)* and (5.37) *(solid line),* respectively [Linder et al. 78]

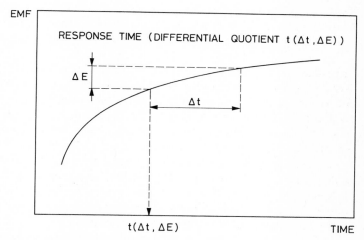

Fig. 5.12. Differential quotient t (Δ t, Δ E) used to describe the response characteristics of a membrane electrode cell assembly [Uemasu and Umezawa 82]

The design of experiments for the determination of response times depends on several factors including the order of magnitude of the expected response time. Slow electrode responses (several seconds to minutes) can easily be followed by simply exchanging two samples or by injecting a standard into the sample solution. In the case of fast responses (< 1s) it is necessary to use arrangements which do not disturb the response time of the electrode or overlap it with the mixing time of the solutions. Different techniques for the observation of short response times have been described: a fast dumping technique [Fleet et al. 74], a flow-through cell [Rangarajan and Rechnitz 75] and an injection technique [Lindner et al. 76; Lindner et al. 78] (see [Lindner et al. 85]). In addition special arrangements for the characterization of the response of microelectrodes have been proposed: a rapid immersion method (Vyskočil and Kříž 72), a flow-through cell [Ujec et al. 79; Ujec et al. 80], an ionophoretic method [Lux and Neher 73] and a method in which the phase boundaries of laminar-flowing electrolytic solutions are crossed [Janus and Lehmenkühler 81 b].

Obviously, response times should also specify the composition of the solutions used, the stirring or flow rate, the temperature and the pretreatment of the membrane electrode.

5.10 Lifetime

With regards to the lifetime of an electrode, completely different demands are imposed on sensors for clinical or electrophysiological applications. In clinical analyses, the solvent polymeric membrane electrodes should be suitable for continuous use in undiluted biological samples (serum, plasma, blood, urine) for at least several months. In contrast, liquid membrane microelectrodes need only have lifetimes of hours or days. Nevertheless, it would be advantageous to have microelectrodes with long lifetimes.

At present, with the current knowledge of membranes and electrode bodies for solvent polymeric membrane sensors, the lifetime of an electrode is not limited by factors such as mechanical defects, electrical shunts, chemical deterioration, or surface contamination, but rather by the loss of the neutral carrier or plasticizer from the membrane into the sample solution. The membrane components usually exhibit a moderate molar mass of about 500 to 1000 g/mol, and so the mobility of these molecules in the membrane is relatively high. As it is not possible to use membrane components of infinitely high lipophilicity [Oesch and Simon 79], a gradual loss of components from the membrane phase has to be taken into account. In plasticized PVC-membranes a loss of the plasticizer down to a level of less than 30 wt.-% plasticizer causes an extremely high membrane resistance (Fig. 4.7) which severely hinders EMF measurements. Similar problems may also result from a reduction in the carrier concentration since this causes a breakdown in the ion selectivity and electrode function of the sensor [Oesch and Simon 80]. Neutral carriers are normally embedded in the membrane at typical concentrations of about 10^{-2} M; these carriers have limiting concentrations of about 10^{-4} M (Fig. 4.3) [Oesch and Simon 80]. It has been shown that the lifetime of solvent polymeric membranes is governed by the rate of loss of the membrane components from the membranes.

In contrast, the lifetime of liquid membrane microelectrodes is in most cases not limited by a similar loss of components. This is mainly due to the special geometry of the electrode bodies (extremely small ratio of the membrane surface to the volume of the membrane solution) and the lipophilicity of the carriers and solvents in-

Table 5.4. Rate constants of the kinetic processes involved in the transfer of membrane components from the membrane phase into the sample solution [Oesch et al. 85 a]

Kinetic process	Rate constants [cm s^{-1}]
a) interfacial exchange reaction	k'
b) diffusion through a stagnant boundary layer in the sample	$\dfrac{D_s}{k\delta_s}$
c) linear diffusion in the sample	$\dfrac{1}{k}\left(\dfrac{D_s}{\pi t}\right)^{1/2}$
d) spherical diffusion in the sample	$\dfrac{D_s}{k\, r_o}$
e) linear diffusion in the membrane	$\left(\dfrac{D_m}{\pi t}\right)^{1/2}$

k' : exchange reaction rate constant of the first order [cm s^{-1}];
k : partition coefficient between membrane and sample;
D_s : diffusion coefficient in the sample [cm^2 s^{-1}];
δ_s : thickness of the stagnant boundary layer in the sample [cm];
r_o : radius of the formally spherical membrane [cm];
D_m : diffusion coefficient in the membrane [cm^2 s^{-1}]
Values of these parameters for several relevant carriers and plasticizers have been given in the literature [Oesch and Simon 80; Oesch et al. 85 a] (see also Table 4.2).

volved. The lifetime of currently available neutral carrier microelectrodes is dictated by mechanical (e. g. breaking, plugging), electrical (e. g. leaking paths) or chemical events (e. g. surface contamination, poisoning of the membrane solution). The origin of these (essentially technical) problems is discussed in Chap. 6 and 7.

Theoretical models used to predict the lifetime of neutral carrier membranes have been discussed [Oesch and Simon 79; Oesch and Simon 80; Oesch et al. 85 a]. These models are based on the kinetics of the loss of components from the membrane phase into the sample solution. An extensive treatment of the mathematical derivation and the assumptions made has already been given [Oesch et al. 80]. Later, an extended but semi-theoretical treatment of kinetic processes was presented [Oesch et al. 85 a]. The latter approach has proved to be especially useful in a discussion of the lifetimes of real membrane electrode systems [Oesch et al. 85 b]. Table 5.4 summarizes the rate constants of the processes that take place and their characteristic parameters.

The process with the smallest rate constant will be rate-controlling. Since the rate constants of processes c) and e) (Table 5.4) are time-dependent, a change in the rate-controlling process during the use of the electrode will occur at the transition time t^{tr} [Oesch et al. 85 a], i.e. the time at which the rate constants of the processes being considered are equal. Two specific examples are discussed below: the continuous use of a Ca^{2+}-selective solvent polymeric membrane in a flow-through system and the use of a Ca^{2+} microelectrode in a relatively large sample volume (e. g. extracellular space).

The first example deals with the incorporation of an available optimized Ca^{2+}-selective membrane into a flow-through electrode system [Ammann et al. 85 b]. The geometrical parameters are: circular membrane surface area $A = 1$ mm^2 ($r_o = 5.6 \cdot 10^{-2}$ cm); membrane volume $V_M = 4$ mm^3; formal boundary layer $\delta_s = 20$ μm (for a laminar flow-rate of the sample of 1.8 ml/h) [Oesch et al. 85 a]. The initial concentration of the neutral carrier ETH 1001 within the membrane is $5 \cdot 10^{-2}$ M [Ammann et al. 85 b]. It can be seen from Fig. 4.3 that a 100-fold decrease in the carrier concentration will cause a breakdown in the electrode. The time required to reach this limit is called t^{lim} [Oesch et al. 85 a]. Obviously, t^{lim} represents the lifetime of the electrode. In the flow-through arrangement described above, the kinetic processes b) and e) (Table 5.4) have to be considered for the efflux of the carrier from the membrane. Using the parameters for the carrier ETH 1001 ($D_s = 10^{-5.5}$ cm^2 s^{-1}; k (membrane/serum) $= 10^{2.9}$; Dm $= 10^{-8.7}$ cm^2 s^{-1} [Oesch et al. 85 a]) in Eq. (5.43) the rate constant of process b) was calculated to be $2 \cdot 10^{-6}$ cm s^{-1}, whereas the rate constant of process e) is time-dependent. A transition time t^{tr} can therefore be defined for the two processes b) and e) (cf. Table 5.4):

$$t^{tr} = \frac{D_m \ \delta_s^2 \ k^2}{\pi \ D_s^2}. \tag{5.43}$$

With the above parameters, t^{tr} is 156 s, i.e. during the first 156 s of contact with the serum the efflux of ETH 1001 will be rate-controlled by the diffusion through the boundary layer in the sample. After this time the rate is controlled by the diffusion of ETH 1001 within the membrane towards the contacting surface. Hence, a concentration gradient will be generated and the concentration at the surface will therefore decrease until it reaches the critical concentration c_m^{lim}, at which point a significant deterioration of the electromotive behaviour will occur (Fig. 4.3). As shown above, the disturbance arises if c_m^{lim} equals 1% of c_m^o, where c_m^o is the initial concentration in the membrane.

$$t^{lim} = \left(\frac{1}{t^{lim,b}} + \frac{1}{t^{lim,e}} \right)^{-1}, \tag{5.44}$$

where

$$t^{lim,b} = \frac{V_m \ \delta_s \ K}{A \ D_s} \ln \frac{c_m^o}{c_m^{lim}}, \tag{5.45}$$

$$t^{lim,e} = t^{tr} \left(\frac{c_m^o}{c_m^{lim}} - 1 \right)^2 \tag{5.46}$$

From Eq. (5.44) it can be calculated that the lifetime t^{lim} is 18 days, i.e. after 18 days of continuous use in serum, the concentration of the carrier at the surface of the membrane will have decreased to the minimum working concentration. However, due to a concentration gradient, the entire membrane still contains 84% of the initial amount of the carrier. If sufficient time is allowed for a re-equilibration of this gradient, the electrode will again exhibit satisfactory behaviour and its use can be extended. Such a re-equilibration can also be brought about by a temporary stop in the flow or simply by a change from serum to an aqueous solution (e. g. calibration solution).

The second example involves a neutral carrier Ca^{2+} microelectrode (Sect. 9.6. 3.3, [Lanter et al. 82]) immersed in an almost unlimited amount of extracellular fluid (serum is assumed). For this arrangement the kinetic processes d) and e) (Table 5.4) have to be considered to describe the efflux of the carrier ETH 1001. The required parameters of this two-phase system membrane solution/sample solution are: radius at the tip of the microelectrode $r_o = 5 \cdot 10^{-5}$ cm, surface area at the tip of the microelectrode $A = 7.8 \cdot 10^{-9}$ cm^2, volume of the membrane solution $V_m = 8 \cdot 10^{-5}$ cm^3 (for $r_o = 5 \cdot 10^{-5}$ cm, filling height $5 \cdot 10^{-1}$ cm, taper angle of the micropipette $= 1.4°$), diffusion coefficient of ETH 1001 in the sample $D_s = 10^{-5.5}$ cm^2 s^{-1}, diffusion coefficient of ETH 1001 in the membrane solution (o-nitrophenyl-n-octyl ether) $D_m = 10^{-6.3}$ cm^2 s^{-1} and partition coefficient between membrane solution and serum $k = 10^{2.9}$. At the transition time t^{tr}, the rate constants of the two processes d) and e) are equal and therefore,

$$t^{tr} = \frac{D_m \, r_o^2 \, k^2}{\pi \, D_s^2}. \tag{5.47}$$

With the above parameters, t^{tr} is 25 s, i.e. for the first 25 s of contact the efflux of ETH 1001 will be rate-controlled by the spherical diffusion in the sample. Thereafter, the diffusion of ETH 1001 within the membrane toward the sample solution governs the efflux.

Consequently, a concentration gradient will be generated and the surface concentration of the carrier decreases until the critical concentration c_m^{lim} (1% of c_m^o) is reached. Again a loss in electrode performance occurs. Using Eq. (5.48),

$$t^{lim} = \left(\frac{1}{t^{lim,d}} + \frac{1}{t^{lim,e}} \right)^{-1}, \tag{5.48}$$

with

$$t^{lim,d} = \frac{V_m \, r_o \, k}{A \, D_s} \ln \frac{c_m^o}{c_m^{lim}}, \tag{5.49}$$

$$t^{lim,e} = \frac{D_m \, r_o^2 \, k^2}{\pi \, D_s^2} \left(\frac{c_m^o}{c_m^{lim}} - 1 \right)^2, \tag{5.50}$$

the lifetime is estimated to be about 68.6 days. Thus, after about 2 months of continuous contact with a formally infinite volume of serum, the neutral carrier concentration at the membrane surface would have decreased to the minimum working concentration ($c_m^{lim} = 0.01 \, c_m^o$). In practice, however, the volume of the sample solution is usually restricted to the volume of the cell (picoliter range) and the time of contact with the protein-containing sample is several hours at most. For a cell volume of $4.2 \cdot 10^{-9}$ cm^3 (spherical cell with a radius of 10 μm) the equilibrium for the distribution of the carrier between the membrane solution and the cytosol is theoretically attained after about 4 s (with the assumption that the cell wall is not permeable to the extracted carrier). Under these conditions, the efflux of the neutral carrier ETH 1001 should stop after 4 s. Obviously, the lifetime should be much longer under these circumstances than under the previously mentioned condition of an unlimited sample space. An intermediate situation between the two extremes (infinite sample volume and nonpermeable cell wall) is more realistic. Under such conditions, and taking into account the fact that microelectrodes are mainly in contact with aqueous solutions (conditioning, calibration, storage), the efflux of membrane components has very little influence on their lifetimes. However, attention has to be payed to the efflux of membrane components if one is concerned about a contamination of the cell (Sect. 7.6).

5.11 Anion Interference

It has been repeatedly observed that interference by lipophilic sample anions can occur in the cationic response of neutral carrier liquid membrane electrodes [Buck et al. 73; Morf et al. 74a; Morf et al. 74b; Seto et al. 75; Jenny et al. 80b]. Lipophilic anions Y^- are by definition much more soluble in the organic membrane phase than hydrophilic anions such as chloride or phosphate. Therefore, lipophilic sample anions facilitate the following process:

$$I^{z+} (aq) + z Y^- (aq) + n L (m) \rightleftharpoons IL_n^{z+} (m) + z Y^- (m) \tag{5.51}$$

As a result, the concentration of permeating anions within the carrier membrane is dramatically increased. In extreme cases this will lead to a loss in cation permselectivity. If such lipophilic anions are present in the membrane phase, the concentration of free neutral carrier at the membrane surface will decrease because of the formation of complexes. Finally, an anionic response function will be induced by the presence of charged complex cations IL_n^{z+} [Morf and Simon 78; Morf 81]. In carrier membranes without additives, the unfavourable transition from a cationic to an anionic response is especially likely when the membrane exhibits a high dielectric constant (polar plasticizer, i.e. high extraction constants for the equilibrium in (5.51)) (see Fig. 5.13) [Morf et al. 74a]. Consequently, nonpolar membrane phases can be used to reduce the interference from sample anions. This effect can easily be exploited in membranes exhibiting a selectivity for monovalent cations since these membranes also exhibit their optimal selectivities at low dielectric constants (see Sect. 4.1.2). However, for membrane electrodes with selectivities for divalent cations it is often not possible to reduce the anion interference by using nonpolar plasticizers. In such cases the loss of divalent/monovalent selectivity would be prohibitive. Another way of reducing or eliminating anion interference has been proposed [Morf et al. 74a; Morf et al. 74b; Morf and Simon 78]. The permanent incorporation of lipophilic anionic sites R^-, for example tetraphenylborate, into the membrane phase would, to a large extent, prevent the uptake of sample anions and

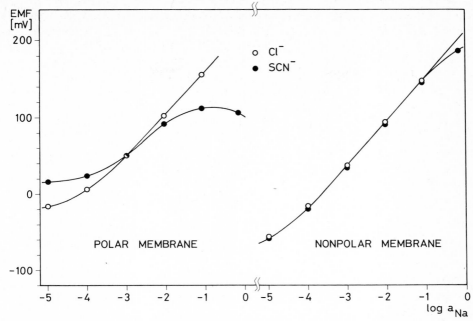

Fig. 5.13. EMF response of PVC membrane electrodes based on a neutral carrier (N,N,N′,N′-tetra-benzyl-3,6-dioxaoctane diamide) to Na^+ activities in the presence of hydrophilic (Cl^-) or lipophilic (SCN^-) ions [Morf 81]. *Left:* polar membrane solvent (*o*-nitrophenyl-n-octyl ether). *Right:* non-polar membrane solvent (dibenzyl ether)

therefore drastically reduce their interference. The addition of salts containing lipophilic anions reduces the free-carrier concentration in the membrane:

$$c_L = c_L^{tot} - nc_{IL_n} = c_L^{tot} - \frac{n}{z_i} c_R, \qquad (5.52)$$

where

c_L : concentration of the free carrier within the membrane;
c_L^{tot} : total concentration of the carrier within the membrane;
c_{IL_n} : concentration of the carrier complexes within the membrane;
n : number of carrier molecules L per complex;
c_R : mean concentration of lipophilic anions R^- within the membrane.

A total consumption of the free carrier L as complexes $IL_n^{z_i}$ is indicated by the conversion of the cationic response into an anionic response. Consequently, in order to obtain a Nernstian cationic electrode function, equation (5.53) must be fulfilled:

$$c_L^{tot} > \frac{n}{z_i} c_R. \qquad (5.53)$$

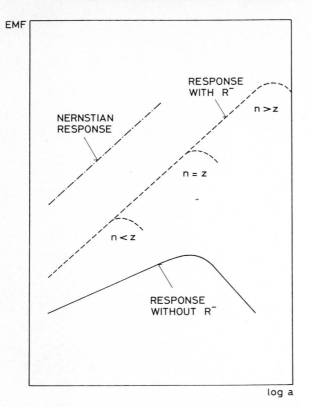

Fig. 5.14. Predicted influence of the membrane additive R^- (lipophilic anionic sites) on the EMF response of neutral carrier liquid membrane electrodes. For a given combination of a cation I^{z+} and an interfering anion Y^-, the complex stoichiometry (number n of carrier molecules L per complex) is varied [Morf and Simon 78]

A detailed theoretical study of the EMF response of neutral carrier-based membranes modified with anionic sites R^- (i.e. cation-exchangers) yields the results schematically presented in Fig. 5.14 [Morf and Simon 78; Morf 81].

Accordingly, the slope of a disturbed EMF response (lower curve, Fig. 5.14) in the analytically useful range of small sample activities approaches that of a Nernstian response if lipophilic anionic sites R^- are added to conventional neutral carrier solvent polymeric membranes. Depending on the stoichiometry n and charge z of the complex betweeen the ion to be measured and the neutral carrier, this linear range can be extended ($n > z$), remain nearly unchanged ($n = z$) or be reduced ($n < z$). Obviously, the addition of R^- is attractive for membrane systems with $n = z$ and especially for $n > z$, where the anion interference is shifted to very high activities. The beneficial effects of R^- have been verified by measurements with Na^+-selective (Fig. 5.15) [Morf et al. 74a], K^+-selective and Ca^{2+}-selective carrier membrane electrodes [Morf et al. 74b]. The observed response curves may be used as a rigorous check for the stoichiometry of the complexes within the membrane. Thus, the

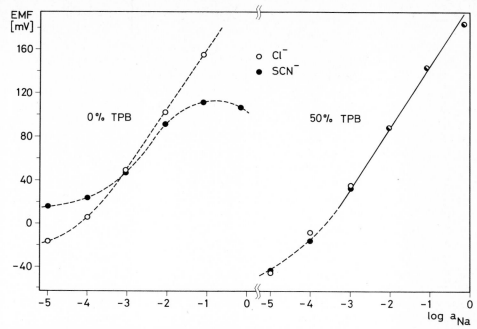

Fig. 5.15. EMF response of neutral carrier membranes as a function of Na$^+$ activities in the presence of different anions [Morf et al. 74a]. *Left:* membrane without lipophilic anionic sites R$^-$. *Right:* membrane with tetraphenylborate (equivalent to 50 mol.-% of the carrier concentration). Na$^+$ carrier: N,N,N′,N′-tetrabenzyl-3,6-dioxaoctane diamide, membrane solvent: *o*-nitrophenyl-n-octyl ether. *Solid line:* Nernstian response

Na$^+$ carrier discussed in Fig. 5.15 should form 1:2 complexes ($z = 1$ and $n = 2$, i.e. $n > z$ (Fig. 5.14)) within the membrane. The results from ^{13}C NMR studies are in good agreement with this conclusion [Morf et al. 74a].

5.12 Concluding Remarks

The potentiometric cell assemblies discussed above include an ion-selective electrode, a reference electrode, an electronic recording device and the sample solution. The two half-cells of microelectrodes can be mechanically coupled to form a double-barrelled microelectrode. Reference microelectrodes always contain silver/silver chloride half-cells without salt bridges, whereas macro reference electrodes may be of different types (Ag/AgCl, Hg/Hg$_2$Cl$_2$, etc.; with or without a salt bridge).

The potential across a liquid membrane is composed of a diffusion potential within the membrane and an equilibrium potential (Donnan potential) at the phase boundary. A useful general description of permselective membranes is obtained by neglecting the diffusion potential. The membrane potential is then mainly determined by the Donnan potential at the phase boundary of the membrane/sample solution. A theoretical treatment of the membrane potential leads to equations of the Nicolsky-Eisenman type which describe the relationship between the EMF and the logarithm of the measuring ion activity in the presence of interfering ions.

Selectivity factors may be expressed in terms of an ion-exchange equilibrium at the phase boundary of the membrane/sample solution. In an ideal case, the potentiometrically observed selectivity between two ions is equal to the ratio of the stability constants of the corresponding carrier complexes in water.

Both the detection limit and the selectivity factor depend on the experimental conditions. Detection limits measured in metal-buffered solutions may be several orders of magnitude lower than those obtained in unbuffered solutions. Selectivity factors are further influenced by the method of their evaluation.

Potentiometric measurements using the membrane electrode cell assemblies described above always yield mean ion activities, which, after correction of the measured EMF for liquid-junction potentials, can be converted to single-ion activities using nonthermodynamic conventions. Mean ion activities are related to free ion concentrations by extended Debye-Hückel formalisms.

A convenient estimation of the liquid-junction potentials of the reference electrode can be made using the Henderson equation. Highly concentrated equitransferent reference electrolytes (for instance 3 M KCl) minimize changes in the liquid-junction potential. For certain special experiments (in vivo monitoring, measurements in small single cells) other reference electrolytes may be more useful.

High performance flow-through electrode systems yield EMF stabilities and reproducibilities with standard deviations less than $100\ \mu V$. Representative values for measurements with microelectrodes are of the order of 0.2 to 0.5 mV.

Response times of electrochemical cells containing ion-selective membrane electrodes are determined by electrical relaxation processes, ion-exchange processes at the phase boundary of the membrane/sample solution and diffusion processes. The response time for neutral carrier membrane electrodes modified with lipophilic anionic sites is governed by the diffusion of ions in the aqueous layer, while the response time of unmodified electrodes is described by the diffusion processes within the membrane. In the case of microelectrodes with extremely high electrical resistances the time constant of the electrical circuit may control the dynamic behaviour. Various possibilities exist for reducing the response time of neutral carrier membrane electrodes. Often, sophisticated experimental setups have to be used in order to measure the intrinsic response time of the membrane.

The lifetime of neutral carrier microelectrodes is not limited by the loss of membrane components into the sample solution, but is determined by mechanical, electrical or chemical events.

Cation-selective neutral carrier membrane electrodes may lose their permselectivity if lipophilic anions are present in the sample solution. The use of nonpolar membrane solutions or the incorporation of lipophilic anionic sites into the membrane phase can drastically reduce or even eliminate such anion interference.

6 Construction of Liquid Membrane Microelectrodes

There are several excellent contributions describing the technical aspects of micro-electrodes for intracellular recordings [Thomas 78; Purves 81]. The book by R.C. Thomas discusses microelectrodes based on various types of membranes and gives much practical information and advice. R.D. Purves clearly outlines the physical, electrochemical and electronic principles of the microelectrode technique. The text gives an excellent description of the reference microelectrode methodology up to 1981 but does not include a discussion of ion-selective microelectrodes.

During the past few years, many new membrane materials as well as improved techniques for the construction of microelectrodes have been investigated. The progress has been especially pronounced in the field of liquid membrane microelectrodes. The following paragraphs focus on the technical aspects of the construction of neutral carrier-based microelectrodes and include some recent advances in the methods applied. It seems that the multitude of construction techniques initially used, i.e. each laboratory followed its own construction procedures, is now being replaced by generally accepted approaches which have resulted from diligent studies.

6.1 Glass Tubings

Micropipettes for liquid membrane microelectrodes can be fabricated from different types of glass (e.g. soda lime glass, borosilicates, aluminosilicates, quartz glass, etc.) which are used as tubings with diameters in the 1 to 3 mm range. Although they may differ considerably in their physical and chemical properties, almost all of them can be pulled to micropipettes. Nevertheless, some types (e.g. the borosilicate Pyrex) seem to be especially suited for the preparation of microelectrodes with very small diameters.

In attempting to compare and select a certain type of glass for use as a micropipette, the specific resistance, softening point and water resistivity of the glass are of particular importance (see Table 6.1).

The specific resistance of the glass, together with the resultant thickness of the glass wall of the micropipette, have a large influence on the occurrence of electrical leakage pathways at the very tip of the electrode.

The softening point (temperature where the glass deforms rapidly; at a viscosity of $10^{7.6}$ P [Hebert 69]) determines the parameters to be chosen for the pulling of the

Table 6.1. Properties of different types of glass used for pulling micropipettes

Glass	Softening point [°C]	Specific resistance (25 °C) [Ω cm]	Reference
soda lime glass	829	–	[Scholze 77]
borosilicate (Pyrex; Corning 7740)	820	10^{15} 10^{16}	[Hebert 69] [Lavallée and Szabo 69]
aluminosilicate (Corning 1720)	915	$< 10^{17}$	[Hebert 69]
quartz glass	1480	10^{13} (300 °C; 0.04 ppm Na)	[Scholze 77]

pipette (e.g. temperature, force of pulling). Since quartz tubings exhibit very high softening points (Table 6.1) they usually cannot be pulled using commercially available pullers.

The water resistivity of the glass should be as high as possible. Hydration of the glass can alter the geometry of the tip and therefore the electromotive behaviour of the microelectrode (e.g. tip potentials). The thickness of the hydrated layer at the glass surface appears to be a function of the glass composition and can be as small as a few monolayers (about 1 nm) [Eisenman 69]. On the other hand, certain special types of glass, such as the K^+-selective glass, show hydrated layers as large as 1 µm [Eisenman 69]. These layers have to be compared with the thickness of the glass wall at the tip of typical microelectrodes which is about 0.1 µm [Agin 69].

The nature, and consequently the properties, of the hydrated layers depend on the composition of the surrounding solution and on the length of exposure to the solution. Hydrated surfaces of any type of glass in contact with metal electrolytic solutions exhibit ion-exchange selectivity properties. In fact, this phenomenon is exploited in H^+- and Na^+-selective glass membrane electrodes, but represents a drawback for the selection of appropriate insulating pipettes for glass membrane microelectrodes and of inert glass for the preparation of reference and ion-selective liquid membrane microelectrodes. For example, the widely used Pyrex glass exhibits an undesirable response of approximately 18 mV per pH unit [Agin 69]. This ion-exchange property is a major problem encountered with reference microelectrodes (see tip potentials in Sect. 7.7).

The occurrence of electrical shunts through the glass wall at the tip of a micropipette is partly a further result of the properties of the glass. In a recent study on the surface and volume resistivity of Pyrex glass it was claimed that the hydration of Pyrex is unlikely to be responsible for electrical leakages [Coles et al. 85]. Indeed, it was found that the composition of Pyrex glass is only influenced to a depth of about 0.5 nm when the glass is treated for 48 h in hydrochloric acid [Wright et al. 80]. It is considered more likely that the shunts are caused by microcracks in the glass wall near the tip [Coles et al. 85]. Microcracks may develop during the pulling procedure. Consequently, the glass properties and the pulling procedure have become important parameters when attempting to avoid electrical leaks.

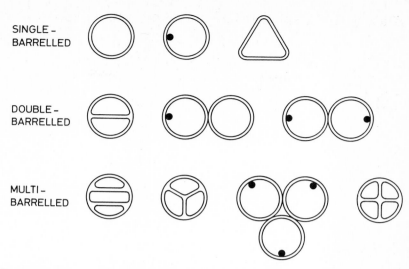

Fig. 6.1. Different shapes of commercially available glass tubings used in the preparation of micro-electrodes. *Black dots:* internal glass filaments

Borosilicates and aluminosilicates are most often used for neutral carrier liquid membrane microelectrodes. The additives B_2O_3 and Al_2O_3 serve as network formers. Both types have advantages for the fabrication of micropipettes. Borosilicates have a distinctly lower softening point which facilitates the pulling of micropipettes, while the aluminosilicates are highly water resistant [Thomas 78]. A common advantage of the two types of glass is their rather high specific resistance (Table 6.1). Both are commercially available in a variety of different configurations (Fig. 6.1).

The shape and thickness of the wall (outer wall as well as internal partition walls) are important parameters. Ultrafine pipettes ($<0.1\,\mu m$) are preferentially prepared from especially thin glass tubings. If theta-like tubings are considered, a thick septum is recommended to avoid an electrical connection between the two barrels. Furthermore, the choice of the appropriate shape of glass tubing will depend on the kind of physiological experiment to be performed (single-, double-, or multibarrelled microelectrodes supply different kinds of information), the filling technique in use (tubings with inner filaments containing sharp angles are much easier to fill) and the type of cell (problems of impalement due to the flexibility of the pipette).

There seems to be no general agreement concerning the importance of cleaning the glass tubing before pulling the micropipettes. Opinions vary from no treatment at all [Thomas 78; Purves 81] to very careful cleaning procedures using strong acids, organic solvents and distilled water [Garcia-Diaz and Armstrong 80; Tsien and Rink 80]. It seems sensible to treat the glass tubing for reasons of cleanliness and for reactivity in the silanization step (Sect. 6.6). However, in an ideal case, the pulled micropipettes should be treated rather than the unpulled tubing. After pulling, 99.98% of the surface near the tip is freshly exposed glass [Deyhimi and Coles 82], so that a previous treatment of the unpulled tubing does not seem to be profitable.

6.2 Micropipettes

A variety of different micropipettes (Fig. 6.2) can be prepared using the available glass tubings (Fig. 6.1). For the fabrication of micropipettes with tip diameters larger than about 0.5 μm any commercially available or home-built puller is adequate. However, for more demanding needs (tip diameters considerably smaller than

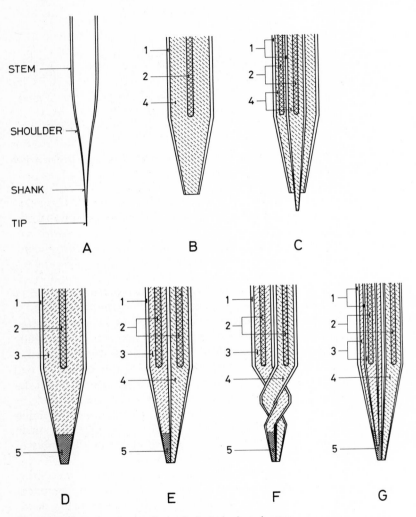

Fig. 6.2. Schematic cross sections of selected micropipettes.
A: customary nomenclature for parts of micropipette; *B:* single-barrelled reference microelectrode; *C:* coaxial reference microelectrode; *D:* single-barrelled liquid membrane microelectrode; *E:* double-barrelled liquid membrane microelectrode (theta-like); *F:* double-barrelled liquid membrane microelectrode (fused and twisted tubings); *G:* coaxial double-barrelled liquid membrane microelectrode (theta-like); *1:* glass pipette; *2:* chlorinated silver wire; *3:* internal filling solution; *4:* reference electrolyte; *5:* membrane solution (with or without PVC)

0.5 µm, multi-barrelled microelectrodes) more sophisticated pulling techniques have to be applied. Various features of the construction of high-performance micropipette pullers have been widely discussed (see e. g. [Brown and Flaming 77; Thomas 78; Purves 81; Flaming and Brown 82]). These studies have evaluated the influence of puller construction on the geometry and performance of microelectrodes. Some of the parameters considered are: the heater coil material (usually 90% platinum and 10% iridium), width (Fig. 6.3) and loop geometry; the pulling strength and pulling acceleration of the solenoid (the pull is achieved by the formation of a temporary magnetic field by passing a current through a wire surrounding an iron core); and the temperature (i. e. the rate of heating and the method of cooling). In fact, the optimal choice of these parameters may vary considerably with the puller construction (e. g. pullers without a cooling step versus airjet pullers; one-stage pullers versus two-stage pullers; single-sided pullers versus two-sided pullers).

To date, the smallest tip diameters are 0.045 µm (Fig. 6.4) (single-barrelled, borosilicate [Brown and Flaming 77]), 0.055 µm (Fig. 6.5) (double-barrelled, theta-like tubing [Brown and Flaming 77]) and 0.05–0.1 µm (single-barrelled, Pyrex tubing with filament [Greger et al. 83; Greger and Schlatter 83]). At present, such ultrafine micropipettes are only useful for reference microelectrodes. Smallest tip diameters of different types of microelectrodes are discussed in Table 2.1.

In puncturing a cell, microelectrodes are only rarely broken. However, impalements may be difficult because of the high flexibility of the shank of the microelectrode (Fig. 6.2, A). The mechanical properties of micropipettes depend on the pull-

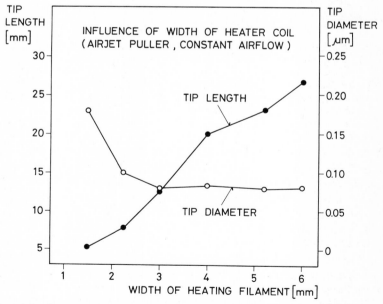

Fig. 6.3. Tip length and tip diameter of micropipettes as a function of the width of the heating filament (trough-type, airjet puller) [Flaming and Brown 82]. Error bars (± 1 std.dev.) are approximately the size of the circles indicated. Redrawn from Flaming and Brown, J. Neurosci. Meth., 1982 [Flaming and Brown 82]

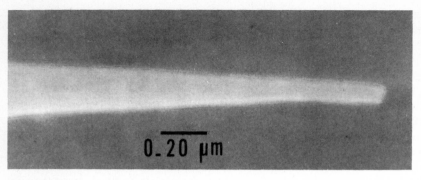

Fig. 6.4. Scanning electron micrograph of a single-barrelled micropipette [Brown and Flaming 77]. Measured tip diameter: 0.075 μm. The true outer diameter of the micropipette is 0.045 μm (i.e. after correction for gold coating). Scale: 0.2 μm. Courtesy of K.T.Brown and D.G.Flaming. From Brown and Flaming, Neuroscience, 1977. Reproduced by copyright permission of Pergamon Press Ltd., Oxford

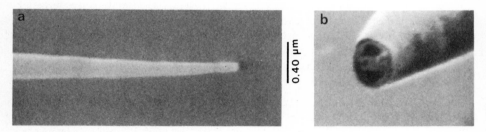

Fig. 6.5. Double-barrelled theta-like micropipette observed by scanning electron microscopy [Brown and Flaming 77]. *right:* theta-like separation of the barrels near the tip (broken to 0.4 μm); *left:* tip of the intact double-barrelled micropipette: 0.085 μm. The true outer diameter of the micropipette is 0.055 μm (i.e. after correction for gold coating). Scale: 0.4 μm. Courtesy of K.T.Brown and D.G.Flaming. From Brown and Flaming, Neuroscience, 1977. Reproduced by copyright permission of Pergamon Press Ltd., Oxford

ing (e.g. length of the shank) and the shape of the tubings. The flexibility of theta-like double-barrelled micropipettes (E in Fig. 6.2, Fig. 6.5) is almost the same in all directions. In contrast, twisted double-barrelled pipettes (F in Fig. 6.2) exhibit the highest flexibility perpendicular to the line joining the centres of the two barrels. Generally, the flexibility of microelectrodes can be minimized by shortening the shank. This is achieved by using airjet pullers in which the heating filaments are rapidly cooled after pulling [Brown and Flaming 77]. The same effect can be achieved by adjusting the geometry and width of the loop of the heating filament (Fig. 6.3). It is possible to produce shank lengths between 6 and 27 mm for constant tip diameters of about 0.1 μm or less [Flaming and Brown 82]. Both long and short shanks have been used in certain cell preparations. For example, fine micropipettes with long shanks are needed in neurophysiological measurements since a considerable depth of tissue has to be penetrated before reaching the cell of interest. On the other hand, micropipettes with short shanks and large taper angles (e.g. 11°

[Chowdhury 69]) are used in patch-clamp techniques and are optimal for liquid membrane microelectrodes because of their low membrane resistances (see Eq. (4.2)).

Specific difficulties are encountered in the pulling of double- or multi-barrelled micropipettes. The septum separating the barrels must not be damaged, as even minute damage to the glass wall between the barrels can result in a mixing of the reference electrolyte and membrane solution. However, the risk of chemical contamination and electrical coupling can be considerably decreased by using theta-like tubings with especially thick septa [Brown and Flaming 77].

6.3 Single- Versus Double- or Multi-Barrelled Microelectrodes

Today, both single- and double-barrelled microelectrodes are widely used (see Table 10.1). Due to technical difficulties triple- (e.g. [Fujimoto and Honda 80; Dresdner and Kline 85; Harvey and Kernan 84b; Harris and Symon 81; Dufau et al. 82; Ullrich et al. 82]) and four-barrelled (e.g. [Silver 76; Kessler et al. 76b]) microelectrodes are only rarely applied. The choice of single- or double-barrelled micropipettes is influenced by many factors including the cell type, cell dimensions and technical considerations.

An intracellularly positioned single-barrelled ion-selective microelectrode, forming part of a cell assembly together with an external bath reference electrode, measures an EMF that is related to the intracellular ion activity, but is superimposed on the cell resting potential. Therefore, in order to evaluate the intracellularly measured EMF relative to a preceding external calibration the superimposed resting potential has to be subtracted from the measured EMF. Consequently, a separate determination of the resting potential of the same cell or of one or several comparable cells is necessary:

$$\Delta E_{cell\ A} = \Delta E_{M,\ cell\ B} + s \log \Delta a_{i,\ cell\ A,} \qquad (6.1)$$

where

$\Delta E_{cell\ A}$	:	difference between the EMF of a calibration solution and the single cell A;
$\Delta E_{M,\ cell\ B}$	:	resting potential measured separately in the single cell B;
s	:	Nernst factor; and
$\Delta a_{i,\ cell\ A}$	:	difference between the activity of the measuring ion I in the calibration solution and the single cell A.

Obviously, Eq. (6.1) allows the determination of the intracellular ion activity of I (see Sect. 7.3). However, as shown in Table 6.2, the resting potentials of different single cells of the same type can vary significantly. These variations are not only due to intercellular deviations. The properties of the microelectrodes (tip diameter, tip potentials, etc.) and the quality of the cell penetration (cell damage, cell contamination) are also factors, which contribute to the uncertainty in the measurement of the ion activity.

Table 6.2. Examples of deviations in resting potentials of single cells of the same type (all resting potentials are given as positive values of n measurements)

Cell	Resting potential [mV]	Remarks	References
sartorius fibre (frog)	78.4 ± 5.3 (n = 1000)		[Ling and Gerard 49]
proximal tubule (bullfrog)	65.9 ± 5.9 (n = 24)	in vivo	[Fujimoto and Honda 80]
sartorius muscle (bullfrog)	68.2 ± 2.3 (n = 8)	in vivo	[Fujimoto and Honda 80]
Purkinje fibre (sheep)	75.5 ± 0.3 (n = 20)	bevelled tip	[Isenberg 79]
	74.1 ± 2.2 (n = 20)	unbevelled tip	[Isenberg 79]
cortical thick ascending limb (rabbit)	72.2 ± 2.0 (n = 54)	basolateral	[Greger et al. 83]
sartorius muscle (frog)	63.1 ± 12.8 (n = 13)	in vivo	[Matsumura et al. 80b]
	65.5 ± 13.1 (n = 10)	in vitro	[Matsumura et al. 80b]
distal tubule (amphibian)	71.0 ± 0.9 (n = 46)	peritubular	[Oberleithner et al. 82a]
	82.9 ± 1.3 (n = 16)	luminal	[Oberleithner et al. 82a]
	9.2 ± 0.6 (n = 17)	transepithelial	[Oberleithner et al. 82a]

Using larger cells it is possible to simultaneously impale the same cell (i.e. A = B in Eq. (6.1)) with an ion-selective and a reference microelectrode (e.g. [de Laat et al. 75; Picard and Dorée 83]). Furthermore, an almost identical situation can be achieved in electrically coupled cells by impaling neighbouring cells with an ion-selective and a reference microelectrode. For example, intracellular ion activities of Purkinje fibres can be evaluated by impaling an ion-selective and a reference microelectrode in different single cells about 200–600 μm apart [Hess et al. 82]. If the resting potential is equally measured by the two microelectrodes the correction for the resting potential is optimal. Under these conditions, the approach with single-barrelled microelectrodes corresponds to measurements with double-barrelled microelectrodes.

The necessity of performing two separate measurements on different cells or of impaling the same cell or electrically coupled cells with two single-barrelled microelectrodes can be circumvented by using double-barrelled microelectrodes:

$$\Delta E_{cell\ A,\ diff} = \Delta E_{cell\ A} - \Delta E_{M,\ cell\ A}$$
$$= s \log \Delta a_{i,\ cell\ A,} \tag{6.2}$$

where

$\triangle$ E$_{cell A, diff}$: net difference between the EMF of a calibration solution and a single cell A (differentiated from the resting potential $\triangle$ E$_{M, cell A}$ of the same single cell A).

Thus, double-barrelled microelectrodes allow a simultaneous recording of changes in both the resting potential and the EMF corresponding to the activity of the ion to be measured. The two potential differences are detected in optimal spatial proximity. This can be an advantage when measurements in excitable tissues are made since locally generated electrical potentials are cancelled out [Somjen 81]. Fast transients such as nerve spike potentials usually do not cancel because the time constants of the dynamic response of the ion-selective microelectrode and the reference microelectrode are different [Somjen 81]. As a result such fast transients remain a problem.

The localization of the tips of fine microelectrodes in tissues can be accomplished by using multi-barrelled microelectrodes. Coloured solutions are ejected from the reference barrel [Frömter et al. 81]. By observing the distribution of this dye with a microscope, the penetrated intercellular space or subcellular compartment can be localized. The procedure is especially helpful if the potential of the organelle to be penetrated is the same as that of the cytosol. The puncture can be observed although the penetration is not indicated by any electrical potential difference.

In polar cells where electrical and activity gradients may exist, it is important to measure the resting potential and the ion activity at almost the same points. Epithelial cells are often too small to accommodate two separate single-barrelled microelectrodes, so that the use of double- or triple-barrelled microelectrodes is highly recommended (see Sect. 8.5).

In physiological experiments double- and multi-barrelled microelectrodes are superior to single-barrelled ones. Unfortunately, the construction of the former is very demanding and intricate, i.e. single-barrelled microelectrodes are often preferred because of the ease of fabrication. Nevertheless, considerable progress in the preparation of multi-barrelled microelectrodes has been achieved. For example, the production of theta-like and other compartmentalized capillaries has been achieved. The filling of such micropipettes has been facilitated by the presence of relatively sharp angles in such tubings which increase the capillary forces. Progress has also been made in silanization procedures. The selective silanization of an ion-selective barrel can now be quite easily carried out [Fujimoto and Honda 80; Zeuthen 80; Deyhimi and Coles 82]. Even if the different barrels are properly pretreated and filled, they can still be electrically interconnected. Such shunts may be due to cracks in the septum or an insufficient thickness and/or strong hydration of the separating glass wall. Appropriate glass tubings and careful pulling minimize these risks. Bevelling may also be helpful. A theoretical treatment of the electrical coupling between three barrels has been made by analyzing the corresponding equivalent circuits [Fujimoto and Honda 80].

6.4 Electrode Bevelling

In principle, a pulled micropipette is ready for use after it has been filled with membrane solution and the internal reference half-cell has been completed. However, it is often recommended that the tips of freshly prepared micropipettes should be broken or bevelled. The reasons for bevelling are a reduction in the electrode resistance, an improvement in the ability to puncture cells and a facilitation in the introduction of current of chemicals into the cell.

During the process of bevelling the circular tip of the micropipette is transformed into a tip shaped like a hypodermic needle (Fig. 6.6). The bevelling is performed on an evenly rotating surface containing the grinding material. The micropipette is held and advanced with great precision against the rotating plate. After treatment over a period of about 1 to 10 min sharp tips with relatively large ellipsoid openings are produced. It is claimed that the optimal angle of the bevelled part lies between 20° and 30° [Brown and Flaming 74; Chang 75; Baldwin 80].

During the bevelling tip-plugging by particles of the abrasive surface can be prevented, for example, by applying pressure on the filling solution [Brown and Flaming 79]. Abrasion products adhering to the outside of the tip can be removed by sonification [Brown and Flaming 79].

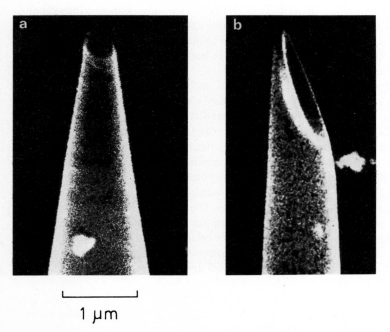

1 μm

Fig. 6.6. Scanning electron micrographs of an unbevelled *(left)* and a bevelled *(right)* microelectrode [Hess et al. 82]. Courtesy of P. Hess, P. Metzger and R. Weingart. From Hess, Metzger and Weingart, J. Physiol., 1982 [Hess et al. 82]. Reproduced by copyright permission of Cambridge University Press, Cambridge

Many different grinding techniques which are both rapid and precise have been proposed [Brown and Flaming 74; Kripke and Ogden 74; Brown and Flaming 75; Chang 75; Tauchi and Kichuchi 77; Ogden et al. 78; Brown and Flaming 79; Baldwin 80] and various bevelling apparatus are commercially available. Various grinding materials (e.g. alumina, diamond dust, ceric oxide) embedded or dispersed in different materials (e.g. polyurethane, saline solution (jet stream), agar discs, dry treatment on alumina particles) are in use [Brown and Flaming 74; Kripke and Ogden 74; Ogden et al. 78]. An alternative to the above grinding techniques is etching by hydrofluoric acid [Muheim 77].

By using 0.05 μm alumina particles embedded in a surface film of polyurethane on glass, micropipettes (Pyrex) with tip diameters smaller than 0.1 μm can be bevelled [Brown and Flaming 74]. Scanning electron micrographs of such tips clearly show that the surface produced is very smooth and that the cutting edge of these bevelled tips is very sharp. Measurements have shown the edge to be about 30 nm thick including the gold coating [Brown and Flaming 74]. There is some controversy concerning the fragility of these very sharp ends. For larger microelectrodes (> 0.5 μm) coarser abrasive surfaces (e.g. 0.3 μm alumina particles and, especially, diamond dust) are used [Brown and Flaming 79].

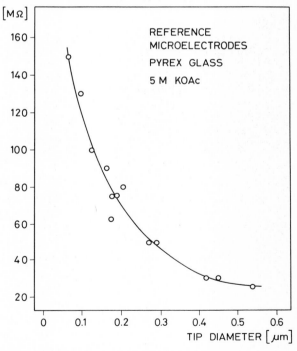

Fig. 6.7. Electrode resistance of bevelled microelectrodes as a function of their tip diameters [Brown and Flaming 74]. Scanning electron micrographs were used to determine the tip diameters at the base of the bevelled edge (for details see [Brown and Flaming 74]). Redrawn from Brown and Flaming, Science, 1974 [Brown and Flaming 74]

Most of the reports on bevelling deal with reference microelectrodes filled with reference electrolyte. The microelectrode resistance is usually continuously monitored [Ogden et al. 78; Brown and Flaming 79] or measured immediately after bevelling [Brown and Flaming 74]. The values of the resistances can then be used as a measure of the tip diameter (Fig. 6.7) [Brown and Flaming 74]. However, such correlations are only valid and useful if the pulling and bevelling occur under ideally constant conditions. It is not possible to exactly correlate the reported resistivity data (e.g. for a 3 M KCl reference electrolyte) with the tip diameters of microelectrodes that have been prepared in different laboratories.

Bevelling offers several advantages. For example, membrane resistances can be reduced by about 50% while keeping the tip diameter of the microelectrode constant. The penetration of cell membranes is clearly facilitated and the recorded potentials are often more stable and reliable [Isenberg 79]. Furthermore, the risk of tip-plugging is reduced. Uncertainties concerning the cell penetration and the generation of less perfect seals by bevelled tip openings seem to be unfounded [Isenberg 79]. In spite of all these advantages, bevelling is not universally employed [Purves 81].

Direct bevelling of microelectrodes already containing a membrane solution has only rarely been described. Normally the micropipettes for ion-selective microelectrodes are bevelled before they are filled with membrane solution [Tsien and Rink 80; Hess et al. 82].

6.5 Tip Observation

Three techniques can be used to estimate or exactly determine the inner and outer tip diameters of microelectrodes:

- light microscopy
- electron microscopy
- resistance measurements.

A fundamental limitation to any microscopic determination is the fact that the radiation of a given wavelength cannot be used to resolve structures smaller than its own wavelength. For a light microscope the resolution is determined by the wavelength of visible light, which ranges from about 0.4 μm to 0.7 μm. Accordingly, the limit of resolution is about 0.4 μm [Latimer 79; Tsien and Rink 80; Wischnitzer 81; Weakly 81]. Many investigators use light microscopy to estimate the size of microelectrode tips. For fairly large tip diameters (>1 μm) this method is accurate enough. However, it is not sufficiently accurate for very fine tips.

The limit of resolution can obviously be reduced by using electrons instead of visible light. Under optimal conditions electron microscopes exhibit resolutions of 0.1 nm. For complex samples such as biological ones, typical resolutions of about 2 nm have to be considered [Wischnitzer 81; Weakly 81]. Conventional transmission electron microscopy yields a two-dimensional picture of the object, whereas three-dimensional images can be obtained by using scanning electron microscopes. Many scanning electron micrographs of microelectrodes have been presented. This method has been very helpful in confirming tip diameters (see Sect. 6.4) and in giving confidence to the reliability of ultrafine microelectrode construction techniques (Figs. 6.4 and 6.5). Scanning electron microscopy has also been used in the transmission mode [Fry 75], which eliminates the need for coating the pipettes with, for example, gold. It is therefore possible to use the same microelectrode after three-dimensional microscopic observation. It further allows an estimation of the thickness of the glass wall [Fry 75].

It is often very difficult to evaluate the tip diameters of microelectrodes from the measured resistances (see [Lewis and Wills 81; Zeuthen 81b]). It was shown by model calculations that the assumption that smaller tip diameters result in higher resistances can be misleading [Purves 81]. The study demonstrated that resistances can depend on parameters such as the reference electrolyte, the sample electrolyte, the thickness of the glass wall and the angle of the taper. Consequently, a relationship between tip diameter and electrical resistance is only informative if a set of microelectrodes has been prepared under almost identical conditions (see Fig. 6.7). Indeed, micropipettes pulled on different days but with exactly the same puller adjustment may exhibit highly reproducible tip diameters (e.g. $0.382 \pm 0.003 \, \mu m$ ($n = 20$), measured by scanning electron microscopy [Lopez et al. 83b]). A study on the influence of the gas flow during the pulling of micropipettes with gas cooling has shown that microelectrodes with the same tip diameter but of different shank lengths (and therefore taper angle) may drastically differ in their resistance: tip diameter $0.049 \, \mu m$, shank length 10 mm: $90 \, M\Omega$ (5 M KOAc); tip diameter $0.048 \, \mu m$, shank length 6 mm: $44 \, M\Omega$ (5 M KOAc) [Brown and Flaming 77]. Furthermore, the resistances of microelectrodes are usually non-ohmic and therefore depend on the conditions of their evaluation [Robinson and Scott 73; Purves 81].

It has been suggested that micropipettes with well-defined tip diameters could be produced by mechanically breaking the tip. It is easy to bring a microelectrode into direct contact with a polished plexiglass rod while observing the tip and its mirror image on the plexiglass surface under a light microscope. When the electrode and rod are in direct contact, a slight tap on the table near the microscope causes the tip to break in steps of about $0.5 \, \mu m$ [Oehme 77]. A more complicated but more reproducible technique has been described for the preparation of predetermined tip diameters in the 3 to 200 μm range [Briano 83]. After calibration by electron microscopy, the bevelling of the micropipettes and a continuous monitoring of their resistances can also lead to well-defined tip diameters (Fig. 6.7) [Brown and Flaming 74].

Finally, another simple method of estimating tip diameters has been proposed in which the movement of the meniscus of the filling solution into the very fine tip is observed [Robinson and Scott 73].

6.6 Silanization

The first liquid membrane microelectrodes were prepared by Orme without any pretreatment of the glass walls, i. e. the organic membrane solution was brought into contact with the native glass surface [Orme 69]. However, glass surfaces are quite hydrophilic and therefore repel organic liquids. Silica surfaces can exhibit about

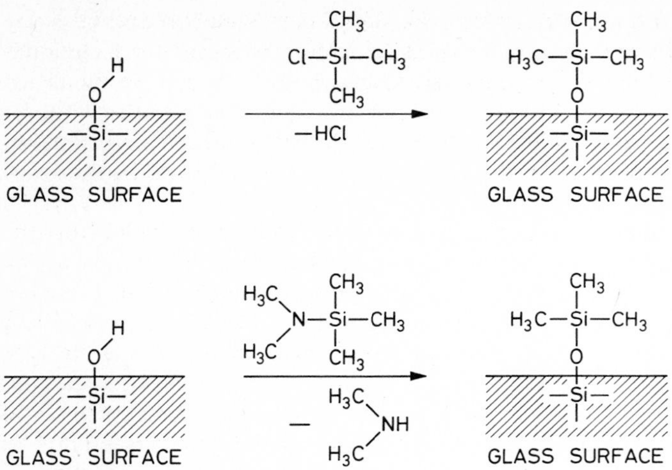

Fig. 6.8. Silanization of hydrophilic glass surfaces with reactive chlorosilanes *(above)* or aminosilanes *(bottom)*

8 μmol/m^2 of hydroxyl groups [Boksányi et al. 76; Deyhimi and Coles 82], which corresponds to about 4.6 free hydroxyl groups per 10 nm^2 of glass [Peri and Hensley 68]. This high density causes the incorporated hydrophobic membrane phase to be easily displaced by the aqueous solution surrounding the microelectrode. In order to avoid disturbances due to this phenomenon (e.g. extremely short lifetimes) the principle of silanization was introduced by Walker in 1971 [Walker 71]. It was already known from earlier studies [Patnode and Wilcock 46; Hertl 68a; Hertl 68b; Blackman 68; Hair and Hertl 69; Hertl and Hair 71] that free hydroxyl groups can be made to react with silicon compounds to yield a lipophilic glass surface (Fig. 6.8). Since 1971, the technique proposed by Walker has been modified and improved considerably. Today, the various techniques can be classified into liquid and vapour methods.

Walker's silanization method was proposed for the treatment of single-barrelled microelectrodes. It was based on the use of liquid silanization reagents [Walker 71]. The technique was later extended to double-barrelled microelectrodes [Lux and Neher 73]. In brief, the microelectrodes are dipped into an organic solution containing a reactive silicon compound (e.g. chlorosilanes, aminosilanes, siloxanes dissolved in carbon tetrachloride). A column (e.g. 200 μm) of this solution is drawn up into and expelled from the shank of the microelectrode several times. During this time, the silicon compound is expected to bind covalently to the free hydroxyl groups of the glass (Fig. 6.8). Afterwards, depending on the reagents and procedures applied, the treated micropipettes are either filled immediately with membrane solution (see e.g. [Lux and Neher 73; Oehme et al. 76; Nicholson et al. 78]) or are baked before the electrode is filled (see e.g. [Walker 71; Brown and Pemberton 76]). A serious drawback to this technique may arise if the procedure is performed in the presence of an aqueous internal filling electrolyte [Lux and Neher 73; Oehme et al. 76]. The reactive silicon compound is usually more reactive with water than with the glass surface, and as a result polymeric silicon compounds are formed (e.g. methylpolysiloxanes [Patnode and Wilcock 46]). The formation of such polymers can plug the very fine tips of the microelectrodes [Zeuthen 80; Tsien and Rink 80].

A second type of liquid silanization makes use of silicon oils [Fujimoto and Kubota 76; Garcia-Diaz and Armstrong 80; Oberleithner et al. 83a]. The silicon polymers do not covalently bind to the glass surface as the highly reactive monomers do. Instead they form a lipophilic layer on the glass wall. Surprisingly, relatively fine micropipettes can be treated with such oils without any risk of plugging the tips (< 1 μm [Garcia-Diaz and Armstrong 80]; 0.5 μm [Fujimoto and Honda 80]). A controlled treatment of the glass surface using defined silicon oils therefore avoids the difficulties which arise from the formation of silicon polymers in the hydrolysis of the reactive monomers.

The most sophisticated silanization technique is based on a vapour phase treatment with reactive silicon compounds. The procedure was first described by Coles and coworkers [Coles and Tsacopoulos 77; Deyhimi and Coles 82; Munoz et al. 83]. It is the result of a detailed study on the influence of the type of silanization reagent and the experimental conditions on the silanization process. The degree of lipophilicity of the glass surface was determined quantitatively by measuring the contact angle of the silanized cylindrical capillaries filled with distilled water. Using this procedure, the hydrophobicity of Pyrex glass after treatment with different chloro- and aminosilanes under various conditions was evaluated [Deyhimi and Coles 82]. The experiments have shown that aminosilanes are more reactive than chlorosilanes, and therefore silanize the glass surface more efficiently. In the case of Pyrex glass, an acid pretreatment of the freshly drawn micropipettes further increases the hydrophobization (optimal conditions found: 5 min in 60% HNO_3 [Deyhimi and Coles 82]). The optimal vapour treatment is achieved when dimethylamino trimethylsilane (Fig. 6.8) is applied to acid-treated and dried micropipettes at 250 °C for 5 min (Fig. 6.9) [Deyhimi and Coles 82]. This aminosilane exhibits several advantages including its relatively high vapour pressure, relative inertness towards hydrolysis and the formation of the non-corrosive by-product dimethylamine [Munoz et al. 83]. The above conditions are optimal for borosilicates. The applications of this technique using other types of glass will produce different hydrophobicities and surface resistivities [Munoz et al. 83].

An arrangement for the silanization of double-barrelled liquid membrane microelectrodes with gaseous reagents has been proposed (Fig. 6.9) [Munoz et al. 83]. Using this set-up it is possible to selectively silanize the ion-selective barrel. Of course, the same equipment can also be used for the silanization of single-barrelled microelectrodes. This method is relatively complex (Fig. 6.9) and time-consuming (only one micropipette is silanized at once). In a simpler procedure [Tsien and Rink 80] the vapour-phase silanization is performed in an oven as illustrated in Figure 6.10. A batch of micropipettes is mounted on a glass plate and covered with an up-turned glass beaker. After predrying the glass (30 min, 150 °C), a small amount of the pure silanization reagent is injected into the volume of the beaker. The silicon compound evaporates immediately and is allowed to react under optimal conditions (30 min, 200 °C; for dimethylamino trimethylsilane). The method is particularly suitable for single-barrelled microelectrodes. However, in using double-bar-

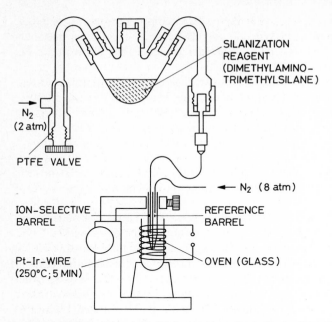

Fig. 6.9. Apparatus for the selective vapour-phase silanization of the ion-selective barrel of a double-barrelled microelectrode [Deyhimi and Coles 82]. A stream of nitrogen carries the vapour of the silanization reagent through the ion-selective barrel. The reference barrel is protected by a stream of nitrogen. The microelectrode is placed in a small oven to adjust the optimal temperature (for details see [Deyhimi and Coles 82]). Redrawn from Deyhimi and Coles, Helv. Chim. Acta, 1982 [Deyhimi and Coles 82]

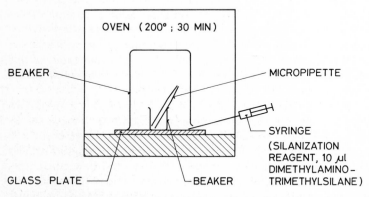

Fig. 6.10. Simple arrangement for the vapour-phase silanization of micropipettes (not drawn to scale)

relled pipettes both barrels will become hydrophobic (but see e.g. [Frömter et al. 81]).

Silanization has several important effects on the properties of liquid membrane microelectrodes. It increases the electrical resistivity of the glass surface and facilitates the uptake of the organic membrane solution. Furthermore, it increases the lifetime of a microelectrode by stabilizing the column of membrane solution within the pipette and/or preventing the hydration of the glass at the very tip [Dagostino and Lee 82]. It is interesting to note that the fabrication of microelectrodes without any silanization is also recommended [Khuri and Agulian 81].

6.7 Electrode Filling

The ease of filling micropipettes with an organic solution is dependent on many factors including the tip diameter, the quality of the silanization of the glass surface, the viscosity of the solution and the filling procedure itself. A large number of different filling methods have been proposed since almost every laboratory has developed its own technique. The various methods can be subdivided into back-filling (filling through the stem) and front-filling (filling through the tip) techniques.

In classical front-filling, a silanized tip is dipped into the organic membrane phase until a column of solution is spontaneously forced into the shank of the microelectrode by the capillary forces [Walker 71]. The internal filling solution is then brought into contact with the membrane solution by filling from the back. This step often only succeeds with great technical expense and skill [Walker 71].

In another version of front-filling, the non-silanized pipette is first filled with an internal filling solution [Lux and Neher 73; Oehme et al. 76]. After liquid silanization (see Sect. 6.6), the membrane solution is carefully withdrawn into the tip up to a height of about 200 μm. Capillaries containing an internal filament (Fig. 6.1) can be filled more easily. This method can also be applied to double-barrelled microelectrodes. The treatment of the ion-selective barrel is the same, but the reference barrel must be protected from a contamination by the membrane solution with a stream of nitrogen [Lux and Neher 73; Oehme et al. 76].

Front-filling procedures usually yield columns with a height of about 200 to 500 μm [Walker 71; Lux and Neher 73; Oehme et al. 76; Lee et al. 80a; Dufau et al. 82]. However, columns as small as 10–150 μm can also be obtained (see e.g. [Sheu and Fozzard 82]). Relative small filling heights decrease the membrane resistance. It is known, however, that the resistance of liquid membrane microelectrodes is only significantly reduced if the height of the column is considerably less than 200 μm [Ujec et al. 80; Tsien and Rink 80] (see [Fujimoto and Kubota 76; Lee and Uhm 81]). Hence, the column heights obtained with front-filling techniques are usually too large to effectively reduce the resistance of the microelectrode. The advantage of short membrane columns can only be realized when coaxial microelectrodes are used (Fig. 6.2, G). With these microelectrodes filling heights of 10–30 μm are feasible, but this requires daunting technical procedures [Ujec et al. 78; Ujec et al. 79; Ujec et al. 80]. Coaxial microelectrodes can only be constructed with tip diameters greater than 1 μm. Using such micropipettes, the membrane resistances of neutral carrier-based microelectrodes can be lowered by more than a factor of 10 [Ujec et al. 81]. Such microelectrodes exhibit extraordinarely short response times (e.g. 4 ms [Pumain et al. 83]).

In order to exploit the properties of poly(vinyl chloride)-containing membrane solutions (see 9.6.3.3) it is necessary to front-fill the micropipettes. In this way, the outer glass wall will be covered by a layer of PVC while a short column (about 5–100 μm) is simultaneously withdrawn into the tip [Marban et al. 80; Rink and Tsien 80]. Because of their high viscosity, PVC-containing membrane solutions must be diluted with tetrahydrofuran for a front-filling [Tsien and Rink 81; Lanter et al. 82].

Filling through the stem (back-filling) is becoming more and more popular with an increase in the use of microelectrodes with very fine tips (< 1 μm) and an ease in filling due to improved silanization procedures. With the help of injection needles [Fujimoto and Kubota 76] or plastic capillaries [Thomas 78; Lanter et al. 80] a small drop of membrane solution is inserted through the stem into the shank of the microelectrode. Usually, the membrane solution then spontaneously flows into the tip. However, if this does not happen vacuum or pressure can be applied. Normally, filling heights of several millimeters are chosen. Consequently, the diameter of the column in the stem is quite large. Thus it is easy to properly bring a capillary or injection needle filled with internal filling solution in to contact with the surface of a membrane solution. With back-filling the probability of the formation of air bubbles at the interface of the membrane and the filling solution is much lower than with front-filling.

6.8 Electrode Shielding

The importance of earthing and shielding the equipment for microelectrode studies is widely recognized [Thomas 78; Purves 81]. Earthing is particularly important for safety reasons and for screening against interferences. Many interference problems arise from capacitances in the input circuit. The input capacitance of the amplifier, the stray capacitance of the input leads and the nonimmersed stem of the microelectrode and the transmural capacitance of the immersed part of the glass shank and stem are the major contributions to the total shunt capacitance [Suzuki et al. 78; Purves 81]. High-quality amplifiers exhibit input capacitances as low as 0.8 pF. The disturbance by the stray capacitance of the connective part is usually minimized by a driven shield (capacitance compensation, capacitance neutralization) [Nastuk and Hodgkin 50; Lux and Neher 73]. However, a shielding of the stem of a microelectrode to avoid disturbances from the transmural capacitance is not widely applied. The transmural capacitance is a function of the length of the immersed part of the microelectrode. Thus, fluctuations in the depth of immersion in the sample solution can cause temporary changes in the measured voltage and a reduction in the response time. For micropipettes of typical internal and external radii and glass as the dielectric medium the transmural capacitance is of the order of 1 pF/mm [Suzuki et al. 78; Thomas 82; Purves 81].

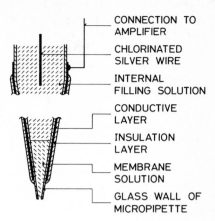

CONNECTION TO
AMPLIFIER

CHLORINATED
SILVER WIRE

INTERNAL
FILLING SOLUTION

CONDUCTIVE
LAYER

INSULATION
LAYER

MEMBRANE
SOLUTION

GLASS WALL OF
MICROPIPETTE

Fig. 6.11. Schematic cross section of a shielded
microelectrode

Several successful shielding techniques have been described to reduce the disturbance by transmural capacitance. These systems were mostly developed for use in excitable cells and employ conventional microelectrodes. Some investigators have shown that such shielding can also be applied to neutral carrier-based microelectrodes [Lewis and Wills 80; Wills and Lewis 80; Tsien and Rink 80; Hess et al. 82]. In a simple shielding procedure the shank of a liquid membrane microelectrode is coated with a conductive silver paint to within 0.5 mm of the tip [Tsien and Rink 80; Hess et al. 82]. A considerably better reduction of the disturbances can be achieved by connecting the conductive layer to the amplifier for capacitance compensation (Fig. 6.11) [Suzuki et al. 78; Lewis and Wills 80; Finkel and Redman 83]. Obviously the conductive part must be coated with an insulating layer in such a case, (Fig. 6.11). Above this, where the upper end is not in contact with the sample solution, the conductive material must be connected to the input by a feedback loop.

6.9 Reference Microelectrodes

For more than 30 years micropipettes filled with aqueous electrolytes have been used as reference microelectrodes (conventional microelectrodes, potential (PD) microelectrodes, indifferent microelectrodes, Ling-Gerard microelectrodes), to introduce a current (stimulation, voltage clamp) and to inject chemicals (ionophoresis, pressure injection). Reference microelectrodes measure the resting potential of a cell. However, it is not possible to unambiguously determine the true resting potential (see also Chap. 8). Small changes in the liquid-junction potentials and the tip potentials must be considered. Even more uncertain are errors due to cell damage (Sect. 7.5), cell contamination (Sect. 7.6) and suspension effects (Sect. 8.1).

The invention of fine reference microelectrodes with tip diameters of less than 1 μm is usually attributed to Ling and Gerard [Ling and Gerard 49]. The pipette preparation, the electrode filling and the theoretical understanding of interfering potentials (liquid-junction potentials (Sect. 5.7), tip potentials (Sect. 7.7)) have been

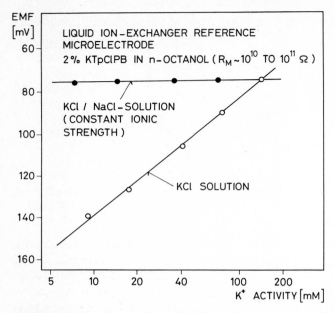

Fig. 6.12. The use of ion-exchanger liquid membrane microelectrodes as reference microelectrodes [Thomas and Cohen 81]. The EMF of the KTpClPB/octanol-based membrane electrode and a 3 M KCl electrode was determined in the solutions indicated. The slope of the response in solutions containing various amounts of KCl but of constant ionic strengths (through addition of NaCl) is 0.9 mV/pK. Redrawn from Thomas and Cohen, Pflügers Arch., 1981 [Thomas and Cohen 81]

continually improved ever since (see [Thomas 78; Purves 81] and this Chapter). Moreover, the contamination of cells by the reference electrolyte has been observed and shall be discussed seriously (see Sect. 7.6).

Several parameters influence the choice of reference electrolytes: the necessity of keeping the changes in the liquid-junction and tip potentials as small as possible, the need for chloride ions in an electrochemically well-defined inner silver/silver chloride half-cell, the risk of cell contamination by electrolyte leakage and the necessity of low microelectrode resistances. Unfortunately, it is impossible for a reference electrolyte to simultaneously guarantee minimal liquid-junction potentials, minimal tip potentials and minimal cell contamination. It seems that 0.5 M KCl or K_2SO_4/KCl solutions are optimal: changes in the liquid-junction potentials are still acceptable and the cell contamination is clearly reduced in comparison to a 3 M KCl solution [Fromm and Schultz 81]. Tip potentials should be small. They can be further minimized by the type of glass, the pretreatment of the micropipette and the filling procedure (Sect. 7.7) [Purves 81].

Today, reference microelectrodes with tip diameters of less than 0.1 μm can be prepared. These allow the measurement of resting potentials of very small cells (e. g. cardiac ventricular muscle (10–15 μm) [Marban et al. 80]; red cells (2 μm × 8 μm) [Lassen and Sten-Knudsen 68]; thick ascending limb cells (1–2 μm in height) [Greger and Schlatter 83]; Leydig cells (10 μm diameter) [Joffre et al. 84]; MDCK-cells (10 μm diameter) [Paulmichl et al. 84] and even of subcellular organelles (mitochondrion, nucleus, Golgi region (see Sect. 8.4)). Such ultrafine reference microelectrodes reduce the risk of cell damage and improve the stability of intracellular recordings.

Attempts have been made to overcome the difficulties inherent to conventional reference microelectrodes by using a liquid membrane electrode as the reference system [Thomas and Cohen 81]. The proposed membrane solution consists of 2 wt.-% potassium tetrakis(p-chlorophenyl) borate dissolved in 1-octanol. Such a membrane electrode exhibits almost the same selectivity for Na^+ and K^+ ions ($K_{NaK}^{Pot} \sim 1$) (Fig. 6.12). Thus, if the sum of the Na^+ and K^+ activities is constant and exactly the same extra- and intracellularly, the liquid membrane electrode should be useful for the measurement of the cell membrane potential. It was found that these requirements are fulfilled in snail neurons, the papillary muscle of guinea-pigs [Thomas and Cohen 81], the vas deferens of guinea-pigs [Aickin and Brading 84; Aickin 84] and the epithelium of rabbit colons [Wills 85]. The liquid membrane reference microelectrode offers clear advantages: the absence of liquid-junction potentials and tip potentials, and the absence of cell contamination due to the reference electrolytes. The relatively rare use of this alternative reference microelectrode is probably due to the following disadvantages: strict demands are imposed on the extra- and intracellular Na^+ and K^+ activities, the resting potential can only be measured accurately if K_{NaK}^{Pot} is exactly equal to one, other species should not interfere (Ca^{2+}, Mg^{2+}, organic cations [Thomas and Cohen 81]), a considerable leakage of 1-octanol has to be expected [Thomas and Cohen 81], and finally, the microelectrode has a relatively high membrane resistance (10^{10}–$10^{11}\Omega$) and is therefore unsuitable for recording short transients.

6.10 Concluding Remarks

Glass tubings of many different types of glass and shapes are available for the preparation of micropipettes. Borosilicate tubings are especially suitable and are most often used nowadays.

Single-, double-, and multi-barrelled micropipettes of different geometry and properties can be pulled from glass tubings. The smallest reported tip diameters are summarized in Table 2.1.

In using single-barrelled ion-selective microelectrodes the resting potential must be evaluated by either impaling the same cell with a reference single-barrelled microelectrode or by carrying out separate measurements in other cells of the same type. Double-barrelled microelectrodes allow a measurement of the resting potential in ideal proximity to the ion-selective barrel. At present, technical and electrical difficulties occasionally prevent the use of double- or multi-barrelled microelectrodes.

The technique of bevelling microelectrode tips has been well investigated. Bevelling can be advantageous for the electrode resistance, the cell impalement, EMF stability and the prevention of tip plugging.

Tip diameters larger than 1 μm can be judged by light microscopy. In order to observe smaller tip diameters electron microscopy has to be applied. Estimations of tip diameters on the basis of membrane resistances are uncertain. The preparation of microelectrodes with predetermined tip diameters is difficult.

Glass surfaces are hydrophilic and must be made lipophilic by silanization in order to obtain high resistances and to facilitate the filling with an organic membrane solution. The application of dimethylamino trimethylsilane in the gaseous phase at about 200 °C produces an excellent hydrophobization of the glass surface. Different arrangements have been developed for carrying out the silanization reaction.

Micropipettes can be filled with membrane solution either through the tip (front-filling) or the stem (back-filling). Front-filling produces relatively short columns. However, they are still too large to cause a significant reduction in the membrane resistances. In order to make use of the advantages of PVC-containing membrane solutions front-filling is preferentially applied. From a technical point of view back-filling is easier and is more suited for extremely fine micropipettes.

The shielding of the stem of a microelectrode can prevent disturbances from transmural capacitance. This is achieved by covering the glass shank with an insulated conducting layer that is connected to the amplifier input.

Intracellular reference microelectrodes measure cell resting potentials. The absolute values are unreliable because of unknown liquid-junction potentials, tip potentials, cell damage and cell contamination. Liquid membrane microelectrodes which should not suffer from these drawbacks have been proposed as reference microelectrodes. However, they exhibit other disadvantages such as selectivity problems and slow electrode responses.

7 Use of Neutral Carrier-Based Liquid Membrane Microelectrodes

7.1 Concepts of Use

Block diagrams and equivalent circuits of microelectrode arrangements have already been extensively discussed in textbooks [Tietze and Schenk 80; Horowitz and Hill 80] and in monographs [Neher 72; Geddes 72; Thomas 78; Purves 81]. Therefore, the electronic details of the measuring equipment will not be repeated here. Figure 7.1 shows a schematic block diagramm of the measuring arrangement used for the characterization of neutral carrier microelectrodes. The set-up allows the simultaneous use of four ion-selective microelectrodes and a reference microelectrode. The operational amplifier is placed on top of the signal source, i. e. directly on the stem of the microelectrode in order to minimize current leakage, capacitive loading and noise pickup.

Extremely low input currents can be used with the aid of a guard. The high impedance elements of the circuit are placed in an aluminium Faraday cage to reduce or eliminate the disturbances from electromagnetic fields. As the set-up has only been used for the characterization of microelectrodes (no cell punctures), differential potentiometry using an additional low-impedance electrode (Pt wire) could be applied in order to improve the EMF stability of the recording. The EMF of each of the four indicator microelectrodes is measured differentially against the same reference microelectrode. EMF tracings are recorded with a four-channel chart recorder.

During physiological experiments the cell preparation is perfused with an electrolytic solution. The bath solution contains an external conventional reference electrode. In order to carry out the ion activity measurements within a single cell, an intracellular reference microelectrode and an intracellular ion-selective microelectrode are also required (Fig. 7.2).

Several different cell assembly arrangements are available. Their use depends on the aim of the study (e. g. number of parameters to be determined) and the experimental conditions (e. g. cell size). The characteristic features of measurements using one single-barrelled or one double-barrelled microelectrode have already been discussed in Sect. 6.3. The unique opportunity of simultaneously measuring additional parameters together with the membrane potential and a given ion activity within the same single cell makes this technique very attractive. Two different methods are commonly used in order to obtain more information than is possible with a simple microelectrode arrangement (Fig. 7.3, upper): in the first the cell is penetrated with several single-barrelled microelectrodes (Fig. 7.3, middle) whereas the second method uses a multi-barrelled microelectrode (Fig. 7.3, lower). With both methods the si-

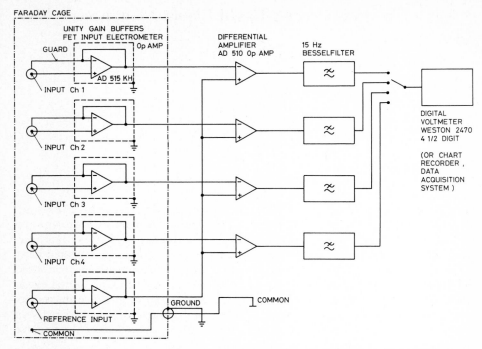

Fig. 7.1. Block diagram of a typical high-impedance measurement system for the differential potentiometric determination of ion activities. Four ion-selective microelectrodes and a reference microelectrode are each connected to an operational amplifier (AD 515 KH; common mode input impedance: 10^{15} Ω; common mode input capacitance: 0.8 pF; input bias current: < 75 fA). The active low-pass Besselfilters (2nd order) exhibit a gain factor of -1. The schematically drawn selector is usually replaced by an analog multiplexer

multaneous monitoring of several ion activities and the membrane potential is possible. However, each method has its own particular features. The specific characteristics of measurements using several single-barrelled microelectrodes are:

– one of the reference microelectrodes can be used for current injection (cell membrane resistance measurement) or to inject electrolytes or other substances (ionophoretic or pressure injection),
– the various parameters are not determined in ideal proximity, i.e. the resting potential and the ion activities are measured at different positions within the cell which, in the case of intracellular ion gradients, makes the interpretation of data difficult (see Sect. 8.5),
– a certain amount of control over the quality of the impalement (Sect. 7.4) and the degree of cell damage (Sect. 7.5) is possible,
– technical difficulties are encountered in impaling the same cell with several microelectrodes.

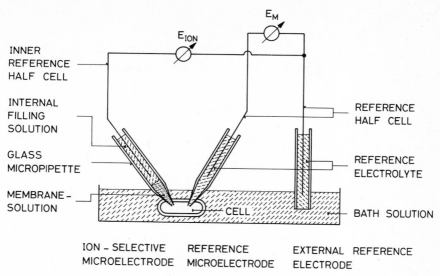

INNER
REFERENCE
HALF CELL

INTERNAL
FILLING
SOLUTION

GLASS
MICROPIPETTE

MEMBRANE –
SOLUTION

E_{ION}

E_M

REFERENCE
HALF CELL

REFERENCE
ELECTROLYTE

BATH SOLUTION

CELL

ION – SELECTIVE   REFERENCE   EXTERNAL REFERENCE
MICROELECTRODE   MICROELECTRODE   ELECTRODE

Fig. 7.2. Schematic representation of a typical microelectrode arrangement for intracellular measurements of ion activity

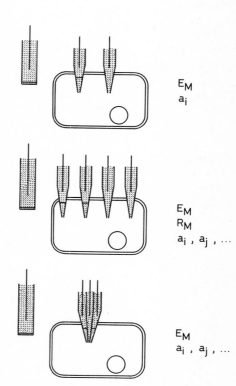

E_M
a_i

E_M
R_M
a_i , a_j , ...

E_M
a_i , a_j , ...

Fig. 7.3. Various arrangements of microelectrodes for intracellular studies. The cell assemblies are completed as shown in Fig. 7.1 and 7.2

The specific characteristics of measurements with multi-barrelled microelectrodes are:

- the cell has to be impaled with only one microelectrode,
- the various ion activities and the resting potential are determined in ideal proximity,
- little control over the quality of the impalement is possible,
- risk of electrical shunt or chemical contamination between the different barrels of the microelectrode must be taken into account,
- technical difficulties exist in the preparation of the microelectrodes.

Impressive examples of both methods exist in the literature: simultaneous penetration of an egg cell [de Laat et al. 75] or a snail neuron [Thomas 77] with five intracellular microelectrodes; simultaneous penetration of a nucleus with an ion-selective and a reference microelectrode and the surrounding cytoplasm with a second reference microelectrode [Civan 78]; penetration of a snail neuron with four microelectrodes for the intracellular calibration of a Li^+-selective neutral carrier microelectrode [Thomas et al. 75]; and the use of triple-barrelled microelectrodes in the proximal tubule of a bullfrog kidney [Fujimoto and Honda 80], on the cortical brain surface [Harris and Symon 81] or in the cerebellar cortex of rats [Ullrich et al. 82].

7.2 Electrode Testing

It is essential to test the microelectrode cell assemblies before performing any physiological experiments. Often, a series of microelectrodes of the same type are prepared a day before the experiments are to be performed. The completed electrodes are stored overnight and tested just before they are to be used in the cells. The performance of microelectrodes with respect to the same membrane solution can be comparable [Sheu and Fozzard 82; Oberleithner et al. 82a; Kurkdjian and Barbier-Brygoo 83; Greger et al. 83; Greger et al. 84a] or quite different [Tsien and Rink 80; Fujimoto and Honda 80; Chapman et al. 83]. In any case, each microelectrode has to be tested individually. After their characterization the best representatives are selected for electrophysiological studies.

The electromotive behaviour of the available neutral carrier microelectrodes has been described in detail (Li^+ [Thomas et al. 75], Na^+ (extracellular) [Ammann and Anker 85], Na^+ (intracellular) [Steiner et al. 79], K^+ [Oehme and Simon 76; Wuhrmann et al. 79], Mg^{2+} [Lanter et al. 80], Ca^{2+} [Oehme et al. 76; Tsien and Rink 81; Lanter et al. 82], H^+ [Ammann et al. 81b]). The electrode function, detection limit, selectivity factors, response time, membrane resistance, EMF stability and reproducibility serve as a guide for the suitability of a microelectrode. However, it is much too time-consuming to reproduce all of these measurements before an intracellular experiment. Hence, by considering the available reference data and carrying out a few simple tests, it is possible to decide whether the sensor can be used for the experiment under question.

In carrying out studies in which special substances such as blockers or drugs (see Sect. 7.8) are involved, the interference of these chemicals has to be carefully evaluated. An adequate test relies on the electrode function of the measuring ion (in its expected activity range) in the presence of a typical concentration of the interfering species.

There are various simple test procedures which allow the selection of the appropriate microelectrodes. Often, the calibration of the microelectrode is sufficient for this purpose. Measurements in calibrated solutions reflect the slope of the electrode function and, in part, the selectivity of a microelectrode. During the calibration it is possible to observe the response characteristics and the EMF stability.

Another useful property of a microelectrode is its membrane resistance, which can be easily measured and is very informative. Two techniques for the determination of membrane resistances are currently in use. In the first method a constant current is applied for a short period of time (usually $\ll 1$ s). Due to the short pulses of current the resultant changes in the voltage, which are dependent on the resistance of the microelectrode, often have to be observed on an oscilloscope. Usually, the current ist adjusted to 1 nA. The resistance [$M\Omega$] of the membrane then corresponds to the observed change in potential in millivolts. The second method relies on a reference resistor in parallel to the microelectrode (voltage divider method). A known voltage is applied to the cell and the attenuation is measured. In the absence of any additional resistance, the membrane resistance corresponds exactly to the known reference resistance when the sum of the actual EMF and the applied voltage is half of the measured value.

7.3 Calibration and Evaluation

The calibration of microelectrode cell assemblies determines to a large extent the accuracy of the measurements. The problems encountered in obtaining appropriate calibrations are most obvious in the direct potentiometric measurements used in clinical analyses, where particularly severe demands are imposed on the measurements in blood and urine samples [Mohan and Bates 75; Meier et al. 80; Anker et al. 84]. In principle, similar requirements are imposed in electrophysiology on the quantitative determination of ionic species in cells. However, many physiological studies do not primarily require accurate absolute values but are interested in changes in ion activities. Thus, small errors in the calibration procedure usually do not prevent a qualitative interpretation of the results. As shown for calcium (Sect. 8.2), different forms of the ion have to be distinguished in physiological solutions (e. g. total concentration, free ion concentration, ion activity). Direct potentiometric measurements with microelectrodes allow the determination of both the ion activity and the ion concentration of the non-bound (free) fraction of the total concentration in the sample. Consequently, two different calibration procedures have to be employed.

In most applications, the microelectrode cells are calibrated using solutions of known concentrations of the ion to be measured and the interfering ions. Such calibration solutions ideally resemble the sample solutions as closely as possible, i.e. the most important interfering ions are considered at their typical physiological concentrations. The ionic strength and the activity coefficients of ions in the calibration and sample solution are then comparable. Furthermore, the contributions of the interfering ions to the EMF will, to a large extent, cancel out and the two solutions will yield similar liquid-junction potentials. Since the calibrations can never exactly simulate a physiological sample (ionic composition, proteins) a small error is inevitable.

By evaluating the slope (s) of the electrode function in such calibration solutions and by using one of these solutions for a one-point calibration (E_{cal}, corresponding to $c_{i,cal}$), the measured electromotive force in the sample (E_S) can be converted to an extracellular sample concentration $c_{S,extra}$ by

$$c_{S,extra} = c_{i,cal} \cdot 10^{(E_S - E_{cal})/s}, \tag{7.1}$$

or to an intracellular sample concentration $c_{S,intra}$ by

$$c_{S,intra} = c_{i,cal} \cdot 10^{(E_S - E_{cal} - E_M)/s}. \tag{7.2}$$

Equation (7.1) shows that for extracellular measurements of free ion concentrations the accuracy is mainly determined by the ability to simulate the sample composition of the calibration solution. A representative example is discussed in Figure 7.4. The changes in potential which occur within the normal physiological range are relatively small and probably below the EMF reproducibility of most microelectrodes. However, some experiments lead to relatively large changes in the ion activities which can easily be followed with microelectrodes. The presence of tip potentials (see Sect. 7.7) may increase the uncertainties in extra- as well as intracellular measurements.

An intracellular measurement of the EMF of a sample (E_S) has to be corrected for the cell membrane potential (E_M) (Eq. (7.2)). Experiments with single-barrelled microelectrodes introduce an uncertainty in E_M which, in most cases, will be the main source of error (see Table 6.2). As already discussed in Sect. 6.3, the use of double-barrelled microelectrodes improves this situation. The choice of calibration solutions for free Ca^{2+} concentrations is especially difficult since solutions at sub-micromolar levels can only be prepared using Ca^{2+} buffers (for a discussion see Sect. 9.6.3.4).

Microelectrode cells are often calibrated for concentration measurements but the results are presented as ion activities. In order to carry out such conversions, an activity coefficient has to be assumed which agrees with the conditions of the sample (see e.g. [Lee and Uhm 81; Greger et al. 83]). Even if the ionic strength of a sample is known approximately, it is difficult to define an exact mean activity coefficient for a complex physiological sample. If the estimated mean activity coefficients are further converted into single-ion activity coefficients non thermodynamic conventions are also involved. Nevertheless, the assumed activity coefficients seem to be acceptable for most situations. The values claimed can be taken to be accurate to within ± 5%. For example, a change in the ionic strength from 120 to 180 mM of a cytosolic solution containing 10 mM Na^+ and 0.1 μM Ca^{2+} leads to a deviation of the mean activity coefficients of Na^+ and Ca^{2+} by 3.4 and 13%, respectively.

Another method of calibrating microelectrode cells relies on the use of activity scales (see [Lev and Armstrong 75; Lewis et al. 78; Reuss and Weinman 79; Wang et

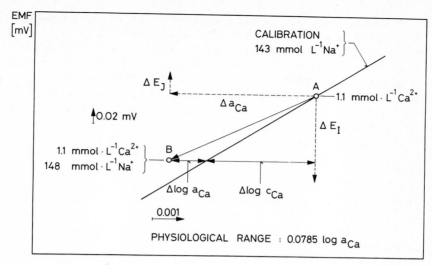

Fig. 7.4. Computed changes in EMF of an ideal Ca^{2+}-selective electrode at a constant Ca^{2+} concentration (1.1 mM) but varying concentrations of Na^+ (from 143 mM (A) to 148 mM (B)). Solution A is assumed to be the one-point calibration solution and is therefore located on the calibration curve with the slope s. The reference electrode contains a 3 M KCl reference electrolyte. Δa_{Ca}: change in Ca^{2+} activity due to an increase in the ionic strength (decrease in the Ca^{2+} activity coefficient). ΔE_I: decrease in the EMF due to a decrease Δa_{Ca} in the Ca^{2+} activity. ΔE_J: change in the liquid-junction potential. The resulting vector AB shows that B is no longer located on the calibration curve through A. $\Delta \log c_{Ca}$: error in calibrating concentration. $\Delta \log a_{Ca}$: error in calibrating activity

al. 84]). Precise values for the mean activity coefficients of single-electrolyte solutions are available. By using such simple calibration solutions, ideal electrode responses can be obtained in the physiological activity range. The ion activity in a sample solution is then obtained by inserting the measured EMF of the sample (E_S) into the calibration curve. The determination is therefore independent on the ionic strength of the sample:

$$a_{S,extra} = a_{i,cal} \cdot 10^{(E_S - E_{cal}) / s} \tag{7.3}$$

and

$$a_{S,intra} = a_{i,cal} \cdot 10^{(E_S - E_{cal} - E_M) / s}. \tag{7.4}$$

The difference ($E_S - E_{cal}$) in Eq. (7.1) to (7.4) includes changes in the liquid-junction potential, which are usually negligible or can be corrected using the Henderson formalism (Eq. (5.35)). Equations (7.3) and (7.4) are only valid if the microelectrode is ideally selective. Of course, if nonideally selective microelectrodes have to be employed interfering ions can be added to the calibration solutions. However, activity coefficients for mixed physiological solutions are not exactly specified (for a discussion of Na^+/Ca^{2+} mixtures see [Butler 68]).

Although calibrations for activities (provided that activity standards are available) yield correct results even if the ionic strength of a sample is not known, they are used less frequently than calibrations for concentrations. The pH scale is the only activity scale commonly used (pH = $-\log a_H$), hence H^+ measurements are always based on activity standards. Indeed, standard pH buffer solutions are widely available. It has been proposed that they should have an ionic strength of 0.16 M in order to minimize the effects of liquid-junction potentials [Sankar and Bates 78; Covington et al. 83].

The calibrations for activities undoubtedly offer advantages. First of all, the activity of an ion is often designated as the physiologically relevant property of an ion [McLean and Hastings 34; Copp 69; Fujimoto and Kubota 76; Robertson and Marshall 79; Solsky 82]. Secondly, the results are less affected by the composition of a sample, and finally, the calibration solutions are very simple in their composition (for the exceptional case of intracellular Ca^{2+} see Sect. 9.6.3.4). Of course, in order to express the ion activities in terms of free ion concentrations the ionic strength of a sample has to be known.

So far, it has been assumed that a microelectrode exhibits ideal selectivity or that the interference can be compensated for by using adequate calibration solutions. These assumptions are valid for most neutral carrier microelectrodes in typical extra- or intracellular environments. However, situations may arise in which corrections for interfering ions become necessary. Corrections according to the Nicolsky-Eisenman formalism can be applied if the selectivity factors are known and if the measuring and interfering ions are determined simultaneously. By solving the Nicolsky-Eisenman equation for the measuring ion I and an interfering ion J (and vice versa) in a mixed calibration solution and an intracellular sample, the intracellular activity of I (and correspondingly of J) is:

$$a_{i,\text{intra}} = \frac{(a_{i,\text{cal}} + K_{ij}^{\text{Pot}}\, a_{j,\text{cal}}) \cdot 10^{\,(E_{i,\text{intra}} - E_{i,\text{cal}})}(a_{j,\text{cal}} + K_{ij}^{\text{Pot}}a_{i,\text{cal}}) \cdot 10^{\,(E_{j,\text{intra}} - E_{j,\text{cal}})\,/\,s_j}}{1 - K_{ij}^{\text{Pot}} \cdot K_{ji}^{\text{Pot}}}. \tag{7.5}$$

Equation (7.5) has been used to interpret the intracellular measurements of Na^+ and K^+ activities using triple-barrelled microelectrodes [Fujimoto and Honda 80] (see also [Lev and Armstrong 75]). Similarly, microprocessor-aided corrections for the interference of Ca^{2+} in the response of Na^+-selective microelectrodes have also been carried out [Boerrigter and Lehmenkühler 84].

7.4 Impalement Criteria

The puncture of a cell can be properly followed by measuring the EMF response of a microelectrode during the impalement. The penetration of a cell by a microelectrode is indicated by an abrupt change in potential. The intracellular reference microelectrode measures the cell resting potential whereas an ion-selective microelectrode measures the decrease in potential corresponding to the sum of the resting potential and the change in ion activity between the extra- and intracellular media

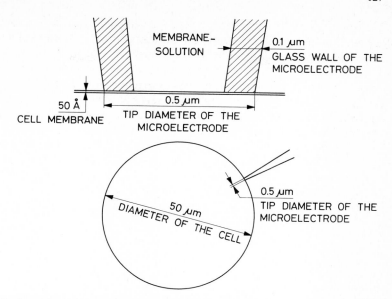

Fig. 7.5. Enlarged schematic representation of the relative dimensions of a microelectrode (tip diameter 0.5 µm), a cell membrane (5 nm) and a single cell (diamter 50 µm); (true to scale)

(both measurements are made with respect to an external bath reference electrode). The total potential increase during the insertion of the indicator microelectrode can be very large (e. g. about 200 mV for Ca^{2+}) or very close to zero (e. g. about 0 mV for K^+, i. e. the resting potential reflects the K^+ gradient [Edzes and Berendsen 75]).

In puncturing a biological membrane with a glass micropipette several criteria should be kept in mind. Figure 7.5 compares the dimensions of a cell, a cell membrane and a microelectrode. The proportions make it clear that the following problems have to be considered:

- cell membrane disruption
- cell damage (Sect. 7.5)
- cell contamination (Sect. 7.6)
- seal of the cell membrane about the microelectrode tip
- generation of electrical shunts during and/or after cell puncture
- exact location of the sensitive microelectrode tip within the cell (bevelled tips)
- changes in tip potentials upon penetration.

Impalement criteria should be chosen so as to ensure the validity of intracellular recordings. Both the impalement procedure and the strictness of the criteria strongly influence the reliability of the collected data. Obviously, the demands vary with the type of cell. In order to judge the quality of an impalement, two different experimental procedures for cell penetration have been proposed. The first method recommends that the cell is first impaled with an ion-selective microelectrode and then, after a stable potential has been established, with the reference microelectrode. Since the intracellular ion activities can be expected to remain constant during the impalement with the reference microelectrode, the nondifferentiated signal should remain stable during the penetration with the second microelectrode. Simultaneously, the nondifferentiated EMF trace of the ion-selective microelectrode allows the observation of short-time variations in the membrane potential.

The second method recommends that the reference microelectrode is inserted first. The subsequent impalement with the ion-selective microelectrode can be followed by the recording of the reference microelectrode. This approach is especially suitable if the ion-selective microelectrode has a larger tip diameter than the reference microelectrode or if the measuring ion (e.g. Ca^{2+}) shows a large electrochemical gradient in the direction of the cell interior [Tsien and Rink 80; Alvarez-Leefmans et al. 81a; Blinks et al. 82].

Various criteria for acceptable impalements have been set:

- the calibrations before and after the intracellular experiment should agree with one another to within a tolerable deviation (e.g. within 1 mV [Reuss and Weinman 79]; within 2 mV [Hansen and Zeuthen 81; Greger et al. 84a]; within 5 mV [Harvey and Kernan 84b]; within 10 mV [Marban et al. 80])
- an instantaneous change in potential should occur during the impalement [Reuss and Weinman 79; Lewis and Wills 81; Giraldez 84] followed by a stable recording (e.g. ± 5 mV during the entire recording [Greger and Schlatter 83])
- the membrane potential should be more negative than a given limiting value [Zeuthen 80; Alvarez-Leefmans et al. 81a]
- a long-lasting depolarization after the impalement with the second microelectrode should not be observed [Alvarez-Leefmans et al. 81a; Thomas 82]
- the EMF signal should show a weak drift during the experiment [Zeuthen 80; Frömter et al. 81; Greger et al. 83], (e.g. < 5 mV/h [Alvarez-Leefmans et al. 81a]; 1 mV/30 s [Reuss and Weinman 79]; ± 1 mV or ± 2 mV/40 s [Higgins et al. 77]
- the reference and the ion-selective barrel should register the same membrane potential [Giraldez 84]
- the changes in the tip potential before and after the intracellular measurement should be limited (< 6 mV Higgins et al. 77]; < 5 mV [Suzuki and Frömter 77])
- during the measurement with an intracellular microelectrode, damage was caused to the membrane with a microneedle and the EMF monitored [Giulian and Diacumakos 77]
- markers can be injected into a cell organelle (e.g. mitochondrion); observation by electron microscopy should show that not a single marker has dispersed into the cytoplasm [Giulian and Diacumakos 77].

For most physiological studies several of the above criteria have to be fulfilled. In addition, even more elaborate tests for valid impalements have been proposed (see Sect. 7.5) [Suzuki and Frömter 77; Lewis et al. 78; Zeuthen 80; Marban et al. 80; Hess et al. 82].

It has been shown that impalements are improved by:

- the use of piezoelectric (or hydraulic) microdrives. These ensure a fast (about 0.1 ms to several milliseconds for a step interval), linear and stepwise (as small as 0.1 μm) motion of the microelectrode [Chen 78; Fromm et al. 80; Sonnhof et al.

82; Zeuthen 82; Greger and Schlatter 83; Schmid and Bohmer 85] (see also [Suzuki and Frömter 77]). Automated impalements using microdrives are available [Brown and Flaming 77; Marshall and Klyce 81].
- the use of bevelled microelectrodes [Isenberg 79; Harvey and Kernan 84b]
- the use of microelectrodes with very small tip diameters [Suzuki and Frömter 77; Marban et al. 80; Greger and Schlatter 83; Harvey and Kernan 84b]
- adequate geometry and/or flexibility of the microelectrodes [Frömter et al. 81; Purves 81; Flaming and Brown 82].

It may be necessary to locate the microelectrode tip within the preparation by means other than electrical observation (e.g. in heterogeneous cell preparations or in subcellular organelles that are are of the same potential as cytosol). A direct microscopic observation of the tip is not always possible because of insufficient resolution. In such cases other techniques can be helpful:

- the position of the tip can be followed optically by using a dye [Frömter et al. 81]. The distribution of the dye indicates the position of the tip within the cell or intercellular environment. Different dyes have been used [Tomita 69; Barrett and Graubard 70; Brown and Flaming 74; Frömter et al. 81]
- the introduction of a current through the reference microelectrode can be used to check whether the ion-selective microelectrode is within the appropriate cell [Reuss and Weinman 79; Alvarez-Leefmans et al. 81a; Harvey and Kernan 84b].
- the administration of blockers influences the EMF signal and therefore can be used to localize the ion-selective microelectrode within the cell (e.g. furosemide [Greger et al. 84a]; amiloride [Lewis et al. 78; Harvey and Kernan 84b]).

7.5 Cell Damage

Intracellular studies with microelectrodes are invasive and inevitably cause some cell damage. Therefore, the results obtained may be subject to some uncertainty. Indeed, if it is considered that typical microelectrode tip diameters are about 100 times larger than the thickness of the cell membrane (see Fig. 7.5), it is suprising that more damage is not inflicted. Tears or incomplete seals about the micropipette may cause a change in the membrane potential and cause an in- or outflow of ions. Different types of damage to the cell membrane last for different lengths of time. Partial destruction of the cell membrane leads to permanent damage and is easily observable. On the other hand, a short-lasting rupture of the membrane followed by an immediate sealing of the cell membrane about the micropipette is only observed as a transient disturbance or may not be detected at all.

Any damage to the cell membrane will be associated with a change in the conductance of the cell membrane. Cell damage is therefore more conspicuous in native cells exhibiting low conductance. Simple calculations have been made which estimate the size and leak resistance of hypothetical pores [Lindemann 75; Higgins et al. 77]. These values give an idea of the size of the damaged area that might influence the native electrical properties of the membrane (Table 7.1).

Table 7.1. Estimation of the diameters of pores filled with extracellular electrolyte which yield shunt resistances equal to that of the native membrane

Cell membrane [reference]	Measured resistance for 1 cm^2 of cell membrane	Membrane resistance for a single cell with an assumed surface of 500 μm^2	Diameter of a pore[a] filled with Ringer solution[b] that yields a shunt resistance equal to the resistance in column 3
Serosal membrane of amphibian urinary bladder epithelia [Frömter and Gebler 77]	7 k$\Omega \cdot$cm^2 (upper limit)	1.4 GΩ	3 nm
Luminal membrane of amphibian urinary bladder epithelia [Frömter and Gebler 77]	65 k$\Omega \cdot$cm^2 (upper limit)	13 GΩ	1 nm
	220 k$\Omega \cdot$cm^2 (amiloride treated)	44 GΩ	0.54 nm

[a] membrane thickness: 10 nm
[b] specific resistance: 100 Ω cm [Lindemann 75]

Leaks with diameters in the order of tens of Å are already sufficiently large to produce a significant distortion of the electrical properties of the cell membrane. It is therefore obvious that the seal of the cell membrane about the micropipette has to be almost perfect. Fortunately, the interaction of the cell membrane with glass micropipettes is very good, a fact that is demonstrated by the very high resistance of the leak between a glass and a membrane (e.g. > 1 GΩ [Bader et al. 82]) and by the very tight pipette-membrane seals of patch-clamp micropipettes (resistances of 10–100 GΩ [Hamill et al. 81]). It is claimed that even better seals can be obtained using silanized glass walls [Aickin and Brading 82; Munoz et al. 83]. The sealing properties of different types of cells are not the same. For example, gall-bladder cells have a strongly convoluted membrane on the lumen side which may be responsible for a marked sensitivity to damage by microelectrodes [Suzuki and Frömter 77].

In spite of the relatively good seals, leakages after punctures with microelectrodes have been repeatedly observed. Cell damage can be shown by:

- recordings of lower than normal resting potentials [Trube et al. 82]
- a rapid collapse of the membrane potential [Greger et al. 83]
- an initially rapid change in potential followed by a drift towards a stable value of the EMF [Lassen et al. 71]. Such a drift can be interpreted as a progressive resealing of the damaged cell about the micropipette (see upper part of Fig. 7.6 and [Higgins et al. 77])
- simultaneous (relatively slow) decrease in the potential and resistance of the cell membrane. This effect indicates a continuous increase in the damaged area (lower part of Fig. 7.6)
- a deviation of the resting potential of the same cell measured by repeated punctures [Lassen et al. 71]

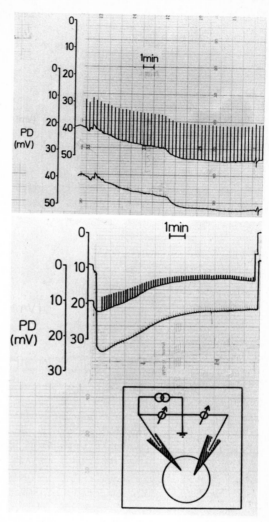

Fig. 7.6. Cell damage after penetration of *Xenopus laevis* oocytes with two single-barrelled micro-electrodes. The membrane potential (PD) and resistance are recorded simultaneously (20 pulses of current (600 ms); shunt resistance: 1 GΩ; see insert). *Upper:* progressive healing, the membrane potential increases from about 40 to about 50 mV after the impalement, simultaneously the membrane resistance increases from about 570 to 680 kΩ, *Lower:* progressive damage, the membrane resistance decreases simultaneously with the potential from about 210 to 52 kΩ. Courtesy of F. Lang

– a shunt in the apical membrane of epithelia analyzed by the voltage divider ratio [Suzuki and Frömter 77; Lewis et al. 78; Zeuthen 80]
– microscopic examination of the insertion of the microelectrode tip into ultrathin sections of tissue [Kessler et al. 77]
– direct intracellular recording of the leaking ion and its effects within the cell [Taylor and Thomas 84]

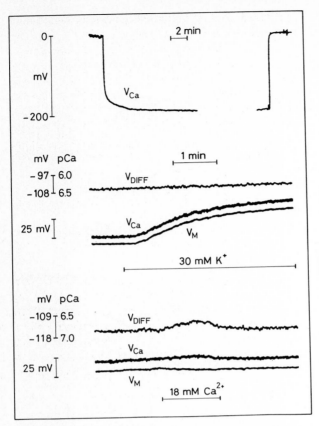

Fig. 7.7. Evidence for the validity of an intracellular Ca^{2+} measurement in heart muscle cells [Marban et al. 80]. Redrawn from Marban, Rink, Tsien and Tsien, Nature, 1980 [Marban et al. 80]

Many of the reported observations of cell damage were made in relatively small cells. Often the experiments were not performed with optimally manufactured microelectrodes.

It is usually assumed that there is no interference from leakages if the measurements satisfy the impalement criteria discussed in Sect. 7.4. Further evidence that the intracellular determinations are correct is based on the following observations:

- a partial and transient depolarization of the membrane potential is induced by an increase in the extracellular K^+ activity. Both the intracellular reference microelectrode and ion-selective microelectrode sense this change in the potential. The differential signal, however, remains constant, indicating that the intracellular ion activity has not been changed due to an increase in the extracellular K^+ activity (for H^+ microelectrodes see [de Hemptinne et al. 82]; for Ca^{2+} microelectrodes see [Marban et al. 80] (Fig. 7.7) and [Coray et al. 80; Weingart and Hess 84]; for Mg^{2+} microelectrodes see [Hess et al. 82])
- the extracellular activity of the ion that is being monitored intracellularly is increased. The stability of the intracellular recording indicates whether or not an in-

flux of the measuring ion has occurred (for Ca^{2+} microelectrodes see [Marban et al. 80] (Fig. 7.7); for Na^+ microelectrodes see [Schümperli et al. 82]).

- repeated impalements of the same cell yield the same results (e.g. 20 punctures of the same cardiac Purkinje fiber yielded a mean membrane potential of -75.5 ± 0.3 mV (std. dev.) [Isenberg 79])
- long-term stability of intracellular recordings is found, e.g. upper part of Fig. 7.7: trace during the impalement of a heart muscle cell with a Ca^{2+} microelectrode and withdrawal after 3 hours of continuous intracellular monitoring [Marban et al. 80]
- change in the membrane potential by stimulations and comparison of the EMF recordings of the ion-selective and reference microelectrodes [Lee and Dagostino 82]
- resistance measurements on epithelia cells (voltage divider ratio) [Lewis and Wills 80]
- good correlations between potentiometric determinations and the results of non-invasive techniques are found (e.g. NMR spectroscopy, distribution of weak acids or bases).

Even if all of the experimental precautions are taken and the validity criteria are fulfilled one cannot be sure that there is no cell damage. For example, both the resting potential and the ion activities may be influenced by leakages or distortions of the cytosolic structures (see Sect. 8.1) and may therefore be reduced to artificial values. However, highly reproducible intracellular measurements on the same type of cell carried out by different investigators indicate an absence of significant errors. Furthermore, progressively improved microelectrode techniques allow preciser punctures to be made (see e.g. [Suzuki and Frömter 77]).

7.6 Cell Contamination

In the course of an experiment using microelectrodes the intracellular fluid can be contaminated for several reasons:

- inflow of extracellular fluid due to cell damage
- outflow of the contents of damaged organelles
- leakage of electrolyte from the reference microelectrode
- leakage of components from the ion-selective membrane solution.

The risk and the degree of cell damage have been discussed in Sects. 7.4 and 7.5. Damage to organelles during the impalement of cells is extremely difficult to assess, but cannot be excluded (see Chap. 8).

During an intracellular measurement the electrolyte of the reference microelectrode and the membrane solution of the ion-selective microelectrode are in direct contact with the cytosol. These highly concentrated solutions containing physiologically important ions, biologically active molecules (ion carriers), organic solvents and lipophilic salts can contaminate the intracellular space. Normally, the volume of the solution in a microelectrode is very large compared to the volume of a cell. For example, reference microelectrodes have volumes of about 10 µl (typical dimensions for this type of microelectrode are: stem inner diameter 0.8 mm, filling height 3 cm, shank 1 cm) while neutral carrier microelectrodes (filling height 5 mm, tip diameter 1 µm, taper angle 1.4°) contain about 80 nl, compared with a cellular volume of 4.2 pl (spherical cell, diameter 20 µm). Immediately after a puncture a steady-state develops between the cytosol and the solutions within the micropipettes. Since the cell membrane does not act as a permeability barrier to the leaking substances, the underlying equilibria are quite complicated. Experiments and calculations indicate that a steady-state is achieved within seconds or minutes in small cells. It should therefore be taken into account that the outflow of components from microelectrodes during an intracellular experiment could contaminate the cell and disturb the physiological event.

The contamination of cells has been mainly discussed for reference microelectrodes. Table 7.2 lists the calculated and measured rates of leakage of electrolytes from reference or injection micropipettes. The leakage rate obviously depends on the tip diameter of the microelectrode and on the concentration of the reference electrolyte. Relatively large micropipettes (~ 1 µm) filled with typical reference electrolytes exhibit leakage rates greater than 100 fmol s^{-1}, while small microelectrodes (~ 0.2 µm) exhibit rates smaller than 10 fmol s^{-1}. The lowest rate given in Table 7.2 is 1 fmol s^{-1} for a 0.5 M KCl reference electrolyte with an electrode resistance of 115 MΩ [Fromm and Schultz 81]. Even if leakage rates are as low as this, a spherical cell with a diameter of 20 µm (volume of 4.2 pl) would show an increase in the KCl

Table 7.2. Measured and calculated rates of electrolyte leakage from intracellular reference microelectrodes

Microelectrode (Tip Diameter/Resistance)		Reference Electrolyte	Rate of Leakage	Reference
Calculated Rates				
0.25 µm	/20 MΩ	3M KCl	60 fmol s^{-1}	[Nastuk and Hodgkin 50]
	/ 5 MΩ	3M KCl	60 fmol s^{-1}	[Coombs et al. 55]
0.2–0.6 µm	/ –	3M KCl	44–136 fmol s^{-1}	[Page et al. 81]
0.6–1.0 µm	/ –	3M KCl	136–230 fmol s^{-1}	[Page et al. 81]
Measured Rates				
–	/50 MΩ	3M AChCl	20 fmol s^{-1}	[Krnjević et al. 63]
–	/25 MΩ	3M AChCl	50–100 fmol s^{-1}	[Krnjević et al. 63]
–	/11 MΩ	3M AChCl	480 fmol s^{-1}	[Krnjević et al. 63]
–	/ 7.5 MΩ	3M KCl	3.2 fmol s^{-1}	[Isenberg 79]
(unbevelled)				
–	/ 2.5 MΩ	3M KCl	11 fmol s^{-1}	[Isenberg 79]
(bevelled)				
–	/ 1.5 MΩ	3M KCl	16 fmol s^{-1}	[Isenberg 79]
(strongly bevelled)				
0.2–0.6 µm	/ –	3M KCl	60 fmol s^{-1}	[Page et al. 81]
0.6–1.0 µm	/ –	3M KCl	150 fmol s^{-1}	[Page et al. 81]
–	/ 16 MΩ	3M KCl	10 fmol s^{-1}	[Fromm and Schultz 81]
–	/ 30 MΩ	3M KCl	5.5 fmol s^{-1}	[Fromm and Schultz 81]
–	/ 62 MΩ	0.5M KCl	1.8 fmol s^{-1}	[Fromm and Schultz 81]
–	/115 MΩ	0.5M KCl	1 fmol s^{-1}	[Fromm and Schultz 81]
0.2–0.3 µm	/50–75 MΩ	1M KCl	4–5 fmol s^{-1}	[Blatt and Slayman 83]

concentration of 14 mM/min. Larger microelectrodes filled with concentrated reference electrolytes show leakage rates of the order of 150 fmol s^{-1} (e.g. tip diameter 0.6-1.0 μm, 3 M KCl reference electrolyte [Page et al. 81]) and would cause an increase of 35 mM/s within the same cell. Consequently, after only 4 s the ionic strength of the cell would be doubled and the osmotic balance significantly disturbed.

Various model calculations have been presented which describe the outflow of reference electrolyte from micropipettes [Krnjević et al. 63; Geisler et al. 72; Purves 79; Page et al. 81]. These models describe the diffusion of ions from a capillary into an unstirred cytosolic volume due to the presence of a concentration gradient. The effect of the hydrostatic pressure within the stem of the microelectrode has also been discussed [Krnjević et al. 63; Isenberg 79]. As shown in Table 7.2 the calculated leakage rates are in good agreement with the experimental values.

Various observations have been reported which indicate that the cell has become contaminated by the leakage of KCl reference electrolyte:

- depolarized membrane potential [Thomas 77; Palmer and Civan 77; Blatt and Slayman 83]
- low cell membrane resistance [Blatt and Slayman 83]
- dependence of the disturbance on the concentration of the KCl reference electrolyte [Blatt and Slayman 83]
- dependence of the disturbance on the composition of the reference electrolyte [Blatt and Slayman 83]
- cell swelling due to the disturbance of the osmotic state of the cell (water uptake) [Nelson et al. 78]
- direct potentiometric measurement of the Cl$^-$ leakage using classical anion-exchanger microelectrodes (Fig. 7.8) [Thomas 78].

Reduction or elimination of KCl leakage from reference microelectrodes can be achieved by:

- using reference electrolytes other than KCl (~0.5 M K$_2$SO$_4$ or Na$_2$SO$_4$ [Blatt and Slayman 83]; 0.6 M K$_2$SO$_4$ [Thomas 77])
- using diluted KCl solutions (0.1 M KCl [Nelson et al. 78]; 0.5 M KCl [Geisler et al. 72; Fromm and Schultz 81; Civan et al. 83; Wills 85]; 0.1 M KCl/0.5 M K$_2$SO$_4$ [Ehrenfeld et al. 85]
- using isotonic cytosol-like solutions as reference electrolytes [Nelson et al. 78]
- using liquid membrane microelectrodes as reference microelectrodes [Thomas and Cohen 81] (see Sect. 6.9).

For many applications, a 0.5 M KCl solution as a reference electrolyte seems to be a good compromise: the leakage rate is reduced, changes in the liquid-junction potential are still tolerable and the tip potentials are not any worse (Sect. 7.7). For measurements in extremely small cells, in which the leakage of a 0.5 M KCl solution is still too high, a physiological reference electrolyte should be used. However, proteins contribute significantly to the anionic charges of a physiological sample. They usually have to be replaced by adequate anions in an artifical physiological reference electrolyte.

KCl LEAKAGE FROM REFERENCE MICROELECTRODES

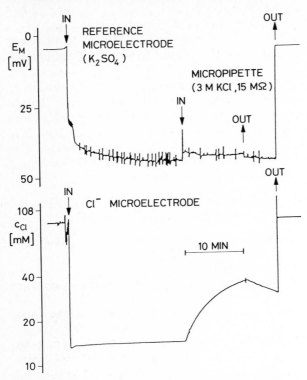

Fig. 7.8. Measurement of the leakage of Cl^- ions from an intracellular micropipette filled with 3 M KCl [Thomas 78]. A snail neuron was penetrated by a reference microelectrode containing a K_2SO_4 solution *(upper trace)* and a classical anion-exchanger (Cl^-) microelectrode *(lower trace)*. During the measurement the cell was also penetrated by a microelectrode filled with KCl *(upper trace)*. The increase in the intracellular concentration of Cl^- was measured with the Cl^--selective microelectrode *(lower trace)*. Redrawn from Thomas, Ion-Sensitive Intracellular Microelectrodes, Academic Press, 1978 [Thomas 78]

To date, the leakage of components from neutral carrier-based membrane solutions into a cell has not been investigated. Among the components of neutral carrier membrane solutions, the ion carriers are physiologically the most active substances. Several carrier molecules used in membrane solutions have proven to be useful in the investigation of the transport of ions across artificial bilayer and natural cell membranes. The compounds are found to selectively change the ionic permeability of these membranes. An almost Nernstian response is observed when the activity of the ion under consideration is varied in the external solution. In artificial bilayer membranes, which exhibit very low conductivities and do not contain a transport system, an extremely low carrier concentration is sufficient to induce ion permeability (Tab. 7.3). Natural cell membranes with their own transport systems require a much higher carrier concentration, in order for the native conductance to be surpassed by the contribution of the ion carrier (Tab. 7.3). If the natural transport sys-

Table 7.3. Effects of natural and synthetic carriers used in membrane solutions on the ion permeability of artifical bilayer and natural cell membranes

Carrier	Concentration	Membrane	Observed effect	Reference
valinomycin	$\sim 10^{-11}$ M	bilayer	decrease in resistance	[Tien 74]
	$5 \cdot 10^{-11}$ M	bilayer	decrease in resistance	[Andreoli et al. 67]
monactin	$5 \cdot 10^{-11}$ M	bilayer	increase in conductivity	[Szabo et al. 73]
monensin	$10^{-6} - 10^{-8}$ M	various cells	change in secretory activities	[Ledger and Tanzer 84]
ETH 1001	$4 \cdot 10^{-11}$ M	mitochondria	Ca^{2+} uptake	[Caroni et al. 77]
	$> 3 \cdot 10^{-5}$ M	bilayer	decrease in resistance	[Vuilleumier et al. 77]
ETH 149	10^{-6} M	bilayer	decrease in resistance	[Margalit and Eisenman 77]

tems are blocked by inhibitors, the effect of the carriers occurs at lower carrier concentrations.

In contrast to reference microelectrodes, where the outflowing electrolytes have been shown to disturb the ionic and/or osmotic state of the cell, ion carriers from ion-selective microelectrodes can influence the transport properties of a cell membrane. On the molecular level, the degree of leakage of a carrier into the intracellular space is essentially limited by the lipophilicity of the carrier molecule (see Sect. 5.10) and the protein content of the aqueous sample solution (see Sect. 4.1.2). The lower the lipophilicity of the carrier and the higher the protein content of the cytosol, the higher is the leakage rate of the carrier. In addition, the geometry of the micropipette will also determine the rate of outflow of carrier molecules. Small tip diameters are advantageous for low leakage rates.

The outflow of the carrier can be maintained at a tolerable level by adjusting the lipophilicity of the neutral carrier molecule. However, it has been shown in Sect. 3.2.4.3 that the lipophilicity of a carrier has its limits. Currently used neutral carriers have distribution coefficients in the two-phase system membrane/aqueous solution in the range of 10^6 to 10^8. However, when the aqueous solution is replaced by a serum with a protein content of about 7%, the distribution coefficients are decreased several orders of magnitude (Fig. 4.5). Accordingly, it can be assumed that most of the neutral carriers discussed so far exhibit distribution coefficients between the membrane solution and the cytosol of the order of 10^3.

In practice, the outflow of neutral carriers should be indicated by changes in the conductance of the cell membrane. The current I_i carried by the ion i depends on the driving force $(V - V_i)$ (V: resting potential [mV]; V_i: equilibrium potential of the ion i according to the Nernst equation [mV]; $V_i = s \log (a_{i,extra}/a_{i,intra})$ and the conductance $g_i [\Omega^{-1}]$ of the membrane for the ion i:

$$I_i = g_i(V - V_i). \qquad (7.6)$$

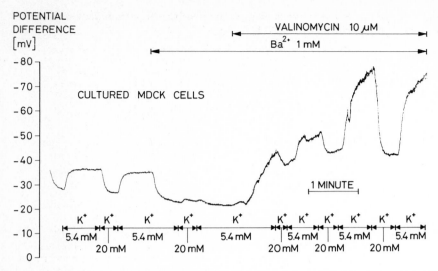

Fig. 7.9. Influence of valinomycin on the K^+ conductance of MDCK cells during the inhibition of K^+ channels with Ba^{2+} [Ammann et al. 85 c]

Under resting conditions, the net electrical current driven by all of the ions i and the pumps p is zero:

$$\sum_i g_i(V-V_i) + \sum_p I_p = 0 \qquad (7.7)$$

The higher the permeability of the cell membrane for the ion i, the higher is the conductance g_i. Usually, the conductance of K^+ is relatively large and the resting potential is almost equal to that of the equilibrium potential of K^+ (e.g. with $K^+_{extra} =$ 4 mM and $K^+_{intra} = 120$ mM : $V_K = -87$ mV). However, if the conductance g_i of another ion i is large, the membrane potential will shift towards the value V_i of this ion i. For example, for $i = Ca^{2+}$, $Ca^{2+}_{extra} = 1.1$ mM and $Ca^{2+}_{intra} = 0.1$ μM, V_{Ca} is +119 mV.

A significant contamination of a cell by the neutral carriers of a membrane solution alters the conductance and potential of the cell membrane. Simultaneously, the dependence of the membrane potential on changes in the extracellular ion activities is altered.

In a preliminary experiment, the influence of valinomycin on the K^+ conductance of cultured Madin Darby canine kidney (MDCK) cells was studied [Ammann et al. 85 c]. It was shown that about half of the conductance of the cell membranes is due to Ba^{2+}-blockable K^+ channels [Paulmichl et al. 84]. Figure 7.9 shows the influence of valinomycin on the K^+ conductance of MDCK cells during treatment with Ba^{2+}. In the absence of valinomycin and Ba^{2+} a stepwise increase in extracellular K^+ (replacing Na^+) from 5.4 to 20 mM leads to a depolarization of about +8 mV. The addition of Ba^{2+} depolarizes the cell by +12 mV to the equilibrium potential determined by ions other than K^+. The response of the K^+ channels towards extracellular changes in K^+ activity vanishes completely. The subsequent perfusion with a Ringer solution containing 10^{-5} M valinomycin during the Ba^{2+} application evokes an almost Nernstian response to extracellular changes between 5.4 and 20 mM K^+ within a minute (experimental: -35 mV; theoretical: 35 mV; 37 °C). Thus, during the complete inhibition of the K^+ channels of MDCK cells, the presence of 10^{-5} M valinomycin in the external perfusion solution induces an ideal K^+ conductance. Future studies will show whether the outflow from intracellular microelectrodes containing valinomycin or other carriers disturbs the properties of cell membranes [Ammann et al. 85 c].

The degree and the course of cell contamination by intracellular ion-selective neutral carrier microelectrodes can be estimated by using the model calculations presented in Sect. 5.10 [Oesch et al. 85 a]. The model describes the diffusion of components from the membrane phase into a spatially limited or unlimited sample volume. The same theoretical approach can be used to describe the flow of components from the perfusion solution into the cell membrane. The experiment shown in Fig. 7.9 can therefore be simulated theoretically.

The time-dependence of the uptake of valinomycin from the perfusion solution into the membrane of a MDCK cell is governed by its diffusion through the adhering stagnant layer surrounding the cell. Using mathematical descriptions for the steady-state diffusion through such a layer, one obtains the following equation:

$$n(t) = A \int_0^t J_\delta dt = \frac{A\ D_S\ c_S^0}{\delta}\ t, \tag{7.8}$$

where

A : surface area of the cell membrane [cm²];
J_δ : flux of valinomycin through the stagnant layer [mol cm^{-2} s^{-1}];
δ : thickness of the stagnant layer [cm];
D_S : diffusion coefficient of valinomycin in the perfusion solution [cm² s^{-1}];
c_S^0 : concentration of valinomycin in the perfusion solution [mol l^{-1}].

The average concentration of valinomycin in the membrane of a MDCK cell is:

$$c_m(t) = \frac{n(t)}{V_m} = \frac{D_S\ c_S^0}{d\ \delta}\ t, \tag{7.9}$$

where

V_m : volume of the cell membrane [cm³]
d : thickness of the cell membrane [cm]

For the small cells under discussion the amount of valinomycin accumulated in the cytosol can be neglected. With the following parameters for experiment shown in Figure 7.9 (assumed dimension of a MDCK cell: $10 \times 10 \times 10$ μm): $V_m = 3.1 \cdot 10^{-12}$ cm³; $\delta = 10^{-2}$ cm; $d = 10^{-6}$ cm; $D_S = 10^{-5.5}$ cm² s^{-1}; $c_S^0 = 10$ μM, one obtains:

$$c_m(t) = 10^{-2.5}\ t\ [\text{mol l}^{-1}]. \tag{7.10}$$

Figure 7.10 illustrates the increase in the concentration of valinomycin in the cell membrane as a function of time for varying thicknesses of the adhering stagnant

CONCENTRATION OF VALINOMYCIN IN CELL MEMBRANE

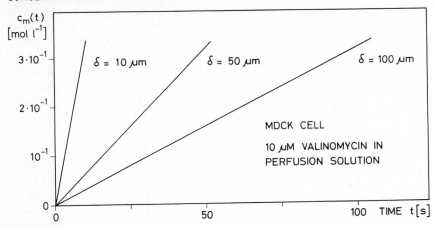

Fig. 7.10. Uptake of valinomycin into a cell membrane from a perfusion solution containing 10 μM valinomycin for varying thicknesses δ of the adhering stagnant aqueous layer (Eq. (7.9)) [Ammann et al. 85 c]

aqueous layer. The experiment shows (Fig. 7.9) that an almost ideal valinomycin-in-duced K^+ conductance is obtained after about 100 s. This is in good agreement with the above calculation which predicts valinomycin concentrations of 3.2 M ($\delta = 10\,\mu m$), $6.3 \cdot 10^{-1}$ M ($\delta = 50\,\mu m$) and $3.2 \cdot 10^{-1}$ M ($\delta = 100\,\mu m$) after 100 s of con-tact with the perfusion solution.

The thickness of the stagnant layer (δ) is uncertain and depends on the stream-ing conditions and the cell preparation. Furthermore, the preparation and use of very dilute aqueous solutions of lipophilic carriers (c_a^0) is problematic for the follow-ing reasons:

– the solubility of the carrier in water is very low (valinomycin: about 10 μM [Kirsch 76]). An exact determination of its concentration in water is difficult. Usually, the aqueous solutions are prepared using ethanolic stock solutions of the carrier
– adsorption of carrier molecules on the glass surfaces of the container walls occurs
– diffusion of carrier molecules into plastic parts (tubings, perfusion chamber) can take place during perfusion.

7.7 Tip Potentials and Electrical Shunts

The microelectrode technique is still subject to troublesome phenomena which are not yet completely understood and which are dependent on the type of microelec-trode used. These interferences are due to the tip potentials of reference microelec-trodes and the resistive or capacitive disturbances of ion-selective neutral carrier microelectrodes. They can often be eliminated or at least reduced to a tolerable level, but are still a drawback to modern microelectrode technology.

Electrical shunt pathways can appear at the tips of neutral carrier microelectrodes. A disturbance is not observed when the shunt resistances through the glass wall and along the glass surface at the tip of a microelectrode are large compared to the electrical resistance of the carrier-based membrane solution. Thus, it is necessary to keep the shunt resistances as high as possible, whereas the membrane resistance should be as low as possible. Membrane resistances are influenced by the composition of the membrane solution (Sect. 4.2.3), the composition of the adjacent solutions and the tip diameter of the microelectrode. On the other hand, the shunt resistances depend on the type of glass, the wall thickness and the degree of silanization of the glass surface. Even after a consideration of these parameters, there are still some neutral carrier-based microelectrodes which are affected by shunt pathways. For example, disturbances have been observed with Na^+- and Ca^{2+}-selective microelectrodes. The presence of shunts results in useless calibration curves and/or erroneous results. However, by modifying the membrane solutions and/or the construction of a microelectrode it is possible to eliminate these problems. The necessary conditions are discussed in the following chapters (Na^+ microelectrode: Sect. 9.3.2.3; Ca^{2+} microelectrode: Sect. 9.6.3.3).

The tip potential is defined to be the measured change in potential when the tip of an intact reference microelectrode is broken [Adrian 56]. Under extreme conditions tip potentials of 100 mV or more can be observed. Some investigators include liquid-junction potentials in the potential difference termed tip potential (since both potential differences are generated at the tip of the microelectrode). However, this does not make any sense because liquid-junction potentials and tip potentials are of quite different physico-chemical origin. Liquid-junction potentials can be kept relatively constant and are theoretically well-defined (Sect. 5.7), whereas tip potentials are not constant and cannot be calculated. Although different models have been proposed to explain tip potentials, their origin is still not completely understood. Most of the theories are based on the fact that a silica surface in contact with an aqueous solution exhibits fixed anionic sites. It is an agreed upon fact that these negative charges play a fundamental role in the formation of the tip potentials. Indeed, tip potentials are drastically influenced by treatment with acids or polyvalent cations. Attempts have been made to rationalize these observations by considering either the glass conductivity or the interaction of a glass surface with an aqueous electrolyte.

One theory describes the tip potentials as interfacial potentials between the glass wall near the tip and the adjacent electrolyte [Agin 69]. If the micropipette is filled with a concentrated electrolyte (e. g. 3 M KCl) changes in the inner-phase boundary potential at the glass wall can be neglected. On the outer surface, however, ion-exchange reactions that depend on the sample composition occur. If the specific resistance of the glass at the tip is sufficiently low, the interfacial potential can act as a voltage divider network. Dry Pyrex glass exhibits a specific resistance of 10^{15} to $10^{16} \, \Omega \, cm$ (see Tabl. 6.1.). Assuming a thickness of 0.1 µm and an effective area of 20 µm^2, the resistance of the dry glass wall at the tip is $2 \cdot 10^{17}$ to $2 \cdot 10^{18} \, \Omega$. Consequently, hydration of the thin glass wall would have to lower the resistance by at least 10 orders of magnitude. It is very unlikely that this occurs. Recently, it has been suggested that an increased conductivity of micropipettes made of Pyrex glass is more likely to arise from microcracks in the glass wall at the extreme tip of the microelectrode [Coles et al. 85].

A different theory is based on the formation of electric double layers (Gouy-Chapman model) [Agin 69; Lavallée and Szabo 69] in which the negative surface charges (Si-O$^-$ groups) are neutralized by the presence of sample cations. The density of the cation layer decreases rapidly with the distance from the surface, yielding the typical profiles of the potential at interfaces. The layer of cations in contact with the surface is immobile while the outer part of the cation cloud is mobile with respect to the rigid part. At the boundary between the fixed and the mobile layer an electrokinetic potential is generated (zeta potential). The zeta potentials are strongly dependent on the type and concentration of the electrolyte [Agin 69]. For example, relatively high concentrations of Th^{4+} ions can reverse the effect of the fixed charges because the strong adsorption of thorium ions causes the diffuse region to be filled with anions instead of cations. It is thought that the tip potentials are a result of the surface conductivity of the cationic layer in the diffuse region of the Gouy-Chapman double layer. The resulting shunt pathway is in parallel to the bulk resistance of the solution which is not in contact with the double layer at the glass wall. The effective width of the double layer, the Debye length, is, however, extremely small at physiological ionic strengths (e.g. 10^{-3} μm for I = 0.1 M [Agin 69]). Hence, the bulk resistance should determine the potential.

It is still uncertain whether one or both of the theories adequately describe the tip potentials. It has been suggested that the double-layers are involved in high-resistivity glass micropipettes (e.g. Pyrex glass), whereas shunts through the glass walls are important in micropipettes manufactured from low-resistivity glass [Lavallée and Szabo 69]. In both cases the tip potentials disappears when the tip of the microelectrode is broken. Thereafter, the conductance of the bulk electrolyte dominates and leads to the well-known diffusion potentials.

Tip potentials are influenced by many parameters:

- the technique used to fill the microelectrode: the boiling method yields considerably larger tip potentials (-12.7 ± 6.2 mV (n = 39), sea water) than the internal-filament method (-0.2 ± 1.4 mV (n = 39), sea water) [Gotow et al. 77]; the alcohol method is worse than the filament or direct-filling method [Okada and Inouye 75; Okada and Inouye 76] (for methods of filling reference microelectrodes see [Purves 81])
- the type of glass used for the micropipette: low-resistivity glass (e.g. NaCaS$_{22.6}$, specific resistance = $6.3.10^{10}$ Ωcm) exhibits significant leak resistances through the glass wall of the micropipette [Lavallée and Szabo 69]
- the sample solution: the tip potentials increase with decreasing ion concentrations in the sample solution [Adrian 56; Küchler et al. 64; Agin 69; Lavallée and Szabo 69; Gotow et al. 77]; the tip potentials depend on the ionic composition of the sample solution (KCl solutions lead to smaller tip potentials than NaCl solutions [Gotow et al. 77]; the pH value of the sample solution [Riemer et al. 74; Gotow et al. 77; Lavallée and Szabo 69], low pH values are advantageous (neutralization of anionic sites on the glass surface) [Lavallée and Szabo 69; Riemer et al. 74; Okada and Inouye 76; Gotow et al. 77])

- the resistance of the microelectrode: certain types of reference microelectrodes that exhibit high resistances were observed to show large tip potentials [Adrian 56; Küchler et al. 64; Purves 81]; a significant correlation was not found [Riemer et al. 74; Gotow et al. 77]
- the temperature: the temperature sensitivity of the tip potentials depends a great deal on the filling method; the sensitivity is low for filament-containing reference microelectrodes [Gotow et al. 77]
- the presence of di- or polyvalent cations: Mg^{2+} and Ca^{2+} [Agin 69; Gotow et al. 77] or Th^{4+} [Agin 69; Lavallée and Szabo 69] may reduce tip potentials
- the time of storage: tip potentials increase with increasing storage time of a microelectrode [Okada and Inouye 76].

It has been deduced from theoretical models and experimental observations that potentials are minimized if:

- large concentration gradients do not exist between the sample solution and the reference electrolyte
- the ionic strengths of the external bath solution and the sample solution are similar
- freshly prepared reference microelectrodes are used
- reference microelectrodes with tip diameters larger than 1 μm are utilized
- glass of high resistivity and relatively low surface charge is used
- the reference microelectrodes are filled by the direct method (with or without filament).

Despite these guide-lines, the magnitude of tip potentials remains uncertain. In principle, tip potentials can be circumvented by:

- using liquid membrane microelectrodes as reference microelectrodes [Thomas and Cohen 81] (for a discussion see Sect. 6.9.)
- blocking the fixed anionic sites on the glass surface. Unfortunately, the masking of the free hydroxyl groups with silanization reagents does not solve this problem since it is almost impossible to completely fill a silanized micropipette with an aqueous electrolyte.

Tip potentials in small reference microelectrodes probably prevent very accurate determinations of resting potentials. However, there are currently no means of completely eliminating these potentials and no theoretical descriptions are available for correcting changes in tip potentials.

7.8 Limitations Due to Selectivity

The selectivity of the microelectrode must be sufficiently good in order to obtain relevant ion activity measurements [Simon et al. 77 a; Simon et al. 78 b; Meier et al. 80; Meier et al. 82]. Using the Nicolsky-Eisenman formalism (Eq. (5.7)) and assuming typical extra- and intracellular activities (Tab. 7.4), the limits of the selectivity factors can be summarized in the equation:

Table 7.4. Extra- and intracellular concentrations and activities [mmol l^{-1}]

Ion	Concentration		Activity[a]	
	Typical	Normal Range	Typical	Normal Range
Extracellular				
H$^+$	$5 \cdot 10^{-5}$	$4.3 \cdot 10^{-5}-\ 5.6 \cdot 10^{-5}$	$3.9 \cdot 10^{-5}$	$3.34 \cdot 10^{-5}-\ 4.35 \cdot 10^{-5}$
Li^{+b}	1	0.7 $-$ 3.0	0.77	0.54 $-$ 2.30
Na$^+$	140	135 -150	105	102 -112
K$^+$	4	3.5 $-$ 5.0	3	2.6 $-$ 3.7
Mg^{2+}	0.6	0.45 $-$ 0.8	0.21	0.16 $-$ 0.28
Ca^{2+}	1.1	1.0 $-$ 1.2	0.37	0.34 $-$ 0.41
Intracellular				
H$^+$	$7 \cdot 10^{-5}$	$6 \cdot 10^{-5}-\ 8 \cdot 10^{-5}$	$5.5 \cdot 10^{-5}$	$4.7 \cdot 10^{-5}-\ 6.2 \cdot 10^{-5}$
Li^{+b}	2	1 $-$ 3	1.5	0.8 $-$ 2.3
Na$^+$	10	5 $-$ 18	7.6	3.8 $-$ 14
K$^+$	120	80 -160	89	61 -116
Mg^{2+}	3	1 $-$ 5	1.1	0.4 $-$ 1.8
Ca^{2+}	$4 \cdot 10^{-4}$	$5 \cdot 10^{-5}-\ 2.5 \cdot 10^{-3}$	$1.4 \cdot 10^{-4}$	$1.7 \cdot 10^{-5}-\ 8.7 \cdot 10^{-4}$

[a] calculated from corresponding concentrations with an ionic strength I = 0.149 M (extracellular) or I = 0.139 M (intracellular)
[b] therapeutic Li$^+$ levels

$$K_{ij,\max}^{Pot} = \frac{a_{i,\min}}{a_{j,\max}^{z_i/z_j}} \frac{p_{ij}}{100}, \tag{7.11}$$

where

$K_{ij,\max}^{Pot}$: highest tolerable value of the selectivity factor;
$a_{i,\min}$: lowest expected activity of the measuring ion I^{z_i};
$a_{j,\max}$: highest expected activity of the interfering ion J^{z_j};
p_{ij} : highest tolerable error in the activity a_i due to interference by a_j (%).

Table 7.5. lists the required log K_{ij}^{Pot} values for extra- and intracellular measurements. The values are given for $p_{ij} = 1$, i.e. for a maximum tolerable error of 1% (worst case). The required selectivity factors are compared with the ion-selectivities of available neutral carrier microelectrodes.

With the exception of the H$^+$-selective and the valinomycin-based K$^+$- selective microelectrodes each sensor is subject to interferences (Tab. 7.5). Therefore, linear calibration curves can not be expected in solutions of typical ion backgrounds. The calibration curves in the vicinity of typical cytosolic compositions have been analyzed according to the equations that govern the behaviour of ion-selective microelectrodes (Nicolsky-Eisenman equation (5.7), Debye-Hückel equation (5.32), Henderson equation (5.35)) [Meier et al. 82]. This was done to demonstrate the possibilities and limitations of the currently used neutral carrier microelectrodes in typical intracellular environments. The results of the computations are presented in a graphical form. The procedure and interpretation of the study are illustrated for a

Table 7.5. Tolerable selectivity factors, log $K^{Pot}_{ij,max}$, for extra- and intracellular determinations (1% error, worst case) of the measuring ion I^{z_i} in the presence of interfering ions J^{z_j} (Eq. (7.11), $p_{ij} = 1$, for concentration ranges see Table 7.4)

I^{z_i}	J^{z_j}	$K^{Pot}_{ij,max}$ Extracellular	$K^{Pot}_{ij,max}$ Intracellular	K^{Pot}_{ij} of Available Microelectrodes (Chap. 9)	
H^+	Na^+	-8.5	-7.5	< -12.7	
	K^+	-7.1	-8.4	$(-9.8)^a$	
	Mg^{2+}	-7.7	-8.0		
	Ca^{2+}	-7.8	-6.3	$(< -11.1)^a$	
Na^+	H^+	4.4	2.8	-1.2^b	-0.8^c
	K^+	-0.6	-3.5	-0.4^b	-2.3^c
	Mg^{2+}	-1.2	-3.1	-3.4^b	-2.4^c
	Ca^{2+}	-1.3	-1.4	-1.3^b	0.2^c
K^+	H^+	2.8	6.0	-0.5^d	-4.4^e
	Na^+	-3.6	0.6	-1.9^d	-3.5^e
	Mg^{2+}	-2.8	0.2	-2.7^d	-5.1^e
	Ca^{2+}	-2.9	1.8	-2.1^d	-4.4^e
Mg^{2+}	H^+	8.9	9.0	2.8	
	Na^+	-3.9	-1.7	-1.1	
	K^+	-0.9	-3.5	-1.4	
	Ca^{2+}	-2.4	0.7	1.1	
Ca^{2+}	H^+	9.3	4.7	-0.1	
	Na^+	-3.6	-6.1	-5.5	
	K^+	-0.6	-7.9	-5.4	
	Mg^{2+}	-1.9	-7.0	< -4.9	
Li^{+f}	H^+	3.1	2.1	-0.5	
	Na^+	-4.3	-3.2	-1.4	
	K^+	-2.8	-4.2	-2.3	
	Mg^{2+}	-3.5	-3.7	-2.7	
	Ca^{2+}	-3.6	-2.1	-0.6	

[a] values obtained with macroelectrodes [Schulthess et al. 81]
[b] microelectrode based on ETH 157 (extracellular application (Sect. 9.3.3.4))
[c] microelectrode based on ETH 227 (intracellular application (Sect. 9.3.3.3))
[d] classical cation-exchanger microelectrode
[e] valinomycin-based microelectrode
[f] for a therapeutical Li^+ concentration of 2 mM (see Sect. 9.2)

sodium-selective microelectrode based on the neutral carrier ETH 227 (without the additive sodium tetraphenylborate) [Steiner et al. 79]. The selectivity factors used were: log $K^{Pot}_{NaH} = -0.9$, log $K^{Pot}_{NaK} = -1.7$, log $K^{Pot}_{NaMg} = -2.7$ and log $K^{Pot}_{NaCa} = -1.7$ [Steiner et al. 79]. The corresponding figures and comments for other currently available microelectrodes are given in Chap. 9. The general aspects of the evaluation of the selectivity are the same. The curvatures of the Nicolsky curves are more or less pronounced depending on the interferences. All of the calculations were performed using published selectivity data [Meier et al. 82].

In Fig. 7.11 a set of calibration curves is shown for the sodium-selective microelectrode and various concentrations of potassium. The area inside the dashed rectangle is given in Fig. 7.12 in which the variability in the concentration of the other

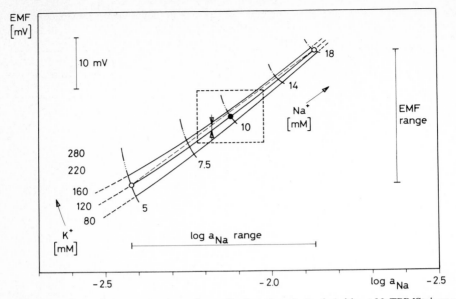

Fig. 7.11. Calibration curves for the sodium-selective microelectrode (without NaTPB [Steiner et al. 79]). Concentration levels were varied from 60–280 mM (K$^+$) and 4–19 mM (Na$^+$) for an otherwise typical background. The calibration curves cannot be approximated by straight lines because of the proximity of the detection limit. Similar figures are obtained for the other microelectrodes discussed in Chap. 9. The central point is marked by a filled circle and the two end points that define the physiological range are indicated by open circles. The vertical distance between the line connecting the end points and the middle calibration curve is termed sag *(arrows)*. The dashed rectangle (± 5 mV, ± 0.1 log a_{Na}) is depicted in Fig. 7.12

interfering ions is also taken into account. Furthermore, the change in the log K_{ij}^{Pot} by ± 0.2 is indicated by circles. From Fig. 7.11 and 7.12 it is evident that the calibration curve has a sub-Nernstian slope due to the interference by potassium, and that the point for 10 mM Na$^+$ is evidently not too far removed from the detection limit (by ~0.6 log a_{Na} units).

A change in the potassium concentration shifts the calibration curve for sodium by the amount indicated; the curvature changes (Fig. 7.11) because both the ionic strength and the interference due to poor selectivity come into play.

The reduction in the activity of sodium due to an increase in the potassium concentration from 120–160 mM is approximately 2.1%. Since the explicit shape of the Debye-Hückel function (activity coefficient $\gamma_\pm$ as a function of ionic strength I) depends to some extent on the ion considered, $d\gamma_\pm/dI$, the sensitivity towards changes in the ionic strength varies from ion to ion. Thus, the above value of -2.1% will be different for Li$^+$ (-1.5%), Mg^{2+} (-7.1%) and Ca^{2+} (-7.5%).

A change in log K_{ij}^{Pot} of ± 0.2 would shift the 10 mM Na$^+$ coordinate as indicated by the half-filled circles (K$_{NaK}^{Pot}$: ◑). The local curvature of the calibration function would also change in a manner somewhat different from that shown in Fig. 7.11 because a change in ionic strength would not be involved.

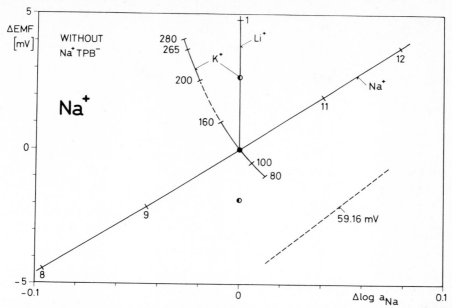

Fig. 7.12. The influence of different factors on the calibration curve of a Na^+-selective microelectrode (without NaTPB [Steiner et al. 79]) in an intracellular environment. The central point (filled circle) represents the response of the ion-selective microelectrode to typical conditions and for average selectivity factors. The (relative) EMF and log activity scales are centered on this point. The long diagonal represents the (Nicolsky) calibration curve through the central point, i.e. only the concentration of the primary ion I is varied. Vertical displacements are the result of an assumed shift in log K_{ij}^{Pot} (± 0.2) because changes in the activities are not involved ($J = K^+$: ●). A change in concentration of any of the interfering species J (background) causes a change in the ionic stength and thus in the activity coefficients of all of the species (oblique curves with marks, cf. Fig. 7.11). Ions not explicitly mentioned give rise to interferences smaller than could be conveniently illustrated. Unless otherwise specified, all concentrations are given in millimoles per liter

Lithium levels as used in the therapy of maniac depressions (~ 1 mM) would strongly interfere with the measurement of sodium ions. In all of the calculations, unless otherwise specified, the lithium concentration was assumed to be 1 µM. For each of the ion-selective microelectrodes discussed in Chap. 9 the following aspects are of importance (explicit numerical values are given below):

1. The apparent slope of the calibration curve for the ion I near the central point, i.e. typical point of its assumed intracellular concentration range (see Fig. 7.11), indicates the sensitivity of the measurement. It is defined as the sum of the local slopes due to the Nicolsky-Eisenman and the Henderson equations (the reference electrode is assumed to contain a 3 M KCl reference electrolyte in order to minimize the influence of the liquid-junction potential).

2. An error in the EMF of 1 mV introduces a bias into the estimated activity of ion I, expressed in terms of percent of the nominal value a_i.

3. Each interfering ion J contributes towards the sum under the logarithm in the Nicolsky-Eisenman equation (Fig. 7.13):

$$\text{contribution} = 100\, k_n / \sum k_n, \tag{7.12}$$

where $k_n = K_{in}^{Pot}\, a_n^{z_i/z_n}$, $K_{ii} \equiv 1$ and the sum $\sum k_n$ is taken over all of the species.

Hence, the Nicolsky-Eisenman equation (5.7) can be reduced to the form

$$EMF = E_o + s \log(a_i + c), \tag{7.13}$$

where for the purposes of this argument c can be regarded to be a constant. If a_i is slightly increased, the EMF will follow in a non-linear manner (curvature close to the detection limit). On the other hand if a given EMF is interpreted according to the Nernst equation (c = zero), a_i will seem to be larger than is actually the case. These effects obviously become worse the larger c is, i.e.: the contribution range for the primary ion I moves from 100% to lower values. In Figure 7.13 the relation $a_i / (a_i + c)$ and its variability under the stated conditions can easily be determined.

4. The major interfering ions are considered in intracellular studies.

5. A change of ± 0.2 in the log K_{ij}^{Pot} of the major interfering ions can notably affect the calibration curve. It can be seen in Fig. 7.13 that such a change has an effect on the EMF similar to that of an extreme change in the concentration of the interfering ion in question. The short vertical marks that signify this change are about as far apart as the contribution bars are long.

6. The calibration curve deviates from the linear behaviour (sag, Fig. 7.11) defined by the end points of the physiological range (Tab. 7.4). These end points can be used to define the expected range in terms of EMF (mV) or log activity. The ratio sag/ EMF range and the local slope are intimately related to the distance, in log activity units, of the central point from the IUPAC-recommended detection limit (Guilbault et al. 76).

7. If, after calibration with a typical concentration of any interfering ion J, the concentration of J is varied over its physiological range, the activity of the primary ion I and the EMF will change although the concentration of I remains the same. Part of the apparent activity of the primary ion I is mimicked by the interfering ions. The size of this increment can be gleaned from Fig. 7.12 by placing a horizontal line (measured EMF) through the point of interest and intercepting the original calibration curve, e.g., a change from 120–160 mM K^+ (after calibration) would indicate ~ 10.5 mM Na^+ or about 0.5 mM more than is actually present. A change of $+0.2$ in log K_{NaK}^{Pot} would by the same token mimic an amount of ~ 1.4 mM Na^+ in excess of what is present. Such a change in K_{ij}^{Pot} can occur due to: e.g., faulty calibration, inappropriate conditioning and/or ageing of the microelectrode. These excesses should be compared with the physiological range of sodium which is 5–18 mM (Tab. 7.4). The contribution (Eq. (7.12)) to the EMF varies between 65 and 87% for Na^+ (sodium-selective microelectrode without NaTPB, see Fig. 7.13 [Steiner et al. 79]). The difference between these numbers and 100% is mainly due to potassium. It becomes even larger if lithium therapy is practiced. The additional sodium mimicked by lithium and potassium can be corrected. The disadvantage is a more cumbersome calibration procedure and more sophisticated data reduction techniques.

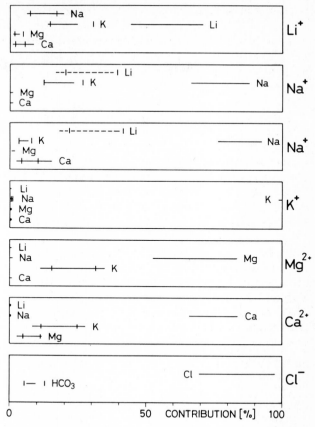

Fig. 7.13. Contributions (Eq. (7.12)) of the individual ions towards the sum under the logarithm in the Nicolsky-Eisenman equation for available microelectrodes. Li$^+$: see Sect. 9.2 [Thomas et al. 75]; Na$^+$ (without NaTPB), Na$^+$ (with NaTPB): see Sect. 9.3.3.3 [Steiner et al. 79]; K$^+$ (classical ion-exchanger): see Sect. 9.4.2. [Oehme and Simon 76]; Mg^{2+}: see Sect. 9.5.4.2 [Lanter et al. 80]; Ca^{2+}: see Sect. 9.6.3.3 [Lanter et al. 82]; Cl$^-$ (classical anion-exchanger): see Sect. 9.7. The range bars were obtained by individually varying the concentrations of each ion over the physiological range as given in Table 7.4., while keeping the other concentrations at their typical levels. For typical concentrations the log K_{ij}^{Pot} were individually varied by ± 0.2; this is indicated by short vertical marks on the appropriate range bar. The lithium contribution (dashed) was calculated for the therapeutic concentration range (1–3 mM Li$^+$ [ten Bruggencate et al. 81; Amdisen 75], while the other contributions were calculated for 1 μM Li$^+$

Good examples of microelectrodes virtually free from interference are the tridodecylamine-based H$^+$ microelectrode and the valinomycin-based potassium microelectrode (see Chap. 9), which is decidedly better than the classical ion-exchanger electrode shown in Fig. 7.13.

8. Error propagation yields an estimate of the permissible uncertainties of the concentrations and the selectivity factors of the interfering species and of the measured EMF if a given uncertainty in the concentration of species I is to be attained. An numerical example is given here.

It is assumed that a sodium-selective microelectrode exhibits the following contributions (Fig. 7.13): sodium 81%, potassium 19%; and that the concentration of sodium in the cell in question is to be determined with a reproducibility of $\pm 5\%$, i.e. ± 0.5 mM. If all of the factors except c_K, K_{NaK}^{Pot} and $\triangle$ EMF are under tight control, three limiting cases can be distinguished:

a) if $\triangle c_K = \triangle K_{NaK}^{Pot} = 0$ then $| \triangle EMF | \leqslant 1.0$ mV can be tolerated;

b) if $\triangle EMF = \triangle c_K = 0$ then $| \triangle \log K_{NaK}^{Pot} | \leqslant 0.08$ can be tolerated; this is approximately equal to the reproducibility of K_{ij}^{Pot} determinations under routine conditions.

c) if $\triangle EMF = \triangle K_{NaK}^{Pot} = 0$ then $| \triangle c_K | \leqslant 25$ mM K^+ can be tolerated; for a comparison, at 120 mM K^+ and $\triangle EMF = \pm 1$ mV a potassium-selective microelectrode could measure potassium to within ± 4.8 mM.

Clearly, none of the conditions stipulated above will strictly be valid. For $\triangle EMF = \pm 0.5$ mV, $\triangle c_K = \pm 2.5$ mM K^+ and $\triangle \log K_{NaK}^{Pot} = \pm 0.1$, the sodium activity a_{Na} is found to be reproducible to within $\pm 6.6\%$.

The calculations presented above can be used to evaluate the performance of any ion-selective microelectrode. The procedure shows that a good estimate of the contribution of interfering ions to the measured EMF is possible, provided that potentiometric selectivity factors and the concentrations of the interfering ions are known. These data are usually available for the hydrogen ion and the physiologically important alkali and alkaline-earth ions. However, other ions and molecules as well as physiologically relevant electrolytes are often involved in experiments. For example, ions such as Mn^{2+}, Co^{2+} and La^{3+} (Ca^{2+}-channel blockers) and Ba^{2+} (K^+-channel blocker) may be present in millimolar concentrations. Figure 7.14 illustrates a few selected examples of compounds often employed in physiological studies. The substances are used in quite different concentrations depending on the experiment. Typical concentrations or representative concentration ranges are given in Fig. 7.14. However, it is not easy to predict the intensity of the interference these compounds will produce. If the substances are added to the bath solution while the ion activities are being measured intracellularly, the risk of interference is drastically reduced. The lipophilicity, the charge and the kind of functional groups that make up the molecule all contribute to the degree of interference. In principle, the influence of these compounds on the EMF response of the measuring ion should be evaluated in each case. So far, only few such tests have been carried out. A careful study has been made on the sensitivity of the classical ion-exchanger K^+ microelectrode to some biologically active substances [Křiž and Syková 81]. The effect of the various concentrations of these compounds (e.g. 10^{-7} to 10^{-3} M) on the EMF, measured in solutions of constant K^+ concentration, was evaluated. It was shown that certain substances interfere with K^+ activity measurements when used at typical concentrations. To date, similar studies using neutral carrier-based microelectrodes have not been carried out. It is known, however, that certain compounds can disturb the sensors. For example, the metabolic blockers dinitrophenol and carbonyl-cyanide-m-chlorophenylhydrazone (CCCP) seriously interfere with the response of the H^+-selective neutral carrier microelectrode (see Sect. 9.1.4.2.4).

ACETAZOLAMIDE
(CARBONIC ANHYDRASE
INHYDRASE , $5 \cdot 10^{-3}$ M)

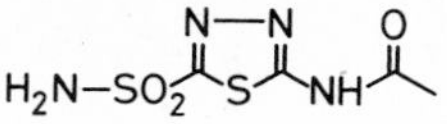

AMILORIDE
(BLOCKER OF Na^+ / H^+ -
COUNTERTRANSPORT ,
10^{-4} – 10^{-3} M)

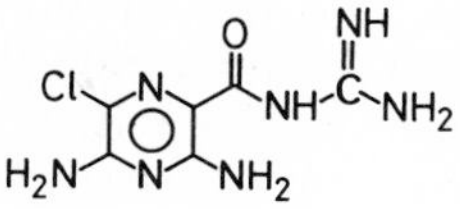

FUROSEMIDE
(DIURETIC , BLOCKER OF
Na^+ / Cl^- COTRANSPORT ,
$5 \cdot 10^{-5}$ M)

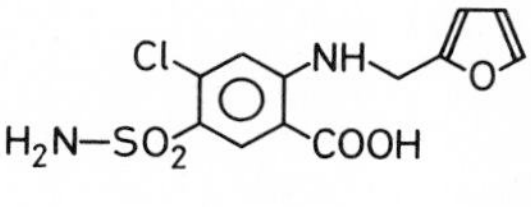

DIDS
(BLOCKER OF
ANION TRANSPORT ,
$5 \cdot 10^{-5}$ – 10^{-4} M)

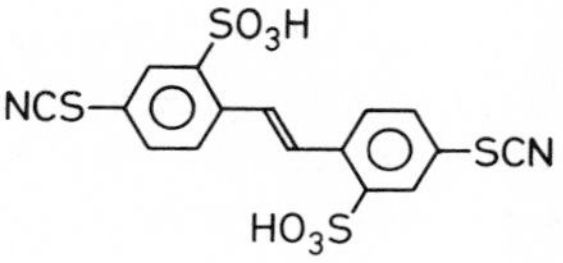

OUABAIN
(NaK-ATPase INHIBITOR ,
CARDIAC GLYCOSIDE ,
10^{-7} – 10^{-3} M)

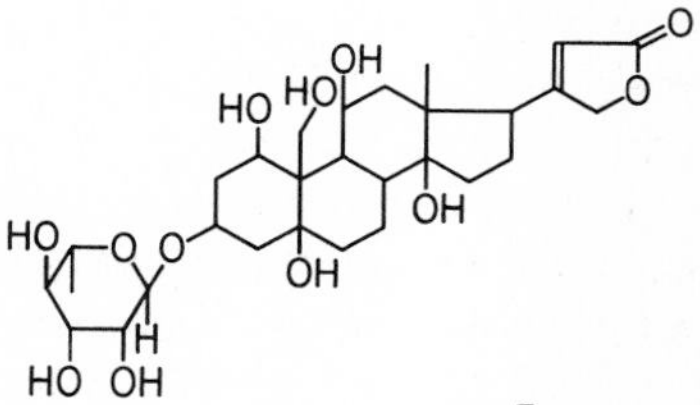

TETRODOTOXIN
(NEUROTOXIN , BLOCKER
OF Na^+ CHANNELS
(ACTION POTENTIALS),
10^{-7} M)

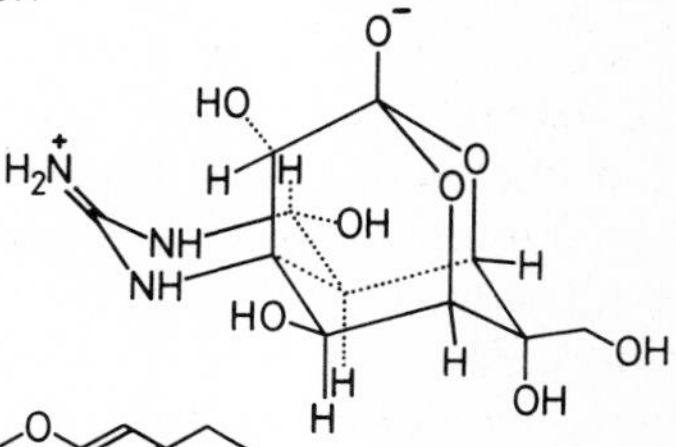

BICUCULLINE
(INDUCTION OF
EPILEPTIC SEIZURES ,
1.2 mg / kg BODY WEIGHT)

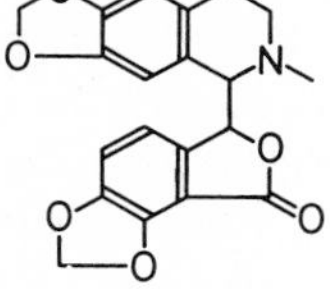

Fig. 7.14. Constitution, function and typical concentrations of a few selected substances frequently used in physiological experiments

7.9 Concluding Remarks

Different ways of using microelectrodes are possible. The number of simultaneously accessible parameters (membrane potential, cell membrane resistance, ion activities) depends on the number of single-barrelled microelectrodes employed or on the number of barrels that make up a multi-barrelled microelectrode. The features of experiments with several single-barrelled microelectrodes are different from those using a single multi-barrelled microelectrode.

Often, the calibration procedure itself is sufficient for testing and selecting suitable microelectrodes for a particular physiological study. Depending on the procedure either ion activities or free ion concentrations can be obtained potentiometrically. In order to measure free ion concentrations it is necessary to take into account the interfering sample ions by adding them to the calibration solution (constant ionic strength). The calibration for ion activities is simpler and does not require a constant ionic strength. In general, activity standards are available (National Bureau of Standards). However, they are not available for intracellular Ca^{2+} measurements at submicromolar levels.

Impalement criteria for intracellular experiments have been discussed. Techniques for the improvement of punctures and the localization of the microelectrode tip within the sample have been proposed.

The evaluation of the degree of cell damage after a puncture with a microelectrode is difficult. Extremely good seals of the cell membrane about the micropipette are necessary to avoid a leakage of electrolytes. Phenomena indicating cell damage are known. Criteria have been proposed to determine whether or not measurements are being carried out on undamaged cells.

Leakage rates of reference electrolyte (3 M KCl) from reference microelectrodes into the cytosol have been measured and calculated. Small cells can be seriously contaminated within short periods of time (seconds or minutes). Reference microelectrodes containing 0.5 M KCl or a physiological electrolyte are a good compromise with respect to tolerable leakage rates, liquid-junction potentials and tip potentials. A preliminary study of the cell contamination by components from neutral carrier microelectrodes shows that the ion carriers of membrane solutions can influence the properties of the cell membrane.

Tip potentials and electrical shunts at the tips of microelectrodes can seriously disturb the potentiometric measurements. Electrical shunts are usually observable and, in most cases, can be eliminated. The origin of tip potentials is uncertain. They can often be kept to a tolerable level but are nevertheless difficult to identify. In addition theoretical models which allow a mathematical correction for changes in the tip potential are not available.

Most of the available neutral carrier microelectrodes are prone to some interference from other physiologically relevant ions. The degree of interference can be evaluated. Consequently, it is possible to predict the suitability of a microelectrode for a given experiment. If other substances such as blockers or drugs are involved, their interference on the electrode response of the ion-selective microelectrode has to be evaluated.

8 General Aspects of Intracellular Measurements of Ions

8.1 The Intracellular Space as the Measuring Sample

The intracellular space is an extremely complex sample (Fig. 8.1) [Alberts et al. 83]. An idea of the complexity of a typical eucaryotic cell, the liver hepatocyte, is given below:

- the cytosol accounts for only about half (54%) of the total volume of a cell
- mitochondria (22%), rough (9%) and smooth endoplasmic reticulum cisternae including Golgi cisternae (6%), nucleus (6%), peroxisomes (1%) and lysosomes (1%) are important organelles
- a single cell contains about 1700 mitochondria, 300–400 peroxisomes and lysosomes, and 10^7 ribosomes
- most of the total membrane area is not made up of the plasma membrane itself (2%), but is made up of the internal membranes (e.g. rough endoplasmic reticulum membrane (35%), smooth endoplasmic reticulum membrane (16%), mitochondrial outer (7%) and inner (32%) membrane)
- a single cell contains about 20% proteins, corresponding to about 10^{10} protein molecules of approximately 10^4 different types
- the cytosol consists of a dense network of protein filaments, the cytoskeleton.

In addition, the cells may contain lipid droplets (0.2 to 5 μm in diameter) and glycogen granules (~30 nm in diameter) in which carbohydrates are stored. Plant cells also contain vacuoles and chloroplasts (~5 μm in diameter). Vacuoles usually occupy more than 50% (in extreme cases 95%) of the cell volume and are separated from the cytosol by a single membrane called the tonoplast. In addition plant cells contain a cell wall that is much thicker and more rigid than the plasma membrane.

   The three-dimensional picture (Fig. 8.2) which was drawn from many electron micrographs [Krstić 76] gives an excellent impression of the compact and complex intracellular space of a single cell. The diameters of the spherical vesicles shown leaving the Golgi complex (Fig. 8.2, lower right) are typically about 50 nm, which corresponds to the smallest available micropipettes (see Sect. 6.2). The tip diameters of neutral carrier microelectrodes are usually about 10 times larger. The tip size is then comparable to the width of the mitochondria (Fig. 8.2, middle left). There are at least two questions which should be considered: can the microelectrode tip be accurately introduced into the cytosol, and, are the properties of the cytosol uniform throughout, as in normal electrolytes? The latter problem is further complicated by the following facts.

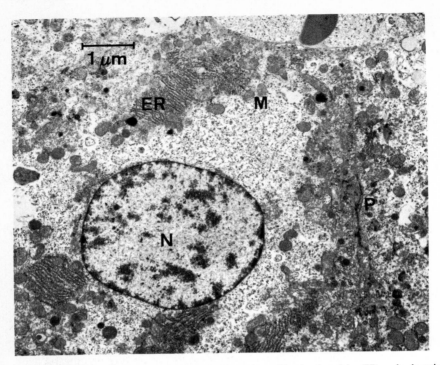

Fig. 8.1. Electron micrograph of a liver cell. *N:* nucleus, *M:* mitochondria, *ER:* endoplasmic reticulum, *P:* plasma membrane. Courtesy of E. Wehrli

Cells contain an extremely high percentage of protein (typically about 20%, erythrocytes: 36.8% [Pauly 73]). Other macromolecules such as nucleic acids and polysaccharides are also present in substantial amounts. The proteins form a large surface carrying many charged and hydrophobic groups. Thus, the cytosol has to be treated as a concentrated and highly organized polyelectrolyte. By expanding the theoretical treatment of polyelectrolytes, an informative description of the cellular fluids can be obtained [Berendsen 67; Oosawa 71]. It has been shown that a high percentage of protein influences the state of the ions and water. The presence of many charged groups produces strong electrical fields which affect the surrounding counterions and solvent molecules. Hence, the following aspects must be considered with regards to studies involving microelectrodes:

– the structure of water
– the solvation properties of the structured water
– the binding of ions
– the activity of ions in dense protein solutions
– the diffusion of ions in structured media.

Proteins will bind water by forming hydrogen bridges to the amide carbonyl groups or by accommodating water molecules in their inner network. Therefore, the assumption that the total intracellular water is free and completely available as a

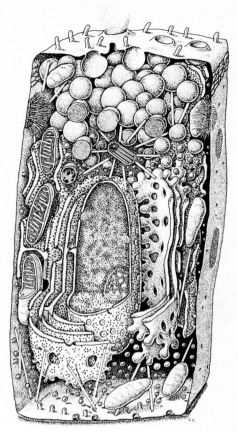

Fig.8.2. Three-dimensional view of a section through an epithelial cell illustrating the high density of protein filaments and internal membrane-bound organelles [Krstić 76]. Courtesy of R. V. Krstić. From R. V. Krstić, Ultrastruktur der Säugetierzelle, Springer 1976. Reproduced by copyright permission of Springer, Berlin Heidelberg New York

solvent for the electrolytes is unrealistic. At least two states of intracellular water must be distinguished. The hydrogen-bonded water is tightly adsorbed to the proteins (e. g. 0.2–0.3 g H_2O/g protein, which corresponds to about 1 molecule of H_2O per polar group of the protein [Pauly 73]). This bound water has a different structure from normal bulk water and therefore exhibits different solvation properties for ions (it is probably osmotically inactive for every solute [Fulton 82]). It is thought that the structure consists of polarized multilayers which form hydrogen bonds less easily and less strongly than bulk water [Ling et al. 73]. As the distance from a protein molecule increases, the remaining water molecules gradually exhibit the properties of normal water.

An estimation of the spatial relationships between the particles in the very dense protein solutions of erythrocytes has been made [Pauly 73]. The fraction of the volume occupied by hydrated hemoglobin molecules is calculated to be 0.37 (for 0.3 g of water bound to 1 g of protein and an assumed diameter of 62 Å). If the residual volume (0.63) of free cell water is assumed to be distributed homogeneously over the surface of the hemoglobin molecules, an 18 Å layer results. Thus, the diameter of hydrated alkali or alkaline-earth metal cations [Marcus 77] is only 3 to 4 times

DENSITY OF PROTEINS IN ERYTHROCYTES

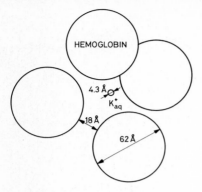

Fig. 8.3. Schematic representation of the spatial relationships between hemoglobin molecules, hydrated K^+ ions and the water layer between the proteins in the cytoplasm of erythrocytes. An elemental cell (true to scale) would countain about 30 ions per hemoglobin molecule [Pauly 73]. Redrawn from Pauly, Biophysik, 1973 [Pauly 73]

smaller than the average thickness of the water layer between the protein molecules (Fig. 8.3). The results of this dense packing on the mobilities, diffusion coefficients and conductivities have been discussed elsewhere [Pauly 73]. In the crowded cytoplasm, protein molecules do not only exert an influence over the ions and the water but also interact with other molecules. This has several important physiological consequences, such as altering enzyme activities [Minton and Wilf 81].

There are two major theories describing the cellular function [Hazlewood 73]. Both make fundamentally different assumptions about the physical state of the ions and the water in the cells (Tab. 8.1).

It is useful to briefly discuss the two theories with respect to the application of microelectrodes to cells, as both theories reveal important aspects that have to be considered in the measurements of intracellular ions with microelectrodes.

The classical view, the membrane theory, was first formulated by Pfeffer [Pfeffer 1877]:

„Über Aufnahme oder Nichtaufnahme eines gelösten Körpers in das Protoplasma entscheidet eine peripherische Schicht dieses, die Plasmamembran, welche sicher überall da gebildet wird, wo Protoplasma an eine andere wässrige Flüssigkeit stößt." [Pfeffer 1877, p. 235].

Today, the membrane theory is the most widely accepted theory and is discussed in almost every textbook. The model describes the cell as a space containing liquid water surrounded by a membrane. The basic structure of the membrane consists of a lipid bilayer (usally about 50 wt.-%) which serves as a relatively impermeable barrier to the flow of most solutes. Proteins (usually about 50 wt.-%) of different functions are dissolved in the lipid phase. Intracellular ions, proteins and other solutes are assumed to be dissolved in a complex artificial aqueous solution. Pumps that are located in the cell membrane regulate the transport of ions and other species and, consequently, consume energy.

Table 8.1. Some aspects of the membrane theory [Pfeffer 1877] and the association-induction hypothesis [Ling 62]

Aspect	Membrane theory	Association-induction hypothesis
state of protoplasm	dilute solution of salts and macromolecules	highly structured and complex medium
boundary between intra- and extracellular space	integrate cell membrane, lipid bilayer structure	absence of integrate cell membrane
control of ion activities	membrane permeability, active transport	selective binding to macromolecules, cooperative adsorption under control of cardinal sites (e.g. ATP), partition between water of different structures
diffusion of ions	theory relies on experiments that only yield slightly reduced diffusion coefficients for the ions (e.g. a factor of 2 for cytosolic K^+ [Hodgkin and Keynes 53]	theory relies on experiments that yield considerably reduced diffusion coefficients for the ions (e.g. a factor of 8 for cytosolic K^+ [Ling and Ochsenfeld 73])
state of water	80–90% free water, 10–20% structured water (bound to macromolecules)	all of the water exists in a different state than free bulk water, existence of deep polarized multilayers
retention of water	osmotic pressure, osmotic activity is regulated by pumps	regulation by intracellular proteins
role of ATP	fuel for pumps	cardinal site, regulates the protein conformation by adsorption and, consequently, influences the protein-ion-water system
resting potential	membrane potential	phase-boundary potential
microelectrode measurements	ion activities in unstructured cytosolic water	erroneous ion activity in a small cytoplasmic volume where the structural integrity has been destroyed or greatly perturbed

A second view of the living cell was formulated by Ling. His association-induction hypothesis offers a different and conflicting picture of the cell [Ling 62]:

„The cell surface about which we have been speaking corresponds roughly with what conventionally has been regarded as the „cell membrane" and, indeed, if one so chooses, there is no reason not to refer to this cell surface as the „membrane". Only one must keep in mind that this „membrane" is more like, say, the skin of an apple which itself constitutes a phase similar to the bulk phase it encloses than the „plasma membrane" in the conventional sense which separates two essentially similar aqueous phases." [Ling 62, p. 283].

The living cell is seen as a fixed-charge system in which the proteins form a lattice with a chain-to-chain spacing of about 20 Å (see also Fig. 8.3). The surface of this protoplasm is in contact with the extracellular fluid and forms an interface, rather than a membrane. It is thought that the cellular processes are regulated by the structured medium. The ion activities are not controlled by active membrane transports and membrane permeabilities but by the selective binding of ions to the macromolecules and the distribution of these ions between water of different structures (this view is also adopted and extended in the sorption theory of Troshin [Troshin 66]). Thus, the unusual composition of the cytosol is responsible for the distribution of the ions. This provocative theory, proposed in order to overcome inadequacies in the membrane theory [Ling 62], was widely propagated ten years ago [Hazlewood 73] and is still under discussion today [Horowitz and Paine 79; Ling 82; Fulton 82]. Of course, since the potentiometric technique measures ion activities in free aqueous solutions, the theory criticizes microelectrode studies as illustrated below.

„Moreover, this film of fluid is precisely the layer of cytoplasm which the impaling electrode has to displace, in order to reach the inside of a cell. Such cytoplasm, torn from its normal structural framework and pushed against adjacent layers of cytoplasm cannot, in our evaluation, be considered as existing in its normal physiological state; what is far more likely is that it resembles cytoplasm of a well ground-up muscle. As such, the K^+ ion would be liberated from its normal adsorption sites, thereby giving high K^+ ion activity reading on the ion-specific microelectrode." [Ling et al. 73, p. 36–37].

This statement was later criticized because it is to be expected that the ions in a film of damaged cytoplasm will reach an equilibrium with the undisturbed cytoplasm after a short period of time [Edzes and Berendsen 75].

Most of the recent studies present data or at least interpret the results in favour of the membrane theory. They conclude that 80–90% of the intracellular water is mobile (like normal water) and only 10–20% is strongly bound to the macromolecules. It has also been found that only a small fraction of the intracellular Na^+ and K^+ ions are immobilized. The existence of active transmembrane transport processes, especially the ubiquitous Na^+/K^+ ATPase, is undisputed. However, the binding hypothesis also puts forward a molecular theory for active transports in epithelia [Ling et al. 73]. In both theories, the importance of the physico-chemical properties of the protoplasm (including association and induction processes) for the physiological functions of living cells is recognized.

Many techniques have been used to elucidate the state of ions and water in cells: microelectrodes [Lev and Armstrong 75, Edzes and Berendsen 75; Palmer et al. 78; Civan 78], conductometry [Pauly 73; Edzes and Berendsen 75; Foster et al. 76], calorimetry [Pauly 73], potentiometry [Pfister and Pauly 72], reference phase [Horowitz and Paine 79; Horowitz et al. 79] and nuclear magnetic resonance spectroscopy [Edzes and Berendsen 75; Shporer and Civan 77]. Two brief examples which illustrate the problems encountered are given below.

Results of earlier NMR studies strongly supported the binding hypothesis [Cope 65; Ling and Cope 69; Chang et al. 73]. Evidence was found for the presence of structured water and a considerable binding of the Na^+ and K^+ ions. Later on, however, it was observed that the properties of a nuclear quadrupole interaction with the sodium nuclei could have led to an erroneous interpretation of the amount of immobilized Na^+ (for a discussion see [Civan 83]). When the nuclear quadrupole effects were properly considered, only a very small fraction of Na^+ ions ($<1\%$) was found to undergo a strong enough interaction with the intracellular binding sites to become observable by NMR spectroscopy [Shporer and Civan 77].

The simultaneous determination of free and bound forms of ions became possible with the development of the intracellular reference phase method (see Sect. 9.3.1) [Horowitz et al. 79]. In oocytes, two-thirds of the water was found to be structured and inaccessible to solutes such as sucrose. Na^+ and K^+ ions exchange only slowly in this aqueous phase but can freely diffuse in the remaining third of the cytoplasmic water. The two phases are in equilibrium. Apparent activity coefficients (see Sect. 8.5) determined by the reference phase method [Horowitz et al. 79] are in excellent agreement with the values reported using microelectrodes [Palmer et al. 78] (see Fig. 8.14). Thus, both bulk water with freely diffusing ions and bound water with partially immobilized ions (yolk platelets) seem to exist in amphibian oocytes. This result, to some extent, satisfies both views of the cell function.

The high complexity of the cell not only makes a description of its function and state more difficult but also complicates the interpretation of intracellular measurements. The following problems arise in studies using microelectrodes:

1) Localization: is the microelectrode tip after cell puncture always located in a cytosolic medium of high water and ion mobility?
2) Selectivity: does the ion-selective microelectrode exhibit sufficient selectivity towards all of the interfering intracellular ions present?
3) Contamination: is the liquid membrane of the microelectrode affected by proteins or other species in the cellular fluid?
4) Activity coefficient: do ion activity measurements in dense protein solutions have the same validity as the corresponding measurements in pure aqueous electrolytes?
5) Liquid-junction potential: are the liquid-junction potentials of the intracellular reference microelectrode in the dense protein solutions the same as in pure aqueous electrolytes?
6) Suspension effect: are the measurements affected by suspension effects due to the presence of a high percentage of polyelectrolytes?

In some cases, it is possible to find satisfactory solutions to these problems:

1) The high reproducibility of repetitive measurements in the same cell indicates that the tip is always located in the same kind of sample, most probably in the cytosol as confirmed by microscopic observations. Results obtained with microelectrodes are in agreement with those obtained using other techniques. This further supports the localization of the microelectrode tip in the cytosol.
2) In characterizing the performance of microelectrodes the selectivities towards the most important interfering intracellular ions are always considered (Chap. 9). However, it is impossible to check the selectivity towards all of the intracellular ions. Thus some risk of interference remains. Standard addition of the measuring ion to a cell extract is an excellent test that can be used to determine any interference. Such an experiment was carried out for neutral carrier H^+ microelectrodes in cell sap solutions [Kurkdjian and Barbier-Brygoo 83]. Another useful

approach is an intracellular calibration in which the ion to be measured is injected in the presence of an intracellular microelectrode [Thomas et al. 75].

3) The membrane solution of the microelectrode may be disturbed or poisoned by
   charged species such as amino acids, proteins or lipids as well as neutral species such as lipids or sugars. These substances are extracted (depending on their
   distribution coefficient) into the membrane solution where they can interact with
   the carrier molecules or influence the physico-chemical properties of the membrane phase. Clinical analyses with neutral carrier membrane electrodes in
   whole blood or undiluted urine have shown that solvent polymeric membranes
   are unaffected by biological samples [Meier et al. 77; Jenny et al. 80a; Jenny et
   al. 80b; Meier et al. 80; Anker et al. 81; Ammann et al. 81b; Ammann et al. 82;
   Anker et al. 83a; Anker et al. 83b; Anker et al. 83c; Anker et al. 84]. The influence of lipophilic anions [Boles and Buck 73; Morf et al. 74a; Morf et al. 74b;
   Seto et al. 75; Ryba and Petránek 76; Morf 81], lipids [Aguanno and Ladenson
   82], proteins [Hill et al. 78; Winnefeld and Schröter 81; Rehfeld et al. 84] and
   erythrocytes [Fogh-Anderson 78; Siggaard-Anderson et al. 83] in solvent polymeric membranes has been widely discussed. It has been shown that optimized
   membranes are free from any disturbances. However, a deposition of macromolecules on the surface can occur [Maruizumi et al. 85]. The effect on the performance of the electrode is negligible. This can be deduced from the high lifetimes (several months) of membranes used continuously in clinical analyzers.
   The compositions of these solvent polymeric membranes are very similar to
   those of the corresponding neutral carrier membrane solutions used for microelectrodes (see Chap. 4). It is therefore very likely that the electromotive behaviour of a microelectrode is not altered by the contact with the cytosolic solution.

4) The activity coefficients of ions in aqueous electrolytes may be different in intracellular fluids because of the electrostatic interactions caused by the presence of
   high concentrations of polyelectrolytes. Consequently, ion activities in solutions
   containing large amounts of charged macromolecules (proteins, nucleic acids,
   polymers) were determined separately [Pfister and Pauly 72; Pauly 73; Lev and
   Armstrong 75]. Other than a binding of the ions to the polyelectrolytes, it was
   found that the activity coefficients of NaCl and KCl are not dramatically affected by the presence of the macromolecules (up to 45 wt.-% protein). For instance,
   the activity coefficient of KCl is only about 15% lower in a 0.1 M KCl solution
   with 20 wt.-% protein than in a pure aqueous 0.1 M KCl solution [Pfister and
   Pauly 72]. These results strongly indicate that the potentiometric values obtained
   in cytosolic solutions are in accordance with the theoretical descriptions of normal electrolytes. This validates the intracellular use of microelectrodes.

Additional interesting aspects of electrode measurements in solutions containing
proteins can be derived from clinical analyses. In blood serum only 93% of the
total volume is free plasma water, the remaining 7% are made up of proteins and
lipids. Direct potentiometric measurements with membrane electrodes in undiluted blood or serum samples yield ion concentrations in the free plasma water.
On the other hand, indirect methods (flame photometry, atomic absorption spectrometry, colorimetry) using strongly diluted serum samples yield concentrations

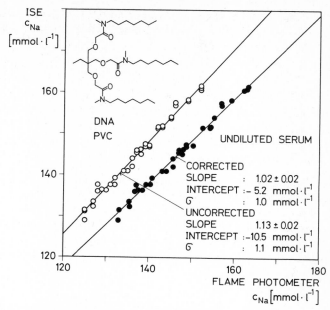

Fig. 8.4. Correlations between measurements of Na$^+$ in blood serum samples using potentiometry (ISE) and flame photometry. 20 serum samples were analyzed twice. ○ : Na$^+$ (ISE) uncorrected for protein and lipid volume; ● : Na$^+$ (ISE) correct for protein and lipid volume [Anker et al. 83 b]

corresponding to the total volume of the serum. The results from the two approaches indeed differ by about 7%, whereby the potentiometric values are the higher ones throughout (Fig. 8.4). Although the cytosol is different in composition from blood there are many similarities. From the above results it can be concluded that microelectrodes only detect ions in the free cytosolic water and that the structured or bound water, i.e. the fraction of water in the vicinity of proteins, is not accessible to microelectrode measurements. It is possible that the tip of the microelectrode is located in crystalline-like regions of protein such as yolk platelets. However, the excellent reproducibility of results from punctures of oocytes suggests that the yolk fraction is not penetrated.

5) The liquid-junction equivalence assumption [Edzes and Berendsen 75], i.e. the assumption that the liquid-junction potential at the reference microelectrode is equally influenced by an aqueous calibration solution and the cytosol, must be fulfilled. In simple electrolyte solutions the liquid-junction potential is adequately described by the Henderson formalism (Sect. 5.7). The usefulness of the Henderson equation (Eq. (5.35)) for the calculation of liquid-junction potentials of reference microelectrodes immersed in cellular fluids has been demonstrated by the simultaneous measurements made with two intracellular reference microelectrodes [Palmer and Civan 77]. The first reference microelectrode was always filled with a 3 M KCl solution, whereas the second one was filled with different electrolytes (0.5 M KCl, 0.1 M KCl, 1.3 M Na$_2$SO$_4$ or 2 M MgCl$_2$). The changes in the liquid-junction potentials of various pairs of microelectrodes were recorded in both an external aqueous saline solution and in the cytosol of salivary

gland cells. The two sets of values were in excellent agreement indicating that the intracellularly observed liquid-junction potentials are still described by the Henderson equation. However, the experiments do not yield any information concerning the absolute values of these potentials. Thus one cannot be certain that the absolute values measured in cytosol may be different from those obtained for pure aqueous solutions (see below).

6) In order to determine intracellular ion activities (see Eq. (7.3)), an exact knowledge of the resting potential of the cell is essential. The influence of the presence of polyelectrolytes (suspension effect) on the measurements of cell membrane potentials is discussed below.

The membrane theory and the association-induction theory offer different theoretical descriptions of the resting potential. In the membrane theory the resting potential is the manifestation of a real membrane potential that can be subdivided into an external phase-boundary potential, a diffusion potential and a second phase-boundary potential facing the intracellular space. The resting potential (E_M) is usually described by the Goldman-Hodgkin-Katz equation:

$$E_M = \frac{RT}{F} \ln \frac{P_K [K]_i + P_{Na}[Na]_i + P_{Cl} [Cl]_e}{P_K [K]_e + P_{Na} [Na]_e + P_{Cl} [Cl]_i},$$ (8.1)

where P is the ion permeability and i and e refer to intracellular and extracellular concentrations, respectively. In contrast, the binding hypothesis treats the entire cell as a fixed-charge system in direct contact with an extracellular solution. Thus, the resting potential is expressed as a surface potential and the permeabilities are replaced by the equilibrium constants of adsorption of the respective ions on the surface of the fixed-charge system [Ling 62; Ling 82]:

$$E_M = E_o - \frac{RT}{F} \ln (K_K [K]_e + K_{Na} [Na]_e),$$ (8.2)

where K_K and K_{Na} are the adsorption constants for K^+ and Na^+, respectively, on the anionic sites of the cell surface.

Despite these differences in the theories the experimentally measured resting potential is still defined as the recorded sudden change in potential during the penetration with the intracellular microelectrode. Although the recordings are stable and reproducible, the interpretation of the change in EMF in terms of the true resting potential is difficult [Tasaki and Singer 68]. The difficulties are due to the presence of high concentrations of proteins. It is well known that electrode studies in colloidal polyelectrolyte suspensions are problematic [Pallmann 30; Jenny et al. 50, Eriksson 51; Tasaki and Singer 68; Brezinski 83]. The abnormal electrode behaviour observed in such systems is attributed to suspension effects. They were first documented by Pallmann for pH measurements (Pallmann effect, Fig. 8.5) [Pallmann 30]. In order to evaluate the influence of suspensions, potentiometric pH determinations were performed in many different colloidal systems. The pH values of these suspensions were found to differ drastically from those of the corresponding filtrates. This phenomenon was termed the suspension effect and was explained by

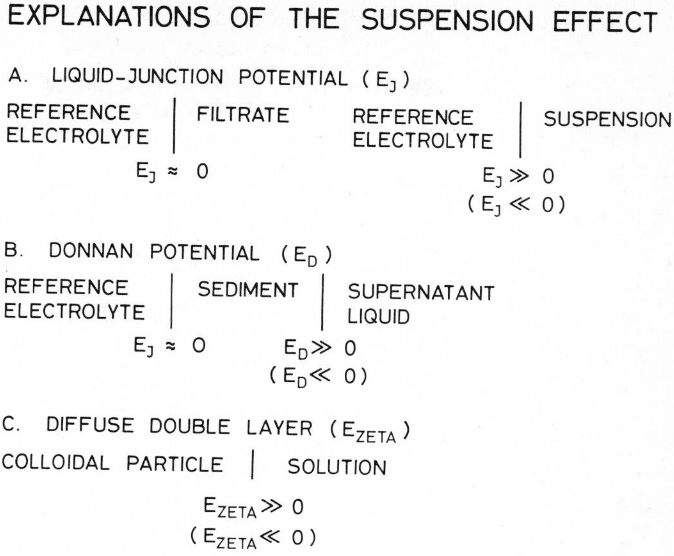

EXPLANATIONS OF THE SUSPENSION EFFECT

A. LIQUID–JUNCTION POTENTIAL (E_J)

REFERENCE ELECTROLYTE		FILTRATE	REFERENCE ELECTROLYTE		SUSPENSION

$$E_J \approx 0$$

$$E_J \gg 0$$
$$(E_J \ll 0)$$

B. DONNAN POTENTIAL (E_D)

REFERENCE ELECTROLYTE		SEDIMENT		SUPERNATANT LIQUID

$$E_J \approx 0 \qquad E_D \gg 0$$
$$(E_D \ll 0)$$

C. DIFFUSE DOUBLE LAYER (E_{ZETA})

COLLOIDAL PARTICLE		SOLUTION

$$E_{ZETA} \gg 0$$
$$(E_{ZETA} \ll 0)$$

Fig. 8.5. Schematic illustration of various models describing suspension effects in potentiometric cells

the presence of ion swarms about the charged colloidal particles, which act as ion-exchanger sites [Pallmann 30].

Several other theoretical analyses of the suspension effect have been proposed (Fig. 8.5): anomaly in the liquid-junction potential [Jenny et al. 50], formation of a Donnan potential [Eriksson 51; Brezinski 83] or the generation of zeta potentials [Eriksson 51]. Jenny et al. [50] found that the transference number of Cl^- ions is reduced in KCl solutions containing cation-exchangers, i.e. the relative diffusion rates of ions are strongly influenced, thus yielding anomalous liquid-junction potentials (Fig. 8.5., A). Using experimental data an empirical equation for liquid-junction potentials in suspensions was obtained [Jenny et al. 50]:

$$E_J = \frac{RT}{F} \left(\ln \frac{a_2}{a_1} - 0.85 \ln \frac{1 + 6.36 a_2}{1 + 6.36 a_1} \right), \tag{8.3}$$

where a_1 and a_2 are the mean activities of KCl in the free solution of the suspension and the electrolyte of the reference electrode, respectively. The experimental evidence seems to support the theory (Fig. 8.6). It was found that the pH measurements in the two-phase system sediment/supernatant liquid do not depend on the position of the glass membrane electrode (Fig. 8.6, C). In contrast, the position of the diaphragm of the reference electrode in the suspension strongly influences the EMF measurements (Fig. 8.6, D). If the reference electrode of a pH glass-membrane cell assembly is immersed in the supernatant liquid, a pH of 6.0 is measured (Fig. 8.6, A). Within the sediment, a pH of 2.0 is measured (Fig. 8.6, B). The difference in pH of 4 units corresponds exactly to the potential difference of two reference electrodes immersed in each phase (240 mV, Fig. 8.6, D).

SUSPENSION EFFECT

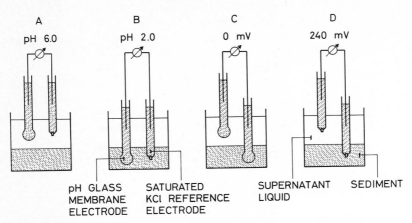

Fig. 8.6. Illustrations of the suspension effect. Redrawn from Jenny, Nielsen, Coleman and Williams, Science, 1950 [Jenny et al. 50]

These unusual potential differences were attributed to anomalous liquid-junction potentials [Jenny et al. 50]. This view neglects a possible Donnan potential between the sediment and the supernatant liquid. Thus, the actual pH values of the two layers should be equal.

On the other hand, a different explanation of the effects observed by Pallmann neglects the large liquid-junction potential differences, but assumes a Donnan-type of potential difference at the boundary between the sediment and the pure liquid solution (Fig. 8.5, B) (Teorell-Meyer-Sievers theory [Eriksson 51; Brezinski 83]). In this view, the sediment behaves as an ion-exchanger with fixed negative charges, preferentially extracting cations from the solution. A Donnan potential (E_D) develops between the two sides due to the asymmetric distribution of ions:

$$E_D = \frac{RT}{F} \ln \frac{a'}{a''}, \tag{8.4}$$

where a' and a'' are the activities of an ion in the sediment and in the bulk solution, respectively. It was found that the potentials calculated on the basis of this theory also fit the experimental data reported by Jenny et al. [Eriksson 51]. Thus, the measurements with the four cell assemblies shown in Fig. 8.6 (A–D) can be rationalized in the following manner: Cell assembly A: the H^+ activity a''_H of the supernatant liquid is measured (pH 6.0):

$$E(A) = E_o + \frac{RT}{F} \ln a''_H. \tag{8.5}$$

Cell assembly B: the H^+ activity a'_H of the sediment is measured (pH 2.0):

$$E(B) = E_o + \frac{RT}{F} \ln a'_H. \tag{8.6}$$

Cell assembly C: the two glass membrane electrodes measure the ratio of the two H^+ activities, a_H'' and a_H'. The Donnan potential between the two phases is given by the reciprocal ratio of the H^+ activities, i.e. the two contributions cancel out and the cell assembly shows an EMF = 0 mV:

$$E(C) = \frac{RT}{F} \ln a_H'' - \frac{RT}{F} \ln a_H' + \frac{RT}{F} \ln \frac{a_H'}{a_H''}. \tag{8.7}$$

Cell assembly D: assuming that the liquid-junction potentials of the two reference electrodes are equal, the two reference potentials E_o are also equal and therefore cancel. Thus, this cell assembly only measures the Donnan potential (240 mV) between the supernatant liquid (pH 6.0) and the sediment (pH 2.0):

$$E(D) = E_o - E_o + \frac{RT}{F} \ln \frac{a_H'}{a_H''}. \tag{8.8}$$

The theory of diffuse double layers of ions at a charged surface has also been used to explain the suspension effects (Fig. 8.5, C) [Eriksson 51].

The occurrence of unusual potential differences in colloidal polyelectrolytes is undisputed but there is no general agreement about their origin. The suspension effect has also been observed in experiments with reference microelectrodes [Tasaki and Singer 68]. KCl-filled microelectrodes were positioned in the polyelectrolyte sediments and in the supernatant liquids. The polyelectrolyte chosen to simulate biological conditions was polyglutamic acid. The potential differences between the microelectrodes were found to be a function of the electrolyte concentration. In dilute solutions potential differences in the range of resting potentials (~100 mV) were obtained (Fig. 8.7) [Tasaki and Singer 68]. The effect increased with increasing polyelectrolyte concentrations.

It is possible that the potential differences due to suspension effects may be superimposed on the intrinsic resting potential of the cell. Thus, in measurements with common electrode arrangements (see Fig. 7.2) the intracellularly recorded potential (ion-selective microelectrode relative to external reference electrode) has to be corrected for the membrane potential (intracellular reference microelectrode relative to external reference electrode) (see Eq. (7.4)). Before the impalement, the reference microelectrode/external bath reference electrode measures an electromotive force E_{extra} in the external electrolyte solution:

$$E_{extra} = E_o + E_{J,e} + E_{J,m} + E_{TP}, \tag{8.9}$$

where

E_o : sum of constant potential differences;
$E_{J,e}$: liquid-junction potential of the external reference electrode;
$E_{J,m}$: liquid-junction potential of the microelectrode;
E_{TP} : tip potential of the microelectrode.

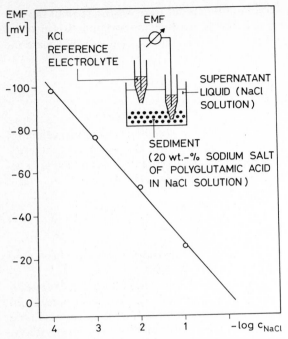

Fig. 8.7. Potential difference between KCl-filled reference microelectrodes as a function of the NaCl concentration in the presence of a polyelectrolyte. Redrawn from Tasaki and Singer, Ann. New York Acad. Sci., 1968 [Tasaki and Singer 68]

After insertion of the reference microelectrode an electromotive force E_{intra} is measured:

$$E_{intra} = E_o + E_{J,e} + E'_{J,m} + E'_{TP} + E_{SE} + E_{RP}, \tag{8.10}$$

where
E_{SE} : potential difference due to suspension effects;
E_{RP} : intrinsic resting potential of the cell.

The measured membrane potential (E_M) is given by

$$\begin{aligned} E_M &= E_{intra} - E_{extra} \\ &= (E'_{J,m} - E_{J,m}) + (E'_{TP} - E_{TP}) + E_{SE} + E_{RP}. \end{aligned} \tag{8.11}$$

E_M coincides with the true cellular resting potential E_{RP} only if the liquid-junction potentials and the tip potentials are the same before and after impalement and there are no potential differences due to the high polyelectrolyte concentrations (suspension effect E_{SE}).

The differences between the liquid-junction potentials $E'_{J,m}$ and $E_{J,m}$ are small assuming that the effects of the polyelectrolytes are contained in E_{SE}. When concentrated equitransferent reference electrolytes are employed the changes in the liquid-junction potential should not exceed 1–2 mV (see Sect. 5.7) corresponding to an error of 4–8% and 8–16% in the determined activity of monovalent and divalent ions, respectively. Changes in the tip potential ($E'_{TP} - E_{TP}$) can give rise to ambiguities in the resting potential measurements. However, studies on the nature of the tip potentials have shown that carefully prepared micropipettes should exhibit relatively constant tip potentials of less than 2–5 mV (see Sect. 7.7).

Assuming that suspension effects are involved in intracellular measurements, they can either lead to erroneous membrane potentials or, in principle, account for the membrane potential itself. It is interesting to consider the following cases (theory of the function of a cell / description of the suspension effects):

1) Membrane Theory / Donnan Potentials: The polyelectrolyte is not in direct contact with the external aqueous phase. Therefore, the Donnan potential is not superimposed on the intrinsic resting potential of the cell membrane. The intracellular ion activities are influenced by the anions of the polyelectrolyte; a fact, that is taken into account in the Goldman-Hodgkin-Katz equation (8.1). The microelectrode measurements are correct.

2) Membrane Theory / Liquid-Junction Potentials: Assuming that the resting potential is operationally defined as the potential difference between the two reference microelectrodes, suspension effects on the liquid-junction potential of the intracellular microelectrode give rise to erroneous resting potentials.

3) Association-Induction Hypothesis / Donnan Potentials: The theory proposes that the intrinsic resting potential of the cell is described by a Donnan potential between the aqueous phase and the polyelectrolyte. The microelectrode measurements are correct.

4) Association-Induction Hypothesis / Liquid-Junction Potentials: The resting potential (which is a Donnan potential) is erroneous due to changes in the liquid-junction potential of the intracellular reference microelectrode.

8.2 Ion Activity, Concentration of Free Ions, Total Concentration

Intracellular ions are not distributed uniformly within a cell and are present in more than one form in the cytosol. Subcellular compartmentation in organelles may further contribute to a non-homogeneous distribution (Sect. 8.5). The various forms an ion may be found in can clearly be illustrated for calcium, since its distribution within a cell is especially varied. Binding to various cellular components and storage in specific cellular compartments are both involved. In addition, different transport systems exist in the plasma membrane and the internal membranes in order to maintain the extremely low cytosolic Ca^{2+} levels. The regulation of intracellular calcium and the exceptionally important physiological role of Ca^{2+} have already been described in detail elsewhere [Carafoli and Crompton 78; Carafoli et al. 82; Carafoli 82; Campbell 83].

CELL MEMBRANE

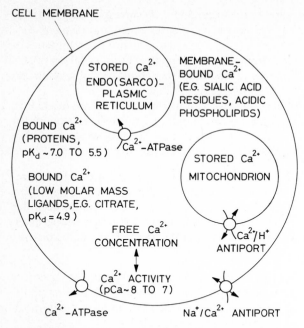

Fig. 8.8. Distribution of total calcium within a single cell. The scheme is not meant to represent a certain type of cell. Instead generalized examples of binding sites, sequestering organelles and transport systems are given

The following rough subdivision for intracellular calcium is possible and is, in principle, valid for any other ion:

- total calcium concentration
- concentration of stored calcium
- concentration of bound calcium
- concentration of free Ca^{2+}
- Ca^{2+} activity.

In order to differentiate between these terms it is helpful to describe the different states of calcium within a single cell (Fig. 8.8):

- stored calcium (in calcium pools such as the mitochondrion or the endoplasmic reticulum)
- bound calcium (e.g. to proteins, to complexing agents of relatively low molar mass or to the surface of the cell membrane)
- free ionized Ca^{2+}.

The free Ca^{2+} concentration in the cytosol of unstimulated cells is about 10^{-8} to 10^{-7} mol l^{-1}. The total amount of cellular calcium may be as high as 1 mmol per kg of cell [Kretsinger 79]. The extremely small amount of free Ca^{2+} is partly due to the binding to various complexing agents. The Ca^{2+} affinities of these agents, however,

Table 8.2. Quantities measured by the most relevant intracellular techniques

Quantity	Technique
ion activity[a]	microelectrodes
free-ion concentration	microelectrodes spectrophotometry fluorometry nuclear magnetic resonance luminescence gelatine reference phase null-point for permeabilized cells
total concentration	X-ray microanalysis flame photometry atomic absorption spectrometry

[a] intracellular pH measurements represent an exception. Various techniques (NMR, spectrophotometry, distribution of labelled weak acids or bases) yield the H^+ activity (definition of the pH scale)

are not strong enough (most pK_d values > 7.0) to account for the extremely low Ca^{2+} levels (pCa < 7.5). Thus, other mechanisms must be involved in the regulation of the cytosolic activity of Ca^{2+}. It has been shown that significant amounts of Ca^{2+} can be sequestered by organelles acting as calcium pools (endo(sarco)plasmic reticulum and mitochondria). Further regulation of intracellular Ca^{2+} is achieved by several transport systems in the plasma and internal membranes. For example, mitochondria exhibit an energy-linked uptake of cytosolic Ca^{2+}. Usually, the mitochondrial calcium can be rapidly ejected into the cytosol, i.e. the mitochondria act as effective regulators of the amount of free Ca^{2+} in the cytosol [Carafoli 74].

It is obvious that the sum of the calcium fractions must be equal to the total intracellular calcium. By subtracting the stored and bound calcium from the total calcium, the free cytosolic Ca^{2+} concentration is obtained. The free Ca^{2+} fraction can then be treated as a simple aqueous calcium solution, i.e. depending on the ionic strength of the cytosol the free Ca^{2+} concentration is related to the Ca^{2+} activity according to the single-ion activity coefficient (Sect. 5.6). It has been shown in Sect. 7.3 that microelectrodes allow the direct determination of the free-ion concentration or the ion activity, depending on the calibration procedure. The uniqueness of the microelectrode approach for a direct measurement of ion activities is expressed in Table 8.2.

8.3 In vitro and in vivo Measurements

Microelectrodes can be used to make both in vitro and in vivo measurements. Under these two experimental conditions the cell preparation has very different environments:

in vitro: in glass, i.e. the cell is isolated from the system in which it is found in na-
            ture
in vivo: in the living organism, i.e. the cell is observed under the genuine condi-
            tions of living systems.

Both experimental conditions are used to describe naturally occurring processes. In
vivo experiments are carried out in the living organism, whereas in vitro studies are
carried out on cells isolated from an organism and kept functioning only by using a
suitable artificial environment. Thus, in vivo and in vitro cells do not share life but a
state, which was referred to as the living state [Ling 62].

An advantage of in vitro studies is the possibility of studying the complex behaviour of cells under the strictly defined conditions of an artificial environment.
The composition and concentrations of the electrolytes and other solutes can be
chosen as required for the particular experiment. Accurate control of the experimental parameters is therefore possible in in vitro experiments. Physiological events
can be observed by adding or removing specific species, such as ions, inhibitors or
hormones. Nevertheless, from time to time in vitro observations must be checked
against the behaviour of the cells in their natural environment (in vivo).

In a few cases, the results obtained in vitro are directly transferable to in vivo
conditions. However, in many situations the two approaches cannot be directly
compared. Despite this, electrophysiological studies can be successfully carried out
on both in vitro and in vivo preparations. However, the physiological conclusions
drawn must take into account the experimental conditions used. Table 8.3 illustrates
the large variety of fundamentally different preparations of the same organ that can
be used for the study of cellular processes. The approaches used for tubular transport studies in the kidney range from in vivo measurements in unanesthetized human beings to in vitro studies on isolated membrane vesicles [Greger et al. 81]. The
former approach is the classical clearance measurement. The results from such
studies usually do not localize or explain the mode of transport. On the other hand,
vesicle experiments provide the most information about the molecular mechanism
of transport across kidney epithelia in vitro. As shown in Tab. 8.3 many intermediate
approaches are available for microelectrode studies.

In vivo studies using microelectrodes are always invasive, i.e. the tissue has to be
exposed and perfused with physiological electrolytes. However, the in vivo approach removes many of the technical difficulties of isolation and manipulation of
cell preparations. Therefore, it can be expected that in vivo measurements closely
describe the true physiological states of living systems.

The puncture of a single cell in vivo may be intricate but is possible. In contrast,
the preparation of in vitro samples usually involves many steps (Fig. 8.9). During
the sample preparation, isolation procedures may be used that are carried out under
nonphysiological conditions. For example, procedures such as centrifugation, chromatography or chemical treatment may seriously influence the meaning of subsequent ion measurements.

Therefore, compared with the true physiological state of living cells in vitro
studies are inevitably affected by the isolation procedure and may be subject to unknown errors. However, this does not decrease the value of in vitro experiments for
the elucidation of some important aspects of physiological events.

Table 8.3. Different types of preparations used for the study of physiological processes in the kidney

Preparation	Condition	Example
clearance measurements, conscious human being or animal[a]	in vivo	[Ammann et al. 81 c; Dütsch et al. 85]
perfused kidney, operative[b]	in vivo	[Matsumura et al. 80; Edelman et al. 78]
isolated perfused kidney[b]	in vitro	[Oberleithner et al. 84; Lang et al. 84]
isolated perfused tubule[b]	in vitro	[Greger et al. 84]
isolated cultured kidney cells[b]	in vitro	[Paulmichl et al. 84]
isolated membrane vesicles[c]	in vitro	[Burckhardt and Murer 81; Wright 84]

[a] macroelectrodes
[b] microelectrodes
[c] patch-clamp electrodes, voltage-sensitive dyes

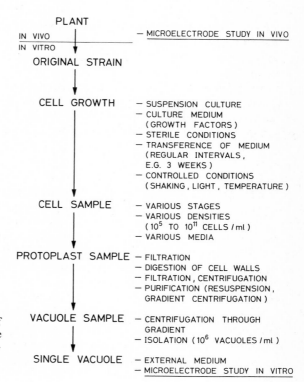

Fig. 8.9. An example of the types of isolation procedures used to prepare a sample for in vitro studies (for details see [Matile 66; Leguay and Guern 75; Alibert et al. 82])

Several factors are known to influence the results (in vitro and in vivo) of intracellular measurements with microelectrodes:

– the composition of the external electrolyte [Suzuki and Frömter 77; Matsumura et al. 80b; Radda et al. 82; Giebisch et al. 81; Schümperli et al. 82; Nicholson and Kraig 81]

Table 8.4. Comparison of the in vitro and in vivo measurements of ion activities using microelectrodes

Ion	Activity	In vitro/ in vivo	Cell	Reference
K^+	56 mM 63 mM	in vivo in vitro	*Necturus,* proximal tubule	[Khuri 79]
K^+	47 mM 73 mM	in vivo in vitro	*Amphiuma,* distal tubule	[Khuri 79] [Oberleithner et al. 80]
K^+	2.6 mM 2.7 mM	in vivo in vitro	Bullfrog, plasma near proximal tubule cell (in vivo); frog serum (in vitro)	[Fujimoto and Kubota 76]

- the partial pressure of CO_2 in the external solution [Zeuthen 78]
- the conditioning time of the perfused preparation [Reuss and Weinman 79]
- the pretreatment of the animal (e. g. season of purchase, temperature [Zeuthen 78]), the electrolyte composition of the living accommodation [Zeuthen 82; Oberleithner et al. 83 b] and the diet of the animals [Lewis and Wills 81].

Despite the different treatments of the cells before the microelectrode studies, the resulting ion activities may be remarkably similar. In certain cases, however, deviations up to 50% have been found (Tab. 8.4). Criteria other than the method of preparation undoubtedly play an important role in these discrepancies.

8.4 Organelles, Single Cells, Cell Population

Eucaryotic cells of higher plants and vertebrates exhibit a remarkable diversity. There are more than 200 different highly specialized types of cells in the human body, linked together by intricate systems of communication, and exhibiting a wide range of size, shape and function [Alberts et al. 83]. Most tissues contain more than one type of cell.

The aim of the microelectrode technique is to unravel (on the single-cell level) the fundamental physiological processes in which ionic species are involved. To this purpose, identified single cells have to be available for the puncture with a microelectrode. Table 8.5 shows that microelectrodes can be used to impale single cells which have been obtained by quite different procedures. Single cells can be directly isolated, cultured, sorted and fractionated into their subcellular components. Directly isolated cells can be fixed for puncture in the form of a single cell (e. g. egg cell) or as a piece of tissue (e. g. strang of Purkinje fibres). In other preparations it is advantageous to work with single cells obtained from cell cultures. Under appropriate conditions (nutrient medium, growth factor) cells can be cultivated in a culture dish. Cultures that are directly preparaed from a tissue or organ are usually heterogeneous (primary culture), but they can be separated into a variety of pure subcul-

Table 8.5. Applicability of microelectrodes and other techniques to various cellular and subcellular samples

Sample	Microelectrodes	Other intracellular techniques
organelle		
– within intact single cell	+	
– within intact cells of a population		NMR
– within intact tissue, organ, animal		
– single organelle from a suspension	+	
– organelle suspension		NMR, distribution of weak acids or bases, trapped indicators
single cell		
– isolated single cell or single cell from a cell suspension or attached culture	+	metallochromic indicatators, photoproteins, fluorescent indicators
– single cell of perfused tissue	+	
– single cell of perfused organ	+	
– single cell of entire animal	+	
population		NMR, trapped indicators,
– cell or organelle suspension		distribution of weak acids or bases, metallochromic indicators, null point for permeabilized cells
– tissue, organ, entire animal		NMR, distribution of DMO

tures (secondary cultures). Cultures prepared from isolated cells are homogeneous from the beginning. A large variety of pure cell lines, strains and cultures are available commercially or from scientific laboratories. Single cells can also be studied using microspectrophotometric and luminescence techniques (Tab. 8.5).

Other intracellular techniques use entire cell populations as the sample (10^5 to 10^{11} cells/ml). Single cells may be separated from tissues using proteolytic enzymes (e.g. collagenase), chelating agents (e.g. EDTA) or ultrasonic waves. The resulting suspension contains a mixture of different types of cells. Various procedures, such as sedimentation or centrifugation are available for the separation of the cells. Flow cytometry is being increasingly applied for the recognition and purification of cell subpopulations from a mixed suspension [Kruth 82; Pinkel 82]. The sorting of cells in flow cytometry is carried out by the tiny regular droplets of solvent that are formed by vibrating the flow chamber. The droplets, each containing a single cell, are then focussed into the center of the stream by a flowing sheath fluid. A laser beam is used to detect the specific scattering, absorbance or fluorescence of the various single cells. Droplets of a selected, identified type of cell are electrically charged and separated in an electric field. Sorting rates of several thousand cells per second are achievable. The sorting of cells is of special importance to methods which use cell populations as the sample (Tab. 8.5, see also Sect. 8.6). Of course, purified cell suspensions or cell cultures (e.g. attached cultures [Werrlein 81]) supply good preparations for single-cell studies with microelectrodes.

Measurements in subcellular organelles can be carried out within intact single cells, intact cells of populations, isolated single organelles or populations of organ-

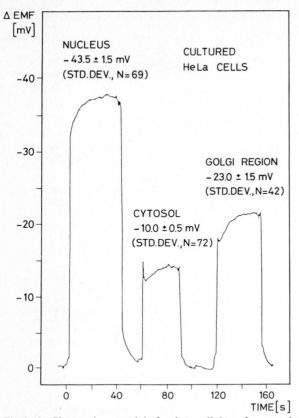

Fig. 8.10. Changes in potential of an intracellular reference microelectrode during the impalements into the nucleus, cytosol and Golgi region of a single cell. Each puncture was achieved by a new penetration of the plasma membrane. Redrawn from Giulian and Diacumakos, J. Cell. Biol., 1977 [Giulian and Diacumakos 77]

elles. Potentiometry using microelectrodes is the only method which can provide results on a single organelle within an intact single cell. However, successful punctures demand greater experimental skills and favourable cell preparations. The ability to impale specific regions within a cell was demonstrated by the injection of markers (e. g. procion brown, cobalt chloride, thorium dioxide) into defined regions of a cell and the subsequent visualization using electron microscopy [Giulian and Diacumakos 77]. Using this technique, it was shown that the internal structures (nucleus, mitochondrion, Golgi apparatus) of intact cells can be penetrated by microelectrodes. Characteristic potential differences were observed for each compartment (Fig. 8.10, [Giulian and Diacumakos 77]). A few other similar studies have been reported (see Table 8.6). In most cases the potential differences across the internal membranes were measured with reference microelectrodes. Ion-selective microelectrodes have only rarely been used for measurements in organelles (Table 8.6).

^{31}P NMR can provide information on the pH of the intracellular compartments in the dense suspensions of single cells (Fig. 8.11, [Martin et al. 82]). However, spec-

Table 8.6. Examples of the use of microelectrodes for measurements in cell organelles

Microelectrode	Cell	Organelle	Localization	Reference
reference microelectrode	salivary glands of *Drosophila*	nucleus	optical (light microscope)	[Loewenstein and Kanno 63]
reference microelectrode	oocytes of *Xenopus laevis*	nucleus	optical (light microscope)	[Loewenstein and Kanno 63]
reference microelectrode	liver of mice	mitochondrion	optical (light microscope)	[Maloff et al. 77]
reference microelectrode	*Drosophila*	mitochondrion	optical (light microscope)	[Tupper and Tedeschi 69]
reference microelectrode	HeLa cells	nucleus, mitochondrion, Golgi region	optical (micro-injection of markers, electron microscope)	[Giulian and Diacumakos 77]
Na^+, K^+, Cl^- microelectrodes	salivary glands of *Chironomus*	nucleus	optical (light microscope)	[Palmer and Civan 77]
K^+ microelectrode	salivary glands of *Chironomus*	nucleus	optical (light microscope)	[Wuhrmann et al. 79]
H^+ microelectrode	*Acer pseudoplatanus*	vacuole	optical (light microscope)	[Kurkdjian and Barbier-Brygoo 83]

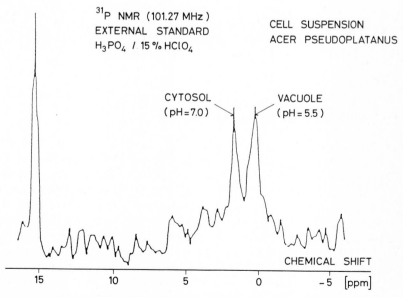

Fig. 8.11. ^{31}P NMR spectrum of a suspension of *Acer pseudoplatanus* cells (the intracellular pH of the cytosol and the vacuoles is indicated). Redrawn from Martin, Bligny, Rebeille, Douce, Leguay, Mathieu and Guern, Plant Physiol., 1982 [Martin et al. 82]

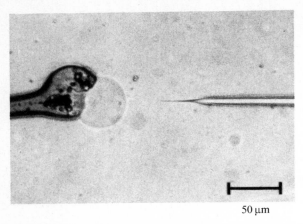

50 μm

Fig. 8.12. The puncture of a single vacuole with a micropipette. The vacuole is fixed with the aid of a suction pipette *(on the left)*. (Courtesy of H. Barbier-Brygoo and A. Kurkdjian)

tra with high resolutions are needed and the compartments must exhibit clearly different pH values (see Fig. 8.11). In addition, the external medium must be free of phosphate during the measurement of the spectrum. Otherwise the signals due to the phosphate of the cytosol and the other compartments may be masked by the signal from the external phosphate [Martin et al. 82].

Other techniques require a pure population of organelles as the sample (Tab. 8.5). Cells can be disrupted using various techniques (e.g. ultrasonic vibration, osmotic shock). The suspensions of the subcellular components can be separated by standard means (ultracentrifugation). Single organelles from such suspensions have also been used for microelectrode studies (e.g. vacuoles (Fig. 8.12) [Barbier 81; Kurkdjian and Barbier-Brygoo 83], mitochondria [Tupper and Tedeschi 69]).

8.5 Cell Compartmentation

Ion activity measurements with microelectrodes are extremely localized. The detection volume is usually at least 1000 times smaller than the volume of the cell. Obviously, any heterogeneity in the sample will lead to space-dependent results. The nonuniform distribution of ions within a cell is a result of cell compartmentation. Strictly speaking, compartmentation should exclusively refer to a nonuniform distribution of diffusible ions between two regions of a cell that are separated by a diffusion barrier (e.g. membrane) [Civan 78]. However, in a discussion of the various aspects of intracellular ion measurements, it is more meaningful to use an extended view of the term compartmentation (see Fig. 8.13).

Intercellular compartmentation arises from tissues containing different types of cells. Microelectrode studies are not affected by this kind of heterogeneity since the measurements are carried out on single identified cells. In contrast, intracellular

INTERCELLULAR COMPARTMENTATION

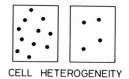

CELL HETEROGENEITY

INTRACELLULAR COMPARTMENTATION

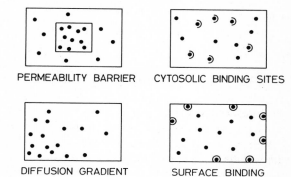

PERMEABILITY BARRIER   CYTOSOLIC BINDING SITES

Fig. 8.13. Origin of intercellular and intracellular compartmentation. Modified from Sies, in Metabolic Compartmentation (H. Sies (Ed.)), 1982 [Sies 82]

DIFFUSION GRADIENT   SURFACE BINDING

compartmentation may influence the results of such measurements. Different mechanisms lead to subcellular compartmentation (Fig. 8.13). Organelles are surrounded by internal membranes which act as permeability barriers for ions and other molecules. As a result, the ion activities in the cytosol and in the organelles may be considerably different. Temporary compartmentation can also take place due to relatively slow diffusion processes. In such a case, concentration gradients will expand from, for example, a stimulation source. Intracellular macromolecules or small dissolved molecules have the ability to bind ions. If the residence time of the ion in the complex is sufficiently long, a localized concentration builds up (Fig. 8.13). Finally, surface groups on the plasma membrane or the internal membranes can also bind ionic species, leading to a compartmentation near the membranes.

The following sections show that microelectrodes can be successfully used to study intracellular compartmentation. On the other hand, it must always be kept in mind that compartmentation can impede the evaluation of extremely local measurements with ion-selective microelectrodes.

Organelles (Permeability Barrier (see Fig. 8.13). The existence of differences in concentration between the cytosol and certain subcellular organelles is undisputed. A heterogeneous distribution is particularly obvious for H^+ and Ca^{2+} ions, whereas the sequestration of Na^+ and K^+ by certain organelles is less pronounced.

Na^+, K^+. Earlier comparisons of total sodium and potassium (as measured by radioactive tracers) with the electrochemically determined Na^+ and K^+ activities

clearly indicated the existence of compartments for these ions [Zeuthen 80]. Recently, Na^+- and K^+-selective microelectrodes have been extensively used to study the compartmentation of these ions [Lev and Armstrong 75; Civan 78]. In these experiments, apparent activity coefficients are introduced. These coefficients are not directly related to the fundamental relationship between concentration and activity (described by the activity coefficients γ_i of the Debye-Hückel formalism (see Sect. 5.6)). They are defined by Eq. (8.12) [Lev and Armstrong 75; Civan 78]:

$$\gamma_i^{app} = a_i / \bar{c}_i, \qquad\qquad\qquad (8.12)$$

where

a_i : local activity of the ion I measured with microelectrodes;

$\bar{c}_i$: average intracellular concentration of the ion I.

A deviation of the apparent activity coefficient γ_i^{app} from the Debye-Hückel activity coefficient γ_i for a comparable aqueous electrolyte indicates a compartmentation of the intracellular space.

The Na^+ (a_{Na}) and K^+ activities (a_K) of the oocytes were measured with microelectrodes. The total amounts of these ions $\bar{c}_{Na}$ and $\bar{c}_K$ were determined by chemical analysis or X-ray microanalysis [de Laat et al. 74; Civan 78]. The ratio a_{Na}/a_K and $\bar{c}_{Na}/\bar{c}_K$ varied considerably during the development of the oocytes (Fig. 8.14) [Palmer et al. 78]. The large differences between γ_{Na}^{app} and γ_K^{app} in the mature oocytes and the activity coefficients of normal electrolytes clearly indicate the formation of subcellular compartments during the maturation of the egg (note that the Debye-Hückel activity coefficients γ_{Na} and γ_K are almost equal in aqueous electrolytes of physiological composition).

Na^+- and K^+-selective microelectrodes have provided other insights into the distribution of ions between the cytosol and the organelles. Direct measurements in the cytosol and the nucleus of the same cell (*Chironomus* salivary gland) have shown that organelles other than the nucleus must be responsible for the observed compartmentation [Civan 78].

Ca^{2+}. The endo(sarco)plasmic reticulum and the mitochondria exhibit storage properties for Ca^{2+}, i.e. they contain much higher concentrations of calcium than the cytosol. Much of the knowledge concerning the Ca^{2+} biochemistry of these compartments is the result of experiments carried out on isolated organelles. The mechanisms for the uptake and release of Ca^{2+} and the regulation of these pathways are partly well understood [Carafoli et al. 82].

To date, these extremely small compartments have not yet been successfully impaled by Ca^{2+}-selective microelectrodes. Alternative methods have been used to measure the free concentration of Ca^{2+}. These methods include the use of the metallochromic dye arsenazo III (suspension of isolated hepatocytes [Bellomo et al. 82]) and the photoprotein aequorin (isolated *Chironomus* salivary gland cells [Rose and Loewenstein 75]). Other techniques that have been used to study the Ca^{2+} loading of organelles are X-ray microanalysis and electron microscopy (calcium-staining, e.g. with potassium antimonate or alizarin red) (see [Nilsson and Coleman 77; Slocum and Roux 83]).

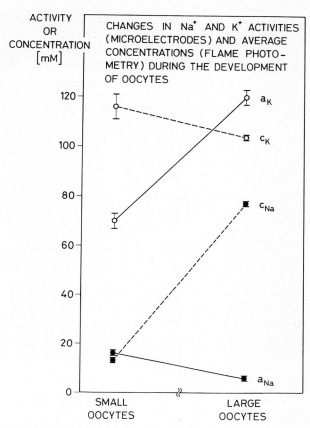

ACTIVITY OR CONCENTRATION [mM]

CHANGES IN Na$^+$ AND K$^+$ ACTIVITIES (MICROELECTRODES) AND AVERAGE CONCENTRATIONS (FLAME PHOTO-METRY) DURING THE DEVELOPMENT OF OOCYTES

Fig. 8.14. Intracellular activities a_{Na} and a_K (microelectrodes) and average intracellular concentrations $\bar{c}_{Na}$ and $\bar{c}_K$ of Na$^+$ and K$^+$ (atomic absorption spectrometry, flame photometry, X-ray microanalysis) in small immature and large mature oocytes [Civan 83]. Redrawn from Civan, Epithelial Ions and Transport, Wiley, 1983 [Civan 83]

Mg^{2+}. The intracellular distribution of magnesium in organelles is poorly understood. This is a result of the lack of reliable intracellular techniques for the determination of magnesium. Mg^{2+} ions are undoubtedly bound to various complexing agents such as the adenine nucleotides (see also Sect. 9.5.3) or enzymes. It also seems that Mg^{2+}, like Ca^{2+}, can be accumulated in the mitochondria [Polimeni and Page 73].

H$^+$. It is well known that an organelle may exhibit an internal pH different from that of the cytosol. For example, the pH within a mitochondrion can be surprisingly high (pH > 8) [Addanki 68; Cohen and Iles 75]; the vacuoles of plant cells are rather acidic compartments (pH 5 to 6) [Kurkdjian and Barbier-Brygoo 83]; and regions of differing pH are expected to exist in thylakoids [Theg and Junge 83]. So far, it has not been possible to penetrate organelles of intact animal cells with H$^+$-selective microelectrodes. However, neutral carrier H$^+$ microelectrodes have been successfully used in plant cells to study the pH of isolated vacuoles [Kurkdjian and Bar-

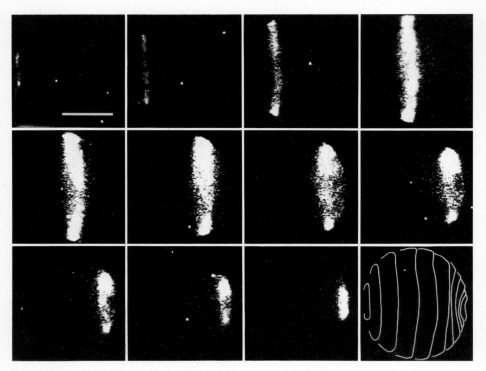

Fig. 8.15. The use of aequorin in the visualization of a Ca^{2+} wave travelling across the cytoplasm of a sperm-activated *Medaka* egg [Gilkey et al. 78]. The successive photographs were taken at intervals of 10 s. The glow spreads in a ring at a linear velocity of about 12 μm/s from the animal to the vegetale pool. Last frame: schematic tracing of the leading edges of the 11 wave fronts. For further details see [Gilkey et al. 78]. Courtesy of L. F. Jaffe. From Gilkey, Jaffe, Ridgway and Reynolds, J. Cell. Biol., 1978 [Gilkey et al. 78]. Reproduced by copyright permission of the Rockefeller University Press

bier-Brygoo 83]. Measurements in vacuoles of intact protoplasts or single plant cells are extremely difficult because of the fragility of the plasmalemma on the one hand, the hardness of the cell wall on the other hand [Kurkdjian and Barbier-Brygoo 83].

Ion-gradients (Diffusion Barrier, see Fig. 8.13). Ions released from a local source, either natural (e. g. transport system of a membrane, internal sources, macromolecules that reversibly bind ions) or artificial (e. g. excitation) create, at least transiently, an activity or concentration gradient within a cell (Fig. 8.13). Such events may result in compartmentation, depending on the time scale. The lifetime of such an intracellular gradient may be very short. If it is less than the analysis time, the gradient will not be detected. Longer lasting gradients are measurable. In this respect, Ca^{2+} is of special interest since its intracellular diffusion is restricted. Several techniques have been employed to study Ca^{2+}-diffusion phenomena in the cytosol.

Spatial nonuniformity in the changes in Ca^{2+} concentration after the excitation of *Limulus* photoreceptors has been observed using scanning microspectrophotometry (with arsenazo III as the indicator [Harary and Brown 84]). The resolution

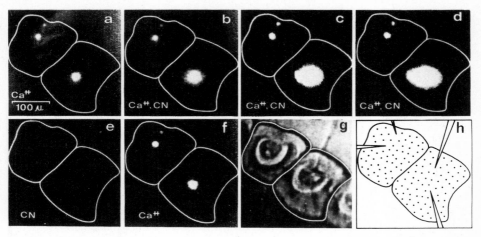

Fig. 8.16a–h. The use of aequorin for the visualization of the restricted diffusion of Ca^{2+} in *Chironomus* salivary gland cells during the local injections of Ca^{2+} from micropipettes (0.2 μm **(h)**) [Rose and Loewenstein 75 b]. A bright-field TV picture of the cells is shown in **(g)**. Leaks of Ca^{2+} are visible in the dark-field TV pictures as luminescent circles (**a** to **f**). **(a)** leaks generated in a medium containing 4 mM Ca^{2+}. **(b** to **d)** 1, 2 and 3 min after the injection of a solution containing 4 mM Ca^{2+} and 2 mM CN^- (mitochondrial inhibitor). **(e)** Ca^{2+}-free medium containing 2 mM CN^-. **(f)** 2 min after returning to a medium as in **(a)**. For further details see [Rose and Loewenstein 75 b]. Courtesy of B. Rose and W. R. Loewenstein. From Rose and Loewenstein, Science, 1975 [Rose and Loewenstein 75 b]. Reproduced by copyright permission of the American Association for the Advancement of Science

of this method is about 2 μm. The single cells were uniformly illuminated while the profiles of the optical transmission were recorded with a linear diode arrangement. Three different diffusion patterns for Ca^{2+} were observed with 20 single cells. The kinetics of the light-induced increase in intracellular Ca^{2+} also vary with location.

The photoprotein aequorin has also been shown to be extremely useful in visualizing Ca^{2+} gradients due to the intensified images produced by its emitted light. Eggs of the fresh water fish, *Medaka,* that were injected with aequorin show a spreading wave of Ca^{2+} during fertilization (Fig. 8.15) [Gilkey et al. 78]. The Ca^{2+} level of the wave, which considerably exceeds the resting free concentration, has been estimated to be about 30 μM. The major source of released Ca^{2+} seems to be the internal granules. The transit time of the wave through the egg (diameter about 1.1 mm) always lies between 2 and 3 minutes [Gilkey et al. 78].

Another impressive study using cells injected with aequorin demonstrated that the diffusion of Ca^{2+} in the cytosol of *Chironomus* salivary glands is constrained [Rose and Loewenstein 75a; Rose and Loewenstein 75b]. The luminescence was observed microscopically with an image intensifier coupled to a television camera and was simultaneously measured using a photomultiplier. Repetitive injections of Ca^{2+} with micropipettes were analyzed under different physiological conditions. The glow was observed to be confined to the area about the tips of the injection micropipettes except when mitochondrial inhibitors were present (Fig. 8.16) [Rose and Loewenstein 75b].

MEASUREMENT OF AN INTRACELLULAR Ca^{2+} GRADIENT
BY A NEUTRAL CARRIER Ca^{2+} MICROELECTRODE

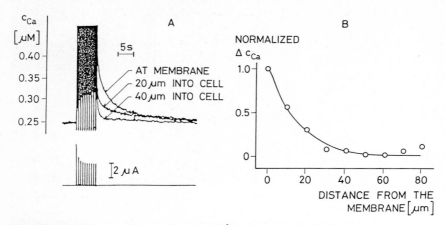

Fig. 8.17 A, B. The use of a neutral carrier Ca^{2+} microelectrode for the measurement of spatial intracellular Ca^{2+} gradients in *Aplysia* neurons following voltage-initiated entry of Ca^{2+} (ten 200 ms pulses from -50 mV to $+30$ mV at 1 s intervals) [Gorman et al. 84]. **A:** recording of the Ca^{2+} deflection at three depths. **B:** the peak amplitude of intracellular Ca^{2+} (measured by the microelectrode immediately following the last pulse) was plotted as a normalized change in the free concentration of Ca^{2+} versus the depth of penetration. Zero distance from the cell membrane was defined as the position of the tip just before being withdrawn from the cell by a 10 μm step (for further details see [Gorman et al. 84]. Redrawn from Gorman, Levy, Nasi and Tillotson, J. Physiol., 1984 [Gorman et al. 84])

As previously mentioned a unique property of microelectrodes is their ability to measure ion activities in the very localized environment of the tip. This property was exploited using Ca^{2+}-selective neutral carrier microelectrodes to detect spatial Ca^{2+} gradients in *Aplysia* neurons [Gorman et al. 84]. The response times of these microelectrodes (tip diameters less than 1 μm) were found to be adequate for such studies in submicromolar Ca^{2+} solutions ($t_{50} = 100$ ms, $t_{90} = 1.5$ s). The influx of Ca^{2+} was generated by optimized voltage-clamp pulses which activated the Ca^{2+} channels. During the stimulation period the response of the microelectrode could not be followed. However a deflection of the EMF signal was observable after the last voltage pulse (Fig. 8.17). The cytosolic Ca^{2+} gradient was analyzed by moving the tip of the Ca^{2+} microelectrode in steps from the neuronal membrane into the cytosolic space [Gorman et al. 84]. Similarly, steep intracellular Ca^{2+} gradients were found in snail neurons using microelectrodes. The tips of the microelectrodes were assumed to be positioned near the membrane where the influx occurs. In this example, the gradient is created by the sequestration of Ca^{2+} [Hofmeier and Lux 81].

It is much more uncertain whether Na^+ and K^+ gradients exist. In vivo studies on ileum epithelial cells in which double-barrelled microelectrodes were advanced stepwise from the lumen through the layers of epithelia cells into the underlying serosal tissues revealed K^+ and Cl^- activity gradients [Zeuthen and Monge 75]. In order to localize the measurements, stains were deposited as the microelectrode was moved stepwise through a cell. A gradient of the electrical potential was recorded at the same time as the activity gradient (mucosal end: depolarized; serosal end: more polarized). The double-barrelled microelectrodes used in these studies had a total tip diameter of less than 0.6 µm and were advanced in steps of 2 µm/s. By using different electrolytes artifacts caused by leaks in the reference electrolyte were shown to be absent [Zeuthen and Monge 75].

Similarly, the intracellular activities of Na^+ and K^+ in the gall bladder epithelium of *Necturus* were found to be different at the mucosal end of the cell which faces the lumen of the intestine ($a_{Na} = 46.3$ mM; $a_K = 64.1$ mM; $a_{Cl} = 100.1$ mM) and at the serosal end of the cell which faces the underlying blood capillaries or lymphatics ($a_{Na} = 12.9$ mM; $a_K = 181.8$ mM; $a_{Cl} = 25.3$ mM) [Zeuthen 76a]. These measurements were carried out with double-barrelled microelectrodes.

Recordings with two reference microelectrodes at different depths in the same cell showed the existence of potential gradients [Zeuthen 76a]. However, evidence for the latter could not be found when these experiments were repeated with a single, very fine (< 0.2 µm) reference microelectrode [Suzuki and Frömter 77]. In these experiments the potential was measured relative to the mucosal solution. It was also found that the electrical gradients could be simulated in the same cell assembly by using reference microelectrodes with large tips. It was concluded that the previously reported gradients were probably due to cell damage or incomplete cell penetration [Suzuki and Frömter 77] (see also comments in [Reuss and Weinman 79]).

Zeuthen refined his measurements of the intracellular gradients of the electrical potential in the epithelial cells of *Necturus* gall bladder using single- (Fig. 8.18) and triple-barrelled microelectrodes [Zeuthen 77]. Two microelectrodes were again located at different depths in the same cell. The observed electrical gradient was found to be independent of the direction of penetration (mucosal or serosal membrane) and disappeared when external Na^+ was replaced by K^+. Both observations were set against electrode artifacts. The existence of a potential gradient in the cells was confirmed by the simultaneous detection of two different potentials at two different positions in the same cell by means of a triple-barrelled microelectrode. The observed gradients are about -0.6 mV/µm in the direction perpendicular to the mucosal membrane [Zeuthen 77].

Similar studies with double-barrelled ion-selective microelectrodes showed the existence of Na^+ and K^+ activity gradients in epithelia [Zeuthen 78]. Recessed-tip glass membrane Na^+ microelectrodes and classical ion-exchanger K^+ microelectrodes were used in these studies. The response time of the Na^+-selective barrel was substantially longer than that of the reference electrode barrel, which rendered the interpretation of the recordings more difficult. On the other hand, the response time of the K^+-selective barrel was short. This made a simultaneous recording of the potential gradient and the K^+ gradient feasible (Fig. 8.19). The probability of artifacts produced by the microelectrode was analyzed in detail. Most but not all of the sources of error could be ruled out [Zeuthen 78]. The validity of the observed gradients is strengthened by the measurement of Na^+- and K^+-concentration gradients in the interspaces of epithelial tissue and in the cytosol of epithelial cells by electron probe X-ray microanalysis [Gupta et al. 76; Gupta and Hall 79].

An intracellular gradient in pH was ascertained by using double-barrelled H^+ microelectrodes. They were moved stepwise from the inner mucosal side (pH 6.7;

INTRACELLULAR GRADIENT OF ELECTRICAL
POTENTIAL MEASURED WITH MICROELECTRODES

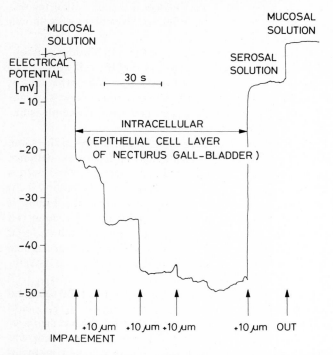

Fig. 8.18. A microelectrode recording of an intracellular electrical potential gradient in *Necturus* gall bladder epithelium. A single-barrelled reference microelectrode was advanced stepwise across the cell from the mucosal side. Redrawn from Zeuthen, J. Membr. Biol., 1977 [Zeuthen 77]

$E_M = -30$ mV) to the serosal membrane (pH 7.1; $E_M = -50$ to -60 mV) [Zeuthen 80].

Binding in the Cytosol (Fig. 8.13). Various intracellular macromolecules and compounds of low molecular mass may act as complexing agents for ions. An ion can be considered to be free if it can move freely within the aqueous phase. The electrostatic interactions can be evaluated on the basis of the Debye-Hückel theory [Edzes and Berendsen 75]. Binding is indicated by deviations in the activity/concentration relationship and by its influence on the diffusion coefficients. Furthermore, binding also exerts an influence on dynamic properties such as the exchange rates of ions with a macromolecular matrix (e.g. proteins). Binding must be clearly distinguished from storage in cellular compartments (see Fig. 8.13).

Many cytosolic binding sites have been identified. Organelles may also contain their own complexing agents. For example, Ca^{2+} binding sites in mitochondria are well documented [Carafoli 74].

As in the case of ion gradients, compartmentation due to binding to cytosolic sites is most clearly evident for Ca^{2+}. The main complexing agents of low molecular

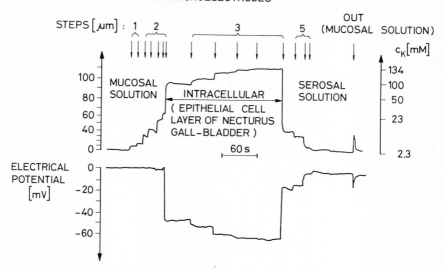

INTRACELLULAR K⁺ ACTIVITY GRADIENT AND ELECTRICAL
POTENTIAL GRADIENT SIMULTANEOUSLY MEASURED WITH
DOUBLE–BARRELLED MICROELECTRODES

Fig. 8.19. A microelectrode recording of intracellular gradients in the K^+ activity and the electrical potential [Zeuthen 78]. A double-barrelled ion-exchanger K^+-microelectrode was advanced stepwise from the mucosal to the serosal side of a *Necturus* gall bladder epithelia cell. The increasing activity in K^+ prior to the impalement was interpreted as being due to a contamination of the space between the electrode tip and the dimpled mucosal membrane with reference electrolyte (NH_4NO_3). Redrawn from Zeuthen, J. Membr. Biol., 1978 [Zeuthen 78]

mass for Ca^{2+} are phosphate, adenine nucleotides, citric, oxalic and other organic acids. The main macromolecular complexing agents are specific Ca^{2+} binding proteins (e. g. calmodulin, troponin C, parvalbumin). The latter usually exhibit a higher affinity for Ca^{2+} ($pK_d \sim 5.5$ to 7) than the small soluble molecules ($pK_d \sim 4$ to 5) [Kretsinger 79]. Therefore, proteins play an important role as cytosolic Ca^{2+} buffers.

The binding of Ca^{2+} to proteins and other complexing agents in blood has been well examined. The fraction of protein-bound calcium amounts to about 40% and that bound to small molecules to about 10% of the total amount of calcium. In agreement with this is the observation that the direct potentiometric measurement of free Ca^{2+} (~ 1.1 mM) amounts to less than half of the total calcium (~ 2.5 mM) measured by atomic absorption spectrometry, colorimetry or indirect potentiometry in acidified samples. Accordingly, blood-bound calcium is seen as a compartment by the macroelectrodes. Bound Ca^{2+} can be quantitatively replaced by H^+ ions. Hence, the acidification (pH 8 to pH 4) of a blood sample by the addition of hydrochloric acid leads to increasing Ca^{2+} activities (Fig. 8.20). In the pH 4 to pH 2 range of the sample the Ca^{2+} activity does not change, demonstrating that the bound calcium is fully replaced at a pH of about 3.5 [Anker et al. 81 b]. Despite the different sizes of macro- and microelectrodes it can be concluded that the latter will also only

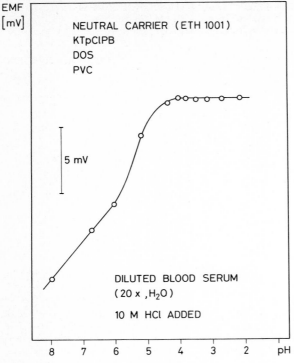

EMF [mV]

NEUTRAL CARRIER (ETH 1001)
KTpClPB
DOS
PVC

5 mV

DILUTED BLOOD SERUM
(20 x , H_2O)

10 M HCl ADDED

8   7   6   5   4   3   2   pH

Fig. 8.20. The displacement of Ca^{2+} from its complexes in diluted blood by H^+ ions as measured by a Ca^{2+}-selective macroelectrode [Anker et al. 81 b]

register the free Ca^{2+} ions. Thus the bound fraction has to be regarded as compartmentalized.

The binding of Na^+ and K^+ is less clear than that of calcium even in blood [Ladenson 77; Coleman and Young 81; Apple et al. 82; Kissel et al. 82]. Despite these uncertainties, the immobilization of Na^+ and K^+ in the cytosol has been studied extensively. It is to be expected that Na^+ and K^+ will undergo weaker interactions with negatively charged sites than Ca^{2+} and Mg^{2+}. It has been determined that very large amounts of Na^+ and K^+ are not accessible to electrochemical studies with microelectrodes. However, ^{23}Na and ^{39}K NMR studies indicate that very little Na^+ or K^+ is immobilized by binding [Civan 83], i.e. other compartmentation mechanisms must be responsible for the discrepancy between local activity and average total concentration (see Sect. 8.1).

Surface Binding (see Fig. 8.13). Compartmentation as a result of the binding of ions to the surface of a cell membrane is well documented for Ca^{2+} and H^+ ions. In vitro studies on artificial bilayer membranes have shown that Ca^{2+} ions are bound to negatively charged phospholipids (for a list of references see [Kretsinger 79]). The interaction of Ca^{2+} with surface sites has also been reported for natural membranes. For example, cardiolipin, a phospholipid of the inner mitochondrial membrane, binds Ca^{2+} with a relatively high affinity ($pK_d \sim 5$) [Carafoli 82]. Surface binding of Ca^{2+} is also shown by the glycocalix, a glycoprotein coat of plasma membranes,

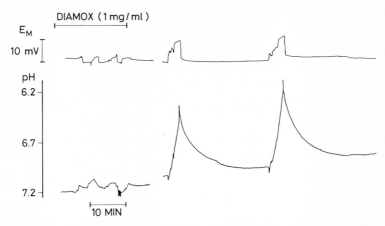

CELL SURFACE MEASUREMENT WITH DOUBLE-BARRELLED
pH-GLASSMICROELECTRODE
(OUTSIDE OF MUCOSAL MEMBRANE OF NECTURUS GALL-BLADDER)

Fig. 8.21. Surface recordings of pH with double-barrelled glass microelectrodes [Zeuthen 80]. The mucosal membrane of *Necturus* gall bladder was touched prior to penetration. The observed drastic acidification of the surface solution between microelectrode tip and apical membrane can be blocked by the carbonic anhydrase blocker diamox *(left trace)*. About 5 to 10 min after removing diamox from the bath solution two contacts with the membrane surface yielded a reversible acidification of about one pH unit *(right trace)*. For more details see [Zeuthen 80]. Redrawn from Zeuthen, Current Topics in Membranes and Transport, Academic Press, 1980 [Zeuthen 80]

which forms a surface with a dense arrangement of negatively charged sites (such as sialic acid residues). The glycocalix may contribute significantly to the Ca^{2+}-buffering about the membrane [Gupta and Hall 79].

The pH near the surface of a cell membrane may well be different from that of the surrounding cytosol. Proteins of the membrane contain various functional groups that can be protonated or deprotonated and that exhibit quite different pK's (for example, carboxyl groups of albumin: ~4; imidazole: 6.9; α-amino groups: 7.8; ε-amino groups: 9.8; phenolic hydroxyl groups: 10.35 [Tanford et al. 55]). The pH of the microenvironment about these groups can be of great significance. For example, the pH in the immediate vicinity of an active site in an enzyme may influence its activity. Indeed, each enzyme exhibits an optimal activity at a different pH.

Because of the size and geometry of microelectrode tips, it is conceivable that one could distinguish between surface activities and bulk cytosolic activities potentiometrically using microelectrodes. Such recordings have indeed been made near the surfaces of cell membranes [Zeuthen 80; de Hemptinne et al. 82; Werrlein 83; Huguenin 85]. In two experiments, drastic changes in the surface pH relative to the external bulk pH were observed (Fig. 8.21) [Zeuthen 80; Werrlein 83]. Both surface measurements were performed with glass microelectrodes. It was found that the pH near the cell membrane was about one pH unit lower. Consequently, the relevant H^+ activity of the extracellular medium acting on the membrane properties probably is not the bulk pH but the surface pH.

For such studies involving the immediate contact of the membrane surface of a microelectrode with the surface of a cell, glass membranes seem more suited than liquid membranes. If the surface of a cell has to be brought into direct contact with a concentrated electrolyte solution (e.g. 3 M KCl) or an organic phase containing an active ion carrier errors may arise.

8.6 Method Specific Properties of Intracellular Measurements of Ions

The different methods used in intracellular ion determinations do not yield coincident results. The differences arise from

- the nature of the cell sample
- the treatment of the cell sample
- the specific features of the intracellular technique.

The strong influence of the nature of the biological sample (Sects. 8.1, 8.2, 8.4, 8.5) and of its treatment before and during the measurement (Sect. 8.3) on the results of intracellular ion determinations has already been demonstrated. The properties of the most popular intracellular techniques are discussed in Chap. 9. Each method involves a different analytical approach and requires a specific technique for the handling of the sample. In principle, it is possible to experimentally obtain any combination of the following parameters:

- ion activity, free-ion concentration, total concentration
- local mean of cell compartment, mean of single cell, mean of cell population
- in vivo, in vitro.

The large number of possible combinations makes a comparison of results difficult. The method-specific character of ion determinations is summarized below (a more refined presentation of some of the features of intracellular techniques can be found in Chap. 9).

Microelectrodes (H^+, Li^+, Na^+, K^+, Mg^{2+}, Ca^{2+}). Ion activities can be measured; extremely local cytosolic ion activities or free-ion concentrations can be evaluated; local measurements in organelles are possible.

NMR Spectroscopy (H^+, Na^+, K^+, Mg^{2+}). Mean free-ion concentrations or pH of extremely large populations of cells (dense suspensions (e.g. 10^{11} cells/ml), organs, entire animal) can be evaluated; in favourable situations the resolution of extra- and intracellular signals or of signals from cell compartments is possible.

Spectrophotometry (H^+, Mg^{2+}, Ca^{2+}). Mean cytosolic free-ion concentrations or pH of cell populations (e.g. 10^7 cells/ml) can be obtained.

Microspectrophotometry, Microfluorometry (H^+, Mg^{2+}, Ca^{2+}). Mean free-ion concentrations or pH of a large cell are measured; scanning mode: cytosolic free-ion concentration of a single cell with a spatial resolution of 2 to 10 μm is obtained.

Photoproteins (Ca^{2+}). Cytosolic free-ion concentration with a spatial resolution of $> 2\,\mu m$.

Weak Acids or Weak Bases (H^+). Mean pH of cytosol and organelles of a cell population is measured; in favourable situations mean pH of single cells are obtained.

Electron-Probe X-Ray Microanalysis (Na, K, Ca, P, S, Cl). Mapping of the total intracellular concentration of an element in a single cell; typical resolution is $\sim 0.2\,\mu m$, optimal resolution is 0.02–$0.03\,\mu m$.

Gelatine Reference Phase (Na^+, K^+). Mean total ion concentrations in the cytosol as well as in the entire intracellular space of large single cells are measured.

Null Point for Permeabilized Cells (H^+, Mg^{2+}, Ca^{2+}). Mean total concentrations in a very dense cell suspension are obtained.

Microelectrodes provide the most localized ion measurements. A few other techniques are available for a relative local mapping of free ion concentrations (luminescence, microspectrophotometry) or for a determination of the elemental concentrations (electron-probe X-ray microanalysis). Most of the methods yield mean concentrations. An excellent critical evaluation of vacuolar pH measurements has shown that mean concentrations obtained by different methods are not necessarily the same [Kurkdjian et al. 85]. The study compares the use of different weak bases (labelled as well as fluorescent), ^{31}P NMR, H^+ microelectrodes and cell-sap measurements. From an analytical point of view, the most important conclusion to be drawn from this work [Kurkdjian et al. 85] is that only measurements on single cells can provide information on the heterogeneity of a cell or the vacuole populations (microelectrodes and microfluorometry with 9-aminoacridine). The significance of the mean pH depends on the technique used. For instance, the buffer capacity of subclasses of an entire population influences the results of the cell-sap technique. The mean pH will be shifted towards the subclass with the highest buffer capacity. Similarly, based on ^{31}P NMR measurements of inorganic phosphate the resulting mean pH will be shifted towards the subclass containing the highest concentration of phosphate. The mean pH's evaluated by the distribution of weak bases (see Sect. 9.1.1) vary drastically (up to 2 pH units) with the type of base.

The large number of intracellular techniques and the variety in the nature of the samples do not allow a complete comparison of the method-specific aspects of intracellular ion determinations to be made. Instead, three techniques shall be compared in three situations: measurements in cells with a heterogeneous population, investigations of cellular compartmentation by organelles and the measurement of ion gradients (Fig. 8.22). The three methods that were selected exhibit widely different spatial responses. They detect ions in a volume of less than $1\,\mu m^3$ about a tip in the case of microelectrodes, in the entire cytosolic space of a single cell in the case of photoproteins and in the total volume of more than a billion cells in the case of NMR measurements. Thus, it is not surprising that differences in the ion measurements are observed, even more so when one considers the complex behaviour of a sample.

Heterogeneous Populations. Tissues as well as isolated cell populations can be heterogeneous. An example of a tissue with structural and functional heterogeneity is the

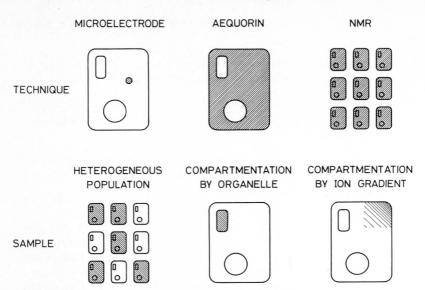

Fig. 8.22. Schematic representation of various parameters with regard to a discussion of method-specific aspects of intracellular ion measurements. The spatial response of various intracellular techniques *(upper)* and the properties of the sample *(lower)* are illustrated

urinary bladder of a toad, which is often used to study transepithelial transport [Civan 78]. The degree of heterogeneity can only be analyzed by using techniques that operate on the single-cell level or by isolating pure populations of the various types of cells (cell sorting (see Sect. 8.4)).

Local measurements with microelectrodes or cytosolic measurements using photoproteins in identified single cells are unaffected by the heterogeneity of tissues or cell suspensions, i. e. physiological processes within a selected single cell can be studied regardless of the variations within the cell population. However, in order to evaluate the degree of heterogeneity of a cell population a fairly large series of measurements on single cells is necessary (Fig. 8.23).

On the other hand, NMR spectroscopy, like other methods relying on cell populations, is confronted with heterogeneity problems. Each signal is usually composed of a superposition of signals from nuclei of the same type distributed throughout the heterogeneous sample. Consequently, the signal describes an envelope of individual signals from equivalent types of cells. Broad peaks therefore indicate a large variation in the ion concentration of a cell population. For instance, the broadness of ^{31}P NMR signals of intact kidneys corresponds to a variation of about 0.2 pH units (pH 7.2 ± 0.2). Experiments on isolated cells originating from the different regions of a kidney have proven that intercellular heterogeneity is responsible for the broad signal observed for the entire organ [Radda et al. 82]. The influence of different subclasses of cells on the shapes of NMR signals is even more pronounced for the liver [Cohen et al. 82 b]. The same difficulties are encountered with cultured cell populations. The ^{31}P NMR spectra of plant cell suspensions exhibit broad signals for vacuolar phosphate. The half-width of the signal corresponds to pH's in the 4.9 to 6.5 range [Kurkdjian et al. 85].

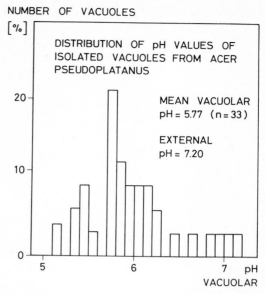

NUMBER OF VACUOLES

Fig. 8.23. The distribution of pH-values of vacuoles isolated from *Acer pseudoplatanus* cells as measured by neutral carrier microelectrodes [Kurkdjian et al. 85]. The percentage of vacuoles of the same pH are presented as a function of the vacuolar pH (external pH: 7.2; mean pH of vacuoles: 5.77 (n = 33)). Redrawn from Kurkdjian, Quiquampoix, Barbier-Brygoo, Péan, Manigault, Guern, in Biochemistry and Function of Vacuolar ATPase in Fungi and Plants (Martin (ed.)), 1985, [Kurkdjian et al. 85]

Since the sensitivity of NMR measurements in cellular samples is poor, it cannot be expected that this method will be useful in the elucidation of heterogeneous samples. First of all, NMR measurements require extremely dense populations or a large volume of tissue material. In an optimal case, the measurements in entire organs allow a spatial resolution of an area with a diameter of about 5 to 10 mm, i.e. in any case the sample will be very heterogeneous. In addition, the resolution of signals corresponding to different ion concentrations is relatively poor. For example, in an ideal case the precision of NMR measurements allows the detection of pH differences of ± 0.05 units [Radda et al. 82].

Compartmentation by an Organelle. It has been proven that certain internal organelles differ in their compositions from the cytosol. Accordingly, it is of interest to determine the ion concentrations of the different compartments. Ideally, the information should come from the same cell.

In principle, microelectrodes can be used to determine the ion activities of different compartments in the same cell. However, in most situations such experiments are hampered by problems such as cell damage and tip localization. Nevertheless, a few examples of successful punctures are known (see Table 8.6). It is often assumed that a local microelectrode recording is representative of the ion activity in the entire compartment. The majority of impalements yield cytosolic values since the probability of impaling a noncytosolic compartment is very small. Indeed, with the aid of a microscope it is possible to introduce the tip of the microelectrode into the cytosol without touching the organelles [Giulian and Diacumakos 77].

The photoprotein aequorin is known to be excluded from organelles after its injection into the cytoplasm (for penetration into the nucleus see [Blinks et al. 82]). Thus, after allowing the aequorin to reach a homogeneous distribution by diffusion the emitted light should correspond to the Ca^{2+} concentration of the total cytosolic phase of a single cell. However, if cells of different Ca^{2+} distributions are studied large errors can result. A hypothetic example of two cells containing the same total amount of free Ca^{2+} but exhibiting a different distribution (homogeneous or restricted to 10% of the cell volume) has been discussed [Blinks et al. 82]. It was shown that the overall emission of light from these two cells will differ by a factor of about 30 due to the nonlinear Ca^{2+}-concentration effect (see Sect. 9.6.1.3.). Besides these difficulties in the quantification of the signals, the method cannot be used for measurements in compartments other than the cytosol.

The most attractive application of the NMR technique to the study of intracellular ions is the measurement of pH by ^{31}P NMR spectroscopy. However, the resolution of the inorganic phosphate signals of different compartments is usually not possible. It also depends strongly on the relative phosphate concentrations in the individual compartments as well as on the pH differences [Cohen et al. 82b]. The best reported resolutions are about ± 0.02 pH units (see e.g. [Jacobus et al. 82]). Typical resolutions of about ± 0.2 pH units are common [Nuccitelli and Deamer 82]. Thus, in analyzing compartmentalized cells each compartment exhibits a ^{31}P NMR signal whose chemical shift is determined by the pH and whose intensity is given by the amount of inorganic phosphate. If the signals are not resolved, the resulting broad signal corresponds to a mean intracellular pH which is a function of the pH, the concentration of the inorganic phosphate of the individual compartments and the heterogeneity of the cell population.

Ion Gradient. A locally induced change in ion concentrations can lead to observable transient gradients. Furthermore, ion-activity gradients may be present in epithelia [Zeuthen 78; Gupta and Hall 79].

Microelectrodes measure ion activities at a single point. They cannot be moved to follow the spatial evolution of a gradient. Only repetitive experiments can map ion gradients. However, the temporal resolution is at best 10 to 100 ms. Two interesting examples have been presented in this chapter: the measurement of ion gradients by the stepwise movement of microelectrodes across a cytosol (Fig. 8.19) and the detection of transient ion gradients after a local stimulation (Fig. 8.17).

Luminescence measurements in cells containing aequorin uniformly distributed in the cytosol are highly suitable for observing ion gradients. The temporal resolution is relatively good (about 10 ms). The examples given in Figs. 8.15 and 8.16 illustrate the impact of this technique for the evaluation of Ca^{2+} gradients.

Nuclear magnetic resonance spectroscopy is unsuitable for the observation of ion gradients within cells. Both the temporal (at least several seconds) and the spatial (mean of populations, limited resolution) properties are insufficient.

8.7 Concluding Remarks

The intracellular space is extremely complex and the corresponding studies are complicated by the presence of large quantities of organelles, high concentrations of proteins and the dynamics of the processes.

In spite of the wide acceptance of the membrane-pump theory, a consideration of the association-induction hypothesis is very challenging and stimulating in a discussion of intracellular methods. Indeed, the controversy between the two theories has given rise to new techniques (e.g. reference phase method) and has evoked a more careful reflection of results and statements. Both views offer valuable criteria for a discussion of intracellular ion measurements.

Liquid-junction potentials, activity coefficients and suspension effects should be given particular consideration due to the special nature of the intracellular sample. To date, suspension effects have not been studied in detail with regard to microelectrode measurements. However, they should be reexamined since the dense intracellular protein solutions can behave like polyelectrolyte suspensions, which can lead to erroneous measurements or false interpretations of the studies.

Neutral carrier microelectrodes are normally used for measurements in the external intercellular space or in the cytosol. In a few cases, the ion activities of relatively large organelles (nuclei) can be measured potentiometrically. Therefore, the microelectrode technique is the only method that can measure ion activities in an organelle of an intact single cell.

The ability to directly measure physiologically important ion activities (instead of concentrations) is unique and should be exploited whenever possible.

Neutral carrier microelectrodes are most frequently applied for the measurement of resting or relatively slowly (seconds to minutes) changing ion activities. Other interesting applications which have been discussed are: the measurement of transient ion gradients after stimulation, the measurement of permanent transepithelial ion gradients and measurements in the microenvironment of a cell surface.

Both in vivo and in vitro measurements are possible. Each approach describes the living cell under markedly different conditions. The kind of cell preparation and its treatment have to be chosen carefully in order to obtain the adequate answer to the physiological problem on hand.

A comparison of method-specific aspects of intracellular ion measurements clearly shows that potentiometry with microelectrodes, like all the other intracellular techniques, is limited by technical difficulties and conceptual problems mainly due to the extraordinary nature of the sample.

9 Neutral Carrier-Based Liquid Membrane Microelectrodes

9.1 Intracellular Measurements of H$^+$

Intracellular H$^+$ activity[1] has been shown to be involved in the regulation of various cell processes. The mechanisms have been discussed in several review articles [Roos and Boron 81; Nuccitelli and Deamer 82]. During the past decade, several improved or novel analytical techniques to assay H$^+$ ions have become available; thus, the amount of research carried out in order to better understand pH-dependent intracellular mechanisms has increased significantly. At present, the methods most frequently used to determine the pH in cells are the distribution of weak acids or weak bases, ^{31}P nuclear magnetic resonance spectroscopy, absorbance or fluorescence spectrophotometry and potentiometry with H$^+$-selective microelectrodes. The different techniques can be classified according to their applicability to a single cell (microfluorometry, microelectrodes) or a population of cells (distribution of weak acids or bases, ^{31}P NMR). The methods applied to single cells are usually invasive (insertion of microelectrodes, microinjection of dyes) but allow a study of the local pH. The methods applied to populations usually involve noninvasive techniques (diffusion of chemicals, magnetic field) which yield the average pH of the cytosol of a vast number of cells. Apart from some similarities, each approach has its own distinct features. To date, none of the methods satisfy all of the demands for an ideal intracellular technique (Chap. 7 and 8).

9.1.1 Distribution of Weak Acids and Weak Bases

This method is based on the distribution of isotopically labelled (e. g. [Kurkdjian and Guern 81; Kurkdjian 82]) or spectrophotometrically detectable (e. g. [Lee et al. 82]) weak acids or bases (the latter approach also belongs to the class of optical techniques (Sect. 9.2.2)). The constitutions of some typical weak acids and bases are shown in Fig. 9.1 (for other compounds see [Roos and Boron 81]).

Assuming that only the electrically neutral form of a weak acid or base is able to permeate the cell membrane, an equilibrium distribution is reached in which the

---

[1] In this text, the symbol H$^+$ designates the hydronium ion. In aqueous solutions, the H$_3$O$^+$ ion is further hydrated to form H$_9$O$_4^+$ ions and more highly hydrated ions.

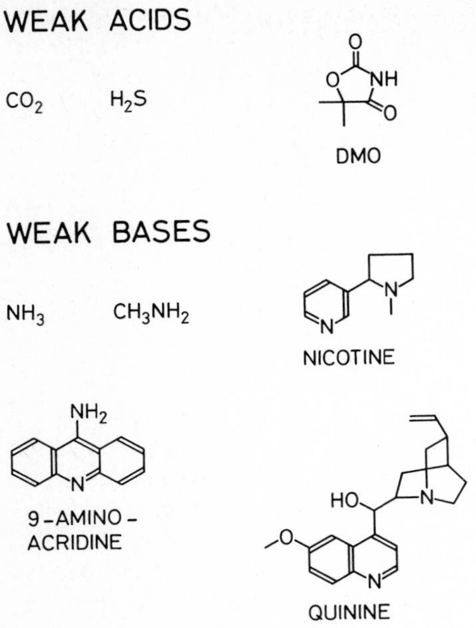

WEAK ACIDS

CO_2 H_2S

DMO

WEAK BASES

NH_3 CH_3NH_2

NICOTINE

NH_2

9–AMINO–
ACRIDINE

QUININE

Fig. 9.1. Representative weak acids and weak bases for the intracellular measurement of pH by the distribution method. DMO: 5,5-dimethyl-2,4-oxazolidine dione

concentrations of the uncharged form (e. g. the protonated weak acid HA) in the extra- (HA_e) and the intracellular (HA_i) fluid are the same:

$$HA_e = HA_i = HA \tag{9.1}$$

$$HA_{tot} = HA + A^- \tag{9.2}$$

If the pK of the acid HA is known and assumed to be the same in the extra- and intracellular medium, then:

$$pH_e = pK + \log \frac{HA_{tot,e} - HA}{HA} \tag{9.3}$$

and

$$pH_i = pK + \log \frac{HA_{tot,i} - HA}{HA}. \tag{9.4}$$

The combination of equations (9.3.) and (9.4.) yields

$$pH_i = pK + \log \left[\frac{HA_{tot,i}}{HA_{tot,e}} (10^{pH_e - pK} + 1) - 1 \right]. \tag{9.5}$$

By analogy, the expression for a weak base HB is:

$$pH_i = pK - \log \left[\frac{HB_{tot,i}}{HB_{tot,e}} (10^{pK - pH_e} + 1) - 1 \right]. \tag{9.6}$$

Thus, by measuring the extracellular pH and the total equilibrium concentrations of the weak acid or base in the extra- and intracellular fluid, the intracellular pH can be obtained. Equations (9.5) and (9.6) have already been discussed in detail [Waddell and Butler 59; Boron and Roos 76; Roos and Boron 81].

The advantages of the distribution method are its applicability to populations of small cells or organelles (e.g. mitochondria [Addanki et al. 68], chloroplasts [Rottenberg et al. 72]) and the simple experimental procedure. One drawback which must be considered is that this method is restricted to a single-point measurement of the pH (destruction of the cells is required) with equilibration times that depend on the surface-to-volume ratio of the cell (organelles: seconds; muscle fibres: 30 min). Further difficulties are caused by an adsorption of indicator molecules on cellular molecules and a transmembrane diffusion of the charged form of the weak acid or base.

An excellent survey of advantages, limitations and sources of errors is given in [Roos and Boron 81] (see also Cohen and Iles 75; Boron and Roos 76; Roos and Keifer 76]). The use of fluorescent amines as weak bases is discussed in [Lee et al. 82].

9.1.2 Optical Techniques

The different optical techniques are all based on the spectrophotometric observation of an optical signal, such as absorbance or fluorescence, after the incorporation of H$^+$-sensitive indicator dyes into cells. Much work has been done to find a suitable method for introducing indicators into cells without damaging or affecting the cell. The most frequently used methods involve: microinjection of dyes [Heiple and Taylor 82], distribution of acids or bases (e.g. fluorescent amines [Lee et al. 82], see also Sect. 9.1.1), in situ synthesis of a dye from a permeant precursor (e.g. 6-carboxyfluorescein diacetate [Thomas et al. 82]) and fluorescent probes which are phagocytized by cells (e.g. fluorescein isothiocyanate dextran [Ohkuma and Poole 78] or fluorescein thiocarbamyl ovalbumin [Heiple and Taylor 82]).

Usually, optical experiments yield information on the average pH of a cell population (e.g. 10^7 cells [Nuccitelli and Deamer 82]). However, recent improvements in optical and electronic devices have promoted the applicability of these techniques to single cells. In particular, quantitative microfluorometry is now widely used to determine the pH of single cells. A schematic representation of a typical microfluorometer is shown in Fig. 9.2 (for a more detailed description see [Heiple and Taylor 82]). To date, the spatial resolution is limited by the beam width to about 10 μm, which is sufficient for certain measurements in a single cell.

An interesting feature of microfluorometry is its use in studying moving cells (e.g. motile amoeba). In such samples the optical path length and/or H$^+$ activity is continuously changing. Thus measurements of the fluorescence at two wavelengths are advantageous [Heiple and Taylor 82]. It is assumed that the protonated (p) and deprotonated form (d) of an injected dye are in equilibrium, each form exhibiting a different absorbance spectrum. The total absorbance of the dye at the two wavelength is described by the Lambert-Beer law:

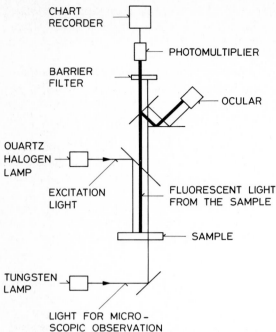

Fig. 9.2. Schematic representation of a microfluorometer

$$A_1 = \varepsilon_{1,p}\, c_p\, l + \varepsilon_{1,d}\, c_d\, l, \tag{9.7}$$

$$A_2 = \varepsilon_{2,p}\, c_p\, l + \varepsilon_{2,d}\, c_d\, l, \tag{9.8}$$

where

A_1	: absorbance at wavelength λ_1;
A_2	: absorbance at wavelength λ_2;
$\varepsilon_{1,p}$; $\varepsilon_{2,p}$	: molar extinction coefficients of the protonated form at λ_1 and λ_2;
$\varepsilon_{1,d}$; $\varepsilon_{2,d}$	: molar extinction coefficients of the deprotonated form at λ_1 and λ_2;
c_p	: concentration of the protonated form;
c_d	: concentration of the deprotonated form;
l	: path length through the sample.

The ratio of the two absorbances and division by $c_d\, l$ yields:

$$\frac{A_1}{A_2} = \frac{\varepsilon_{1,p}\, c_p/c_d + \varepsilon_{1,d}}{\varepsilon_{2,p}\, c_p/c_d + \varepsilon_{2,d}}. \tag{9.9}$$

Using the law of mass action for the acid-base equilibrium (K: equilibrium constant), the term c_p/c_d can be expressed by

$$\frac{c_p}{c_d} = \frac{[H^+]}{K}. \tag{9.10}$$

Substitution of (9.10) in to (9.9) yields:

$$\frac{A_1}{A_2} = \frac{\varepsilon_{1,p} [H^+]/K + \varepsilon_{1,d}}{\varepsilon_{2,p} [H^+]/K + \varepsilon_{2,d}}. \tag{9.11}$$

At any two wavelengths, the absorbance ratio is described by a set of constants and the pH of the sample. If the calibration solutions resemble the sample, it is not necessary that the extinction coefficients and the stability constant are known. H$^+$ activity measurements with the ratio technique are local, fast and reproducible to within ± 0.02 pH units [Heiple and Taylor 82]. However, as in other optical techniques, a calibration for absolute intracellular pH measurements is a problem. Deviations in the calibration curves obtained from in vitro and in vivo measurements have been observed for the fluorescence ratio technique using fluorescein ovalbumin [Heiple and Taylor 82] and trapped indicators such as carboxy fluorescein [Thomas et al. 82].

9.1.3 ^{31}P Nuclear Magnetic Resonance Spectroscopy

The theoretical and technical aspects of nuclear magnetic resonance (NMR) have been extensively discussed in textbooks and will not be treated here. Table 9.1 shows that the properties of the ^{31}P nucleus are very suitable for NMR studies. Known ^{31}P chemical shifts range from about 245 ppm (PF$_2$CH$_3$) to $-$ 461 ppm (P$_4$) (relative to H$_3$PO$_4$), whereas physiologically relevant phosphorus containing compounds cover a range of about 30 ppm (Tab. 9.2).

The utilization of ^{31}P NMR in studies of intracellular pH began in 1973. It is the only approach that is assumed to be totally nondestructive. The measurements are noninvasive since the sample contains naturally occurring ^{31}P nuclei. However, one should not ignore the possible influences of the external magnetic field on the sensitivity of animals and human beings to magnetic fields [Baker 80; Gould 84]. The

Table 9.1. A comparison of the NMR properties of the ^{31}P nucleus with those of the ^{1}H and ^{13}C nuclei [Harris and Mann 78]

Property	^{1}H	^{13}C	^{31}P
natural abundance [%]	99.985	1.108	100
spin quantum number	1/2	1/2	1/2
NMR frequency [MHz] (at 4.7 Tesla)	200.00	50.29	80.96
relative sensitivity^a (relative to ^{1}H)	1.0	$1.8 \cdot 10^{-4}$	$6.6 \cdot 10^{-2}$
standard	(CH$_3$)$_4$Si	(CH$_3$)$_4$Si	85% H$_3$PO$_4$ (external); 88% H$_3$PO$_4$ (external); methylene diphosphonic acid (external); phosphocreatine (internal); glycerophosphoryl choline (internal)

^{a} sensitivity multiplied by the natural abundance

Table 9.2. Ranges of chemical shift for some biologically important phosphorus containing compounds [Shulman 79]

Substance	Chemical shift [ppm]		
88% phosphoric acid (standard)		0	
inorganic phosphate	-3	to	0
sugar phosphates	-5	to	$+1$
phospholipids	-0.5	to	$+1.2$
nucleoside diphosphates	$+6$	to	$+11$
nucleoside triphosphates	$+5$	to	$+23$

role of magnetic fields in the human body is now being increasingly discussed (e. g. evoked neuromagnetic fields [Kaufman et al. 84]).

^{31}P NMR can be used to measure the pH of large cell populations [Ugurbil et al. 79; Gillies et al. 82], cell suspensions [Martin et al. 82], perfused tissues [Gadian et al. 79], organs [Grove et al. 80; Jacobus et al. 82] and animals or human beings [Bore et al. 82]. Information on the metabolism can be obtained simultaneously (e. g. ADP, ATP, phosphocreatine, phosphorus containing sugars). The assignment of a signal is often difficult. It necessitates the extraction and chemical identification of a substance. Furthermore, the sensitivity is relatively poor. Hence, only components that are present at a concentration > 0.3 mM are detectable [Radda et al. 82].

In principle, any NMR signal of a phosphorus containing compound that is sensitive to changes in the pH of a sample can be used to determine the H^+ activity. In practice, however, the signals from inorganic phosphate are most commonly utilized. The observed signals stem from the two conjugates HPO_4^{2-} and $H_2PO_4^-$ which exhibit a pK in the physiological pH range and undergo a fast chemical exchange on the NMR time scale. Therefore, the spectrum consists of a single signal with a chemical shift δ corresponding to the total intracellular inorganic phosphate ($c_{tot} = c_{HPO_4} + c_{H_2PO_4}$). The measured chemical shift reflects the concentration-weighted average of the shifts of the alkaline form (c_{HPO_4}, δ_{HPO_4}) and its conjugate ($c_{H_2PO_4}$, $\delta_{H_2PO_4}$). Through Eqs. (9.12) to (9.14) the dependence of the pH on the chemical shift of the inorganic phosphate in the sample is obtained (9.15).

$$c_{tot}\, \delta = (c_{HPO_4} + c_{H_2PO_4})\delta = \atop c_{HPO_4}\, \delta_{HPO_4} + c_{H_2PO_4}\, \delta_{H_2PO_4}. \tag{9.12}$$

$$\frac{c_{HPO_4}}{c_{H_2PO_4}} = \frac{\delta - \delta_{H_2PO_4}}{\delta_{HPO_4} - \delta}. \tag{9.13}$$

$$pH = pK + \log \frac{c_{HPO_4}}{c_{H_2PO_4}}. \tag{9.14}$$

$$pH = pK + \log \frac{\delta - \delta_{H_2PO_4}}{\delta_{HPO_4} - \delta}. \tag{9.15}$$

Figure 9.3 shows a typical calibration curve obtained in vitro [Garlick et al. 79]. The shape of the curve corresponds to the relationship given in Eq. (9.15).

CHEMICAL SHIFT

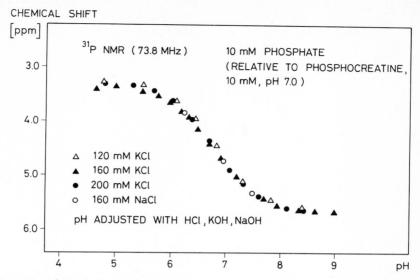

Fig. 9.3. Change in the chemical shift of inorganic phosphate as a function of the pH of the solution at 37 °C [Garlick et al. 79]. The chemical shift is expressed relative to phosphocreatine at pH 7.0. Solutions containing 10 mM inorganic phosphate and 10 mM phosphocreatine were adjusted to various ionic strengths by adding KCl or NaCl and titrated at 37 °C by adding HCl, KOH or NaOH. Symbols: △ 120 mM KCl; ▲ 160 mM KCl; ● 200 mM KCl; ○ 160 mM NaCl. Redrawn from Garlick, Radda and Seeley, Biochem. J., 1979 [Garlick et al. 79]

The NMR spectra obtained are usually simple since the only signals that are observed are those that originate from mobile phosphorus-containing substances with concentrations larger than about 0.3 mM. A representative ^{31}P NMR spectrum of a perfused organ is shown in Fig. 9.4 [Grove et al. 80]. In certain samples the signal from the inorganic phosphate may be disturbed by the presence of other compounds (e.g. yolk proteins [Webb and Nuccitelli 82]). The problem can be circumvented by taking advantage of the different relaxation times of the nuclei involved. The reliability of the intracellular pH measurements using the ^{31}P NMR technique has been discussed in the literature [Shulman 79; Roos and Boron 81; Nuccitelli and Deamer 82]. As shown by Eq. (9.15) it is necessary to know the pK of dihydrogen phosphate. The effects of the phosphate concentration or a complex formation between the inorganic phosphate and the sample cations on the pK are negligible [Gadian et al. 79]. However, changes in the ionic strength of a sample significantly influence the pK and therefore the relative position of the titration curve. Under physiological conditions, a two-fold increase in the ionic strength shifts the curve to higher values by about 0.1 pH units [Gillies et al. 82]. Hence, the composition of the calibration solution and the sample should be as similar as possible.

Usually, 85% H$_3$PO$_4$ is used as a standard for the chemical shift (assignment of zero chemical shift). Due to its chemical reactivity, H$_3$PO$_4$ has to be used as an external standard for measurements in biological samples. A further disadvantage of this standard is the relatively large line-width ($\sim$5 Hz) of the signal. Other compounds (e.g. glycerophosphoryl choline [Navon et al. 77; Webb and Nuccitelli 82]

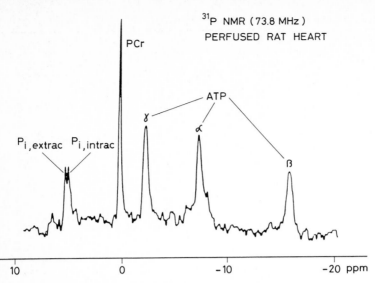

Fig. 9.4. ^{31}P NMR spectrum obtained in 8 min at 73.8 MHz from the perfused heart of a rat [Grove et al. 80]. PCr: phosphocreatine; $P_{i,extrac}$: inorganic phosphate in the perfusion solution; $P_{i,intrac}$: inorganic phosphate in the intracellular space. Redrawn from Grove, Ackerman, Radda and Bore, Proc. Natl. Acad. Sci. USA, 1980 [Grove et al. 80]

or phosphocreatine [Jacobus et al. 82], see Fig. 9.4 and Tab. 9.1) have been proposed as nonreactive and pH-independent internal standards.

In order to detect millimolar amounts of inorganic phosphate in a small sample volume and obtain a suitable signal-to-noise ratio a very large number of cells is required. The methods used to prepare such high density cultures have been discussed [Ugurbil et al. 79; Gillies et al. 82].

For the measurement of changes in pH a typical time resolution of about 1 minute has to be considered [Radda et al. 82]. In some special cases, a continuous measurement of pH with a data collection rate of about 15 s has been reported [Jacobus et al. 82].

The reproducibility of H^+ activity determinations is about ± 0.02 pH units [Nuccitelli and Deamer 82]. The resolution between the signals of inorganic phosphate in different cell compartments depends on the relative concentrations of phosphate. Resolutions as low as 0.02 pH units have been obtained [Jacobus et al. 82]. However, it is usually difficult to attain resolutions of less than 0.2 pH units due to the considerable line-width of cellular signals (e.g. 30–50 Hz at 145 MHz [Gillies et al. 82]). The accuracy of the method is unlikely to be better than 0.1 pH units [Gadian et al. 79; Nuccitelli and Deamer 82].

The necessity of achieving stable physiological conditions for the sample within the spectrometer creates several technical problems. As a result, special arrangements of the sample tube which allow a perfusion and oxygenation of the cells have been developed [Gadian et al. 79; Jacobus et al. 82; Gillies et al. 82]. The modified sample tubings should influence the homogeneity of the magnetic field and the signal-to-noise ratio as little as possible (see the discussion in [Gadian et al. 79]). Fur-

Table 9.3. Course of the development of H$^+$-selective microelectrodes

Membrane	Development	Year, Authors, Reference
solid-state (platinum)	platinum-based hydrogen microelectrode for vacuolar pH measurements	1927, Taylor, Whitaker, [Taylor and Whitaker 27]
glass, solid-state (tungsten)	microelectrode for large cells (giant axons, muscle fibres)	1954, Caldwell, [Caldwell 54]
glass	H$^+$-selective glass capillary inserted into insulating glass capillary is drawn to a micropipette	1961, Kostyuk, Sorokina, [Kostyuk and Sorokina 61]
glass	open-tip type	1964, Lavallée, [Lavallée 64]
glass	protruding-tip type (spear-tip type, exposed-tip type)	1967, Hinke, [Hinke 67]
glass	insulated H$^+$-selective glass	1967, Carter, Rector, Champion, Seldin, [Carter et al. 67]
solid-state (antimony)	glass micropipette coated with antimony	1972, Bicher, Ohki, [Bicher and Ohki 72]
glass	recessed-tip type (tip diameter less than 1 μm)	1974, Thomas, [Thomas 74]
glass	micropipette tip covered with H$^+$-selective glass, double-barrelled	1976, Pucacco, Carter, [Pucacco and Carter 76]
glass	H$^+$-selective glass deposited on microneedle by RF-sputtering	1976, Baumgartl, Shizemitsu, Lübbers, [Baumgartl et al. 76]
glass	H$^+$-selective glass capillary inserted in insulating soda-lime glass capillary, recessed-tip type	1977, Yamaguchi, Stephens, [Yamaguchi and Stephens 77]
glass	double-barrelled recessed-tip type	1979, de Hemptinne, [de Hemptinne 79]
liquid	3-hydroxy-picolinamide as H$^+$ carrier	1979, Erne, Ammann, Simon, [Erne et al. 79a]
solid-state (antimony)	double-barrelled	1980, Matsumura, Kajino, Fujimoto, [Matsumura et al. 80b]
liquid	sodium salt of nigericin as H$^+$ carrier	1980, Matsumura, Aoki, Kajino, Fujimoto, [Matsumura et al. 80a]
liquid	p-octadecyloxy-m-chloro-phenyl-hydrazone mesoxalonitrile (OCPH) as H$^+$ carrier	1981, Harman, Poole-Wilson, [Harman and Poole-Wilson 81]
liquid	tri-n-dodecylamine (TDDA) as H$^+$ carrier	1981, Ammann, Lanter, Steiner, Schulthess, Shijo, Simon, [Ammann et al. 81b]

thermore, some extraordinary arrangements have been achieved by surgically localizing the radio frequency coil on internal organs [Bore et al. 82]. However, for local in vivo studies the noninvasive potential of NMR can be maintained by using external surface coils that allow a local measurement within tissue under consideration [Bendall and Aue 83; Cross et al. 84]. This approach is very promising in the development of medical care using noninvasive techniques.

Finally, the ^{31}P NMR method requires a great deal of expensive instrumentation and an extensive theoretical knowledge.

9.1.4 Potentiometry

9.1.4.1 Survey of H$^+$-Selective Microelectrodes

In the past, many different types of H$^+$-selective microelectrodes have found an application in extra- and intracellular pH studies. The microelectrodes are based on materials such as platinum, tungsten, antimony, glass or liquids containing different H$^+$-selective components (see Tab. 9.3).

The various microelectrodes listed above have been repeatedly discussed with regards to their usefulness in physiological studies [Thomas 74; Matsumura et al.

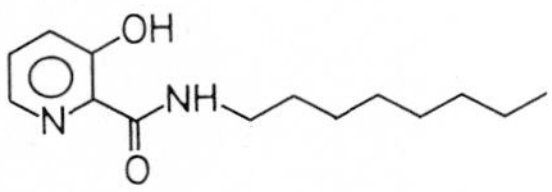

1 (ETH 308)

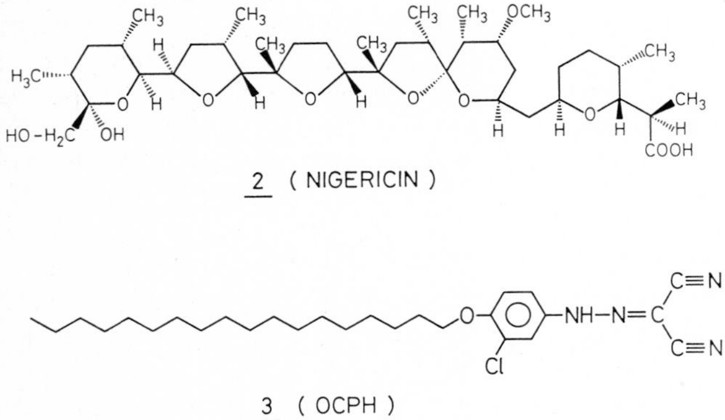

2 (NIGERICIN)

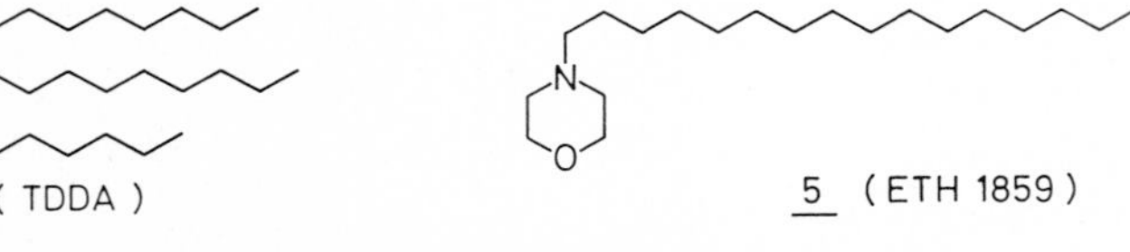

3 (OCPH)

4 (TDDA)

5 (ETH 1859)

Fig. 9.5. H$^+$-selective carriers for liquid membrane microelectrodes

80b; Roos and Boron 81]. The solid-state membrane microelectrodes have only limited applications due to the interference from oxidizing and reducing agents and the incomplete reversibility of the membrane surface reaction [Matsumura et al. 80b]. These difficulties inherent to antimony membranes are not encountered with optimized pH-sensitive glass. The glass membrane microelectrodes exhibit excellent electromotive behaviour. However, their main disadvantages are the high electrical resistances and the serious technical difficulties encountered in the preparation of small and/or double-barrelled microelectrodes. The numerous attempts to produce glass membrane microelectrodes (Tab. 9.3) illustrate the problems of fabrication.

In 1979 progress was made in the development of H$^+$-selective liquid membrane microelectrodes [Erne et al. 79a; Ammann 81; Steiner 82]. Today, neutral carrier microelectrodes based on tri-n-dodecylamine [Ammann et al. 81b] are a good alternative to glass membrane H$^+$ microelectrodes [Thomas 84] and have already replaced them to a large extent.

9.1.4.2 Available Liquid Membrane Microelectrodes

The currently available H$^+$-selective carriers for liquid membrane microelectrodes are shown in Fig. 9.5. Carriers $\underline{1}$–$\underline{3}$ act as electrically charged carriers (see the discussion in [Erne 81]), whereas tri-n-dodecylamine ($\underline{4}$) [Schulthess et al. 81] and $\underline{5}$ behave as neutral carriers for hydrogen ions. Microelectrodes based on $\underline{4}$ [Ammann et al. 81b] have properties comparable to those of glass microelectrodes but are significantly better than the microelectrodes based on charged carriers.

9.1.4.2.1 Microelectrodes Based on 3-Hydroxy-N-octyl-picolinamide ($\underline{1}$)

The antibiotic virginiamycin S has been reported to facilitate the transport of H$^+$ions across lipid membranes [Grell et al. 77]. It has been shown that the 3-hydroxy-picolinyl structural unit is responsible for the H$^+$-carrier properties of the antibiotic [Grell et al. 77]. Indeed, synthetic lipophilic picolinamides can be incorporated into PVC membranes to yield electrodes that exhibit high selectivities for the H$^+$ ion [Erne et al. 79a; Erne et al. 81]. The sensors allow extra- as well as intracellular pH determinations to be made (Fig. 9.6, upper curve) [Erne et al. 79a; Simon et al. 80]. The corresponding microelectrodes maintain this selectivity behaviour [Ammann 81; Steiner 82]. Figure 9.7 illustrates the potential of a microelectrode based on $\underline{1}$ for measurements in the physiological pH range at a typical extracellular background of ions. The same characteristics were obtained in electrolyte solutions with intracellular concentrations of ions [Steiner 82].

Unfortunately, the extremely high electrical resistance of such a membrane (for a tip diameter of about 2.5 µm: ~$10^{12}\,\Omega$) is a limiting factor for practical applications in electrophysiology.

9.1.4.2.2 Microelectrodes Based on the Na$^+$ Salt of the Antibiotic Nigericin ($\underline{2}$)

It is well known that various carboxylic antibiotics incorporated into membranes induce a selectivity for alkali or alkaline-earth cations [Simon et al. 73]. The structurally related carriers nigericin and monensin induce K$^+$/Na$^+$ and Na$^+$/K$^+$ selec-

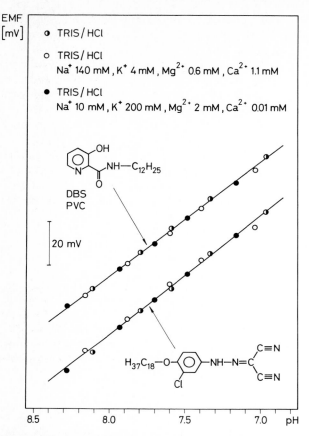

Fig. 9.6. H$^+$ electrode functions of a PVC liquid membrane electrode based on a lipophilic 3-hydroxy-picolinamide *(upper curve)* [Erne et al. 79 a] and the carrier OCPH (3 in Fig. 9.5., *lower curve*) [Le Blanc et al. 76; Coon et al. 76]. The pH-values of the TRIS-buffered solutions without interfering ions (◗), with an extracellular background of ions (○) and with an intracellular background of ions (●) were determined with a glass electrode

tivity (see Sect. 9.3.2.2), respectively [Lutz et al. 70]. Despite their ability to form complexes with metal cations, it was possible to develop a microelectrode based on the Na$^+$ salt of nigericin that is still sufficiently selective towards H$^+$ ions [Matsumura et al. 80a]. In TRIS-buffered solutions (pH 7.6), the response to Na$^+$, K$^+$, Mg^{2+} and Ca^{2+} is almost negligible (2 to 4 mV/decade). The slope of the H$^+$ electrode function is slightly sub-Nernstian (about 53 mV) but is constant in various buffer media. The response time (t_{95}) is typically 2–3 s, the stability of the EMF ± 1.0 mV/h and the reproducibility better than ± 1 mV (0.02 pH units) [Matsumura et al. 80a]. The microelectrodes require a preconditioning time of about 1 week in order for the carrier to attain its maximum H$^+$ selectivity (the underlying mechanisms could not be explained [Matsumura et al. 80a]). Obviously, during an entire week of conditioning the properties and geometry of the microelectrode tip could change due to hydration and dissolution of the glass (Sect. 6.1). Presumably, a partial pro-

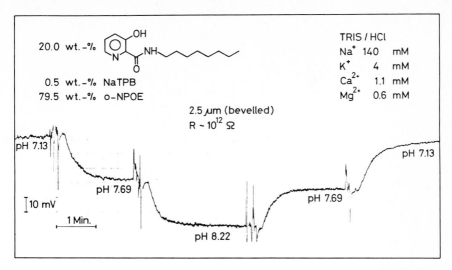

Fig. 9.7. Electrode response of a H$^+$-selective microelectrode based on ETH 308 ($\underline{1}$ in Fig. 9.5) to TRIS-buffered solutions containing typical extracellular concentrations of Na$^+$, K$^+$, Mg^{2+} and Ca^{2+} [Ammann 81]

tonation of the Na$^+$ salt occurs. ^{13}C NMR studies of the acid-base equilibria of synthetic carboxylic carriers in membranes have shown that a 50% protonation of the carboxylic group occurs in the aqueous phase at pH 7.1 in the absence of metal ions [Maj-Zurawska et al. 82].

9.1.4.2.3 Microelectrodes Based on p-Octadecyloxy-m-chloro-phenyl hydrazone-mesoxalonitrile (OCPH, $\underline{3}$)

A pH-responsive membrane based on the H$^+$ carrier OCPH ($\underline{3}$ in Fig. 9.5) incorporated into a block copolymer matrix has been developed [LeBlanc et al. 76; Coon et al. 76]. Macroelectrodes employing such membranes exhibit good EMF characteristics (see Fig. 9.6, lower curve) and have been applied as intravascular catheter electrodes in the monitoring of the pH of blood [LeBlanc et al. 76]. However, this carrier did not induce H$^+$ permselectivity in other polymeric membrane matrices [LeBlanc et al. 76; Erne 81]. Many unsuccessful attempts have been made to develop a membrane solution containing OCPH [Steiner 82], even though the special membrane matrix can be omitted in microelectrodes. Nevertheless, a liquid membrane microelectrode with 0.1‰ OCPH in decanol has been reported [Harman and Poole-Wilson 81]. The use of this microelectrode requires a 3 hour preconditioning period. Furthermore, it is found that many of a series of microelectrodes are defective and that the useful microelectrodes exhibit a small response range of 2 pH units. Until now no reports have been made on the use of microelectrodes based on $\underline{3}$ in physiological studies.

9.1.4.2.4 Microelectrodes Based on Tri-n-dodecylamine ($\underline{4}$)

A comparison of the performance of liquid membrane microelectrodes containing the H$^+$ carriers $\underline{1}$, $\underline{2}$ or $\underline{3}$ with the performance of glass membrane microelectrodes shows that liquid membranes still need some improvement. It was found that lipophilic amines exhibit H^+-carrier properties when incorporated into solvent polymeric membranes [Schulthess et al. 81] (see also [Funck et al. 82]). ^{13}C NMR studies have shown that amines whose conjugate acid in water has a dissociation constant less than 10^{-8} M are predominantly present in the nonprotonated form in organic solvents if they are in equilibrium with aqueous solutions with pH-values above 3 [Schulthess et al. 81]. Such compounds therefore behave as neutral carriers for H^+ ions. The linear range of the pH response depends on the pK of the amine. Thus, the carrier 5, which has a slightly lower pK than 4, exhibits an extended range in acidic solutions.

Solvent polymeric membranes with tri-n-dodecylamine (4 in Fig. 9.5) exhibit selectivities that allow both extra- and intracellular pH measurements. Extremely high selectivities with respect to Na^+, K^+ and Ca^{2+} were evaluated by the fixed interference method. The values of log K_{HM}^{Pot} are -10.4 ($M = Na^+$), -9.8 ($M = K^+$) and < -11.1 ($M = Ca^{2+}$) [Schulthess et al. 81]. The corresponding values obtained with H^+-selective glass membrane electrodes are about -13 for monovalent interfering ions [Buck et al. 74]. The solvent polymeric membranes were found to be very suitable for pH measurements in blood [Anker et al. 83 a; Ammann et al. 85 b]. The values obtained correlate very well with those obtained with glass electrodes [Anker et al. 83 a]. Thus, tri-n-dodecylamine is an attractive compound for use in reliable H^+-selective liquid membrane microelectrodes.

H^+-selective Membrane Solutions. The best performance has been obtained with microelectrodes containing 10 wt.-% of tri-n-dodecylamine, 0.7 wt.-% sodium tetraphenylborate and 89.3 wt.-% o-nitrophenyl-n-octyl ether [Ammann et al. 81 b]. It is necessary to equilibrate this mixture for about 16 hours in a pure atmosphere of CO_2. It is assumed that a partial formation of tri-n-dodecylammonium bicarbonate or carbonate takes place [Muhammed 72], which advantageously influences the electrical resistance of the microelectrodes. The salt formation in the organic phase proceeds very slowly [Muhammed 72], so that the presence of typical levels of CO_2 during physiological studies should not alter the composition of the membrane. On the other hand, if the equilibration time in CO_2 is too long (> 1 day) the microelectrodes may start to deteriorate because of the extensive formation of bicarbonate or carbonate (no free ion carrier). This is in agreement with the theoretical prediction that an ion carrier must be predominantly uncharged [Morf 81]. It has been shown that tri-n-dodecylamine-based membrane solutions which have been equilibrated once in CO_2 for 16 h can be stored for at least 4 months without any further CO_2 treatment [Kurkdjian and Barbier-Brygoo 83] (freshly distilled tri-n-dodecylamine was used). The influence of the purity of the H^+ carrier on the response of the microelectrode has been discussed [Kurkdjian and Barbier-Brygoo 83]. It is also important that a vacuum is not applied during the filling of the micropipettes (e. g. removal of air bubbles), thus avoiding a loss of CO_2 from the membrane phase.

Electrode Function and Selectivity. In Fig. 9.8 the response of a microelectrode to changes in the pH of a sample solution is compared with the behaviour of a glass electrode. There is obviously a nearly theoretical response to hydrogen ion activities in the pH range 5.5–12 (slope of linear regression: 58.0 ± 0.4 mV, 22 °C; theoretical: 58.6 mV), even in the presence of 69 mM Na^+. At pH values below 4 an anion inter-

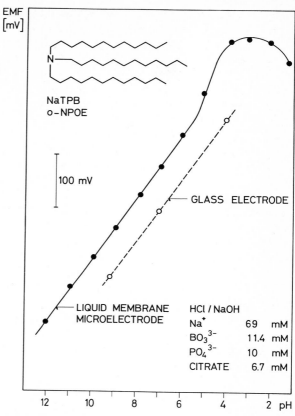

Fig. 9.8. EMF response of a microelectrode based on TDDA to different pH-buffered solutions at a constant background of ions, *upper trace;* pH response of a glass electrode at the same ionic background, *lower trace;* experimental values, dots

ference due to Cl$^-$ leads to EMF's that are too negative [Morf and Simon 78]. The slope in the pH range 4–5.5 has not been satisfactorily explained. It in no way hinders the use of microelectrodes for pH studies in biological systems (see Fig. 9.9).

Comparable electrode functions have been found in cell-sap solutions [Kurkdjian and Barbier-Brygoo 83]. Additional characteristics of the electrode function observed in various buffers have been given in [Kurkdjian and Barbier-Brygoo 83; Aickin 84a; Yoshitomi and Frömter 84; Mutch and Hansen 84; Kafoglis et al. 84; Duffey 84].

Using the Nicolsky-Eisenman formalism to describe the electrode response [Meier et al. 80] and assuming that the error in the activity determination introduced by Na$^+$ is less than 1%, a selectivity factor, $K^{Pot}_{HNa} < 2 \cdot 10^{-13}$, can be derived from Fig. 9.8. An interference from ions in typical intra- and extracellular concentrations cannot be observed in the physiologically relevant pH range (Fig. 9.9). The addition of 24 mM sodium bicarbonate to solutions with an extracellular background (Fig. 9.9, upper curve) does not significantly influence the slope of the electrode response (56.0 ± 0.5 mV).

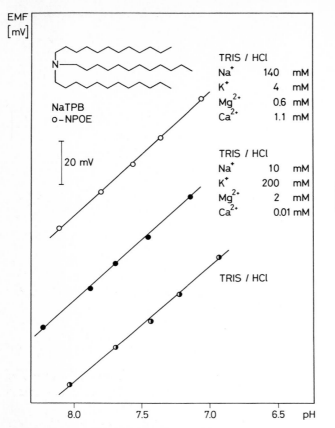

Fig. 9.9. EMF response of a microelectrode based on TDDA to different H$^+$ activities in the physiological range at typical intra- and extracellular backgrounds: H$^+$ response at constant extracellular background *(upper trace)*; H$^+$-response at constant intracellular background *(middle trace)*; H$^+$ response in pure TRIS/HCl buffer solutions *(lower trace)*; experimental values, dots

The addition of 100 mM sucrose to the calibration solutions does not influence the response curve. In the presence of 50 mM bovine serum albumin the slope is slightly reduced [Kurkdjian and Barbier-Brygoo 83]. The pH response is destroyed if uncouplers such as dinitrophenol [Kraig et al. 83] or carbonylcyanide-*m*-chlorophenyl hydrazone (CCCP) [A. Kurkdjian, private communication] are present. The disturbance is probably caused by anionic interference, i.e. the protonated carrier forms ion pairs with the basic form of the uncoupler. The use of other inhibitors such as ouabain or acetazolamide (see Sect. 7.8) does not produce any detectable interferences [Kraig et al. 83].

As already mentioned there are considerable doubts about the interference of CO_2. The formation of salts in the membrane phase due to the presence of CO_2 has been shown to be extremely slow. Permeation of CO_2 through the membrane does not significantly change the pH of the strongly buffered internal filling solution. Further evidence for the absence of CO_2 interference has recently been reported:

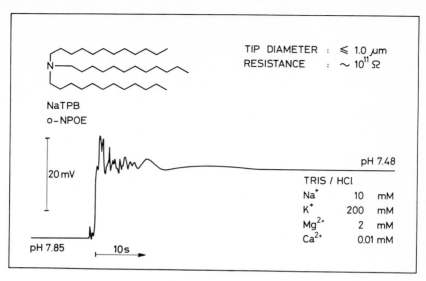

Fig. 9.10. Response of a neutral carrier H$^+$ microelectrode to the injection of 1 M HCl into a buffer solution with a pH of 7.85 and a background of ions as indicated

- Ringer solutions equilibrated with varying partial pressures of CO_2 could be used as perfusion solutions [Yoshitomi and Frömter 84]
- direct comparison of the pH measurements with glass electrodes and the neutral carrier microelectrode during a bubbling of CO_2 into bicarbonate solutions did not indicate any disturbance due to CO_2 [Mutch and Hansen 84]
- the calibration curves obtained from solutions equilibrated with air or with 95% $O_2/5\%$ CO_2 were the same [Duffey 84]
- disturbances due to CO_2 have occasionally been observed, but they can be eliminated by using a CO_2-saturated citrate buffer as the internal filling solution. A comparison of pH measurements involving changes in CO_2 has been performed using recessed-tip glass membrane microelectrodes and neutral carrier microelectrodes. The recordings have shown that both microelectrodes measure the same intracellular pH. Under CO_2-free external conditions, the intracellular pH's agree very well: neutral carrier microelectrode: 7.23 ± 0.05 ($n = 5$); recessed-tip glass microelectrode: 7.23 ± 0.07 ($n = 6$) [Aickin 84a]

Electrode Resistance. Electrical resistances of about 10^{11} Ω (tip diameter 0.8–1.0 μm [Ammann et al. 81 b]; 1.0–1.5 μm [Kurkdjian and Barbier-Brygoo 83]), and 1 to 5 · 10^{12} Ω (tip diameter 0.3–0.6 μm [Kurkdjian and Barbier-Brygoo 83]; <0.2 μm [Yoshitomi and Frömter 84]; <1 μm [Oberleithner et al. 84a]) have been reported for the membranes.

Response Time. In order to evaluate the response time, a microelectrode was conditioned overnight in a buffer solution of pH 7.85 at a typical intracellular background and a change in pH was induced (see Fig. 9.10). After an injection of hydrochloric acid the final pH of the buffer solution was checked with a calibrated glass electrode. The irregularities in the response curve in Fig. 9.10 can be traced to an inadequate mixing of the electrolytes. The 90% response time is < 5 s.

From more sophisticated studies of pH transients in cerebellar microenvironments it was concluded that the response times of H^+ microelectrodes are comparable to those of classical ion-exchanger K^+ microelectrodes, which are known to be very short [Kraig et al. 83].

Furthermore, the following response times have been claimed: 1 s (< 0.2 μm, 10^{12} Ω, driven shield feed-back circuit [Yoshitomi and Frömter 84]); > 15 s (< 0.2 μm, > 2·10^{12} Ω, unshielded microelectrode [Yoshitomi and Frömter 84]); 1 s (< 1 μm, 2·10^{12} Ω [Oberleithner et al. 84 a]); < 1 s (1–3 μm [Mutch and Hansen 84]).

Stability, Reproducibility and Drift in EMF. A typical cell assembly shows an overall drift of 0.6 mV/h over at least 3 days. To test its reproducibility measurements were performed in solutions of pH 7.8 (TRIS/HCl, 140 mM Na^+, 4 mM K^+, 0.6 mM Mg^{2+}, 1.1 mM Ca^{2+}) and pH 7.4 (TRIS/HCl, 140 mM Na^+, 4 mM K^+, 0.6 mM Mg^{2+}, 1.1 mM Ca^{2+}) by switching the solutions every 2 minutes. The standard deviation of the EMF's was 0.23 mV (10 degrees of freedom). Approximately 10 minutes after the preparation of the electrodes (tip > 1 μm) readings with these stabilities and reproducibilities are obtained.

Further information on the reproducibility of EMF's was provided by 21 repetitive intracellular measurements of the pH in ferret ventricles (77 impalements) at a constant buffered extracellular pH [Coray and McGuigan 82]. The mean intracellular pH, which reflects both the physiological variability and the reproducibility of the microelectrode, was 7.31 ± 0.01 (std. dev.) [Coray and McGuigan 82]. The slopes of the H^+ electrode functions were found to be 55.0 ± 1.2 mV (n = 4) before and 54.3 ± 1.0 mV (n = 4) after the intracellular experiments [Duffey 84].

Lifetime. Reports on the lifetime of neutral carrier microelectrodes differ considerably depending on their mode of fabrication. The lifetime appears to be very unpredictable [Thomas 84]. Microelectrodes filled with a CO_2-equilibrated H^+-selective membrane solution can be used immediately and maintain their performance for several days [Kurkdjian and Barbier-Brygoo 83; Kraig et al. 83], up to 3 days [Aickin 84 a] or for 1 day to 3 weeks [Bickel and Cimasoni 85]. The microelectrodes seem to lose their sensitivity with repeated cell impalements [Aickin 84 a].

Intracellular H^+ Activity Studies. Figure 9.11 illustrates the first application of the microelectrode system to the determination of the pH_i in *Xenopus laevis* oocytes [Ammann et al. 81 b]. E_M is the membrane potential measured with a reference microelectrode filled with 3 M KCl against a reference electrode in an extracellular bath (60 mM NaCl, 1.2 mM KCl, 4 mM $CaCl_2$, 10 mM HEPES, titrated with NaOH to pH 7.5, [Na^+]$_{tot}$ = 65 mM); and E_H is the potential difference between the H^+-selective microelectrode and the reference electrode in the extracellular bath (minus E_M). The oocyte was impaled at (a) with the reference microelectrode and at (b) with the ion-selective electrode. At (c) the external bath was replaced by a bicarbonate buffered Holtfreter solution (20 mM NaCl, 1.2 mM KCl, 4 mM $CaCl_2$, 45 mM $NaHCO_3$) equilibrated with CO_2. Point (d) represents the replacement of the Holtfreter solution by the extracellular bath. In order to show that the CO_2 does not in-

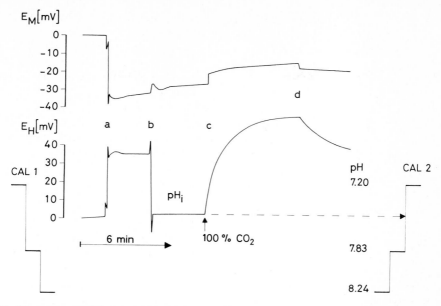

Fig. 9.11. Measurement of the intracellular H$^+$ activity (pH$_i$) in a *Xenopus laevis* oocyte. The tip diameter of the neutral carrier microelectrode was <1 μm [Ammann et al. 81b]

fluence the performance of the electrode, an external calibration was carried out before and after the internal recording. It is obvious that the impalement does not harm the pH-sensitive microelectrode.

9.1.5 Comparison of the Different Methods

The advantages and limitations of the methods used for intracellular H$^+$ activity measurements have been extensively discussed ([Roos and Boron 81; Nuccitelli and Deamer 82; Kurkdjian et al. 85]. Liquid membrane microelectrodes which have only recently been developed were only considered by Kurkdjian et al. The important features of each method are summarized in Table 9.4. Of course, deviations from these general remarks may occur in special applications. Nevertheless, Table 9.4 allows a convenient comparison of the techniques. Unfortunately, there have only been a few contributions which directly compare the different methods. The examples compiled in Tab. 9.5 list method-dependent deviations of measured intracellular pH's. Although most of the comparisons have not been performed using the most recently developed modifications of a technique (e. g. using neutral carrier microelectrodes instead of protruding-tip type glass microelectrodes), the results of the different methods usually agree to better than 0.1 pH units. Of course, it is difficult to decide which method yields the more accurate pH. Many physiologists acknowledge that microelectrodes offer the best approach for obtaining absolute values of pH. Indeed, the calibration procedures of ion-selective microelectrodes seem to be reliable.

Table 9.4. Comparison of methods for the determination of intracellular pH

Method	Method-specific advantages	Method-specific disadvantages	Typical sample	Influence on the cell	Temporal resolution	Spatial resolution	Calibration	Reproducibility [pH-units], std. dev. given	Accuracy [pH-unit]
Distribution of weak acid/base							calibration for the measurement of extracellular pH and the distributed weak acids or bases		
– isotopically labelled	– simple procedure; very small cells and organelles	averaged pH of cytoplasm and organelles; concentration – dependent influence of the indicator on the cell pH, more suited to follow changes in pH since absolute values not very reliable; usually only one measurement per aliquot; adsorption of the indicator to the cell membrane; errors due to penetration of the charged form,	– 0.1 µl cell suspension volume; very small cells	– quasi-noninvasive; destruction of the cell for a single determination	given by equilibration time of the weak acid/base which is dependent on the surface-to-volume ratio of the cell; seconds for organelles; minutes to hours for e.g. muscle fibres; fast changes in pH not detectable	cell population; average pH of cytoplasm and organelles			
– spectrophotometrically detectable	– continuous monitoring		– 0.05 µl cell suspension volume	– quasi-noninvasive, cell can be studied further					
31P NMR	simultaneous measurement of other physiological phosphorus-containing compounds; information on cell compartmentation	large number of cells needed, i.e. high density cultures are necessary (e.g. difficult for vacuoles); expensive instrumentation	method which requires most of material: at least ~1 ml of cell suspension volume (~10^{11} cells per ml); typically 10 ml samples of 20–50% cell suspension; organs; entire animal or human being	noninvasive	dependence on the sample concentration (relation between accumulation time and signal-to-noise ratio); typically several minutes; in special cases of the order of seconds; continuous monitoring possible	cell populations; resolution of pH of different compartments is possible in certain samples	calibration is slightly influenced by changes in ionic strength	±0.02	0.1

Technique	Advantages	Problems / procedure	Applicable sample	Further problems	Response	Resolution / measured quantity	Calibration	Accuracy (pH)	
Optical techniques									
– conventional dyes	– fast response	– adsorption to cytoplasmic molecules; fluorescence quenching; microinjection; influence of the indicator on the cell pH	– 0.1 µl cell suspension volume (e.g. 10^7 cells of 2 µm diameter)	indicator may interact with cell membrane or with species in the cell; in some cases impalement of micropipette for microinjection	very fast response; continuous monitoring; ratio measurements: measuring time is limited by change of the excitation filters (~1 sec)		– calibration for an absolute pH determination is problematic	– ±0.02	
– trapped indicators (carboxyfluorescein)	– fast response; very small cells; no microinjection	– dye leakage; cell-dependent esterase activity	– cell population; average over cytoplasm			– cell population; average pH of cell cytoplasm	– deviations of calibration curves in vitro and in vivo	– ±0.02	
– fluorescein ovalbumin	– fast response; very small cells	– microinjection or stimulation of endocytosis	– 1 cell (mobile) or 1 phagosome, cytoplasmic pH is measured; pH gradients can be observed			– 1 cell or 1 phagosome; resolution limited by beam width (typically ~10 µm)	– deviations of calibration curves in vitro and in vivo	– ±0.02	0.1
Microelectrodes	simultaneous measurement of other ion activities and resting potential; very local measurement	impalement of microelectrode(s)	extruding-tip glass microelectrodes: 1 large, nonmotile cell; recessed-tip glass electrode or liquid membrane electrode: 1 nonmotile cell (>10 µm)	invasive; risk of membrane leakage of membrane materials out of microelectrodes into the cytoplasm and cell wall	dependence on tip diameter and membrane material (membrane resistance); typically 0.5 seconds to several seconds	local pH near microelectrode tip; detection volume is about 1 fl; measurements in cell compartments are possible in certain samples	convenient calibrations in aqueous buffer solutions with typical ion background; absolute pH	< ±0.01	

Table 9.5. Measurement of intracellular pH using different techniques under the same conditions

Cell	Method (intracellular pH, std. dev. given)			Remarks, Reference
muscle fibres of giant barnacle	a) ISE[1]	$(7.26 \pm 0.011$	$(n=6))$	DMO and MA require 30 minutes and 5 hours, respectively, for steady-state distribution. Deviation of DMO and MA with respect to ISE can be explained by an error in the pK_{intra} for DMO and MA, or by diffusion of the ionic form of DMO and MA, or by cell compartmentation [Boron and Roos 76]
	DMO[2]	$(7.20 \pm 0.005$	$(n=30))$	
	b) ISE[1]	$(7.34 \pm 0.025$	$(n=6))$	
	DMO[2]	$(7.29 \pm 0.009$	$(n=27))$	
	c) ISE[1]	$(7.34 \pm 0.010$	$(n=6))$	
	DMO[2]	$(7.27 \pm 0.009$	$(n=22))$	
	d) ISE[1]	$(7.48 \pm 0.003$	$(n=3))$	
	MA[3]	$(7.28 \pm 0.020$	$(n=12))$	
axons of giant squid	e) ISE[1]	$(7.35 \pm 0.006$	$(n=126))$	pH's are not significantly different [Boron and Roos 76]
	DMO[2]	$(7.36 \pm 0.25$	$(n=6))$	
Nitella translucens	f) ISE[1]	$(7.54 \pm 0.15$	$(n=8))$	The low pH's obtained with the antimony-ISE[4] could be due to insulation problems at the tip of the electrode [Spanswick and Miller 77]
	DMO[2]	$(7.42 \pm 0.07$	$(n=7))$	
	ISE[4]	$(6.74 \pm 0.15$	$(n=6))$	
Xenopus eggs	g) ISE[5]	$(7.39 \pm 0.11$	$(n=11))$	Excellent agreement between ISE and ^{31}P NMR; ISE shows faster time resolution than ^{31}P NMR [Webb and Nuccitelli 82]
	^{31}P NMR[6]	$(7.42 \pm 0.04$	$(n=6))$ $(n=6))$	
	h) ISE[5]	$(7.69 \pm 0.05$	$(n=5))$	
	^{31}P NMR[6]	$(7.66 \pm 0.06$	$(n=14))$ $(n=3))$	
	i) ISE[5]	$(7.70 \pm 0.10$		
	^{31}P NMR[6]	$(7.64 \pm 0.02$		
Rat liver	k) ISE[7]	$(7.18 \pm 0.02$	$(n=29))$	The low values of the microelectrodes have been discussed with regard to cell compartmentation [Cohen et al. 82b]
	^{31}P NMR	$(7.25 \pm 0.02$	$(n=7))$	
	DMO	$(7.20 \pm 0.03$	$(n=9))$	

a) $pH_{extra} = 7.20$

b) $pH_{extra} = 7.80$

c) $pH_{extra} = 8.20$

d) $pH_{extra} = 7.50$; 3–5 h of equilibration

e) $pH_{extra} = 7.70$

f) $pH_{extra} = 6.0$

g) unfertilized egg

h) fertilized egg

i) A23187-activated egg

k) portal venous: $pH = 7.4$; $p_{CO_2} = 4.7 - 5.15$ kPa; lactate: 1.9–2.5 mM

1) protruding-tip type glass membrane microelectrode

2) ^{14}C-labelled 5,5-dimethyl-2,4-oxazolidinedione (DMO)

3) ^{14}C-labelled methylamine (MA)

4) plastic insulated antimony membrane microelectrode

5) recessed-tip type glass membrane microelectrode

6) Nicolet Magnetics NT 200 spectrometer (81 MHz)

7) double-barrelled glass membrane microelectrode

9.1.6 Concluding Remarks

After a period of about thirty years, during which solid-state and glass membrane microelectrodes were frequently used for intracellular pH studies, neutral carrier liquid membrane microelectrodes based on tri-n-dodecylamine are now increasingly taking their place. The introduction of highly selective liquid membrane microelectrodes has greatly simplified the fabrication of pH-sensitive microelectrodes and has made the detection of shorter H$^+$ activity transients possible. Small and double-barrelled H$^+$ microelectrodes are now easily accessible.

Alternative techniques for intracellular pH measurements include ^{31}P nuclear magnetic resonance, optical techniques and the distribution of isotopically labelled weak acids or bases. The most important properties of microelectrodes are their ability to simultaneously measure the electrical properties and absolute pH of a cell, and to make local and continuous measurements in single cells.

9.2 Intracellular Measurements of Li$^+$

Typical physiological Li$^+$ activities are in the micromolar range. Potentiometric sensors exhibiting such low detection limits in an extra- or intracellular environment are not yet available. Millimolar concentrations of lithium salts have shown therapeutic results in the treatment of manic-depressive diseases [Ehrlich and Diamond 80]. The importance of clinical administrations of Li$^+$ has given rise to many investigations on the physiological and biochemical role of Li$^+$ ions. On the cellular level, and especially in the nervous system, interest is focussed on the transport of Li$^+$ across cell membranes. The mechanisms of the therapeutic action of Li$^+$ ions are still unclear partly due to difficulties in the determination of Li$^+$ in cells. Techniques such as atomic absorption spectrometry [Janka et al. 80] and flame photometry [Gorkin and Richelson 79] have been applied to determine the total concentration of Li$^+$, but they have only allowed limited physiological studies. The development of a Li$^+$-selective liquid membrane microelectrode based on a neutral carrier has opened up new possibilities for the elucidation of the mechanisms of the therapeutic effects of Li$^+$ ions.

To date, only one type of Li$^+$ microelectrode has been proposed [Thomas et al. 75]. The membrane solution is based on the Li$^+$ carrier ETH 149 (1 in Fig. 9.12). Since 1975, several other synthetic Li$^+$ carriers have been developed (Fig. 9.12). The Li$^+$/Na$^+$ and Li$^+$/K$^+$ selectivities are crucial parameters with regards to the reliability of extra- and intracellular measurements (see Sect. 7.8). In comparison to the microelectrode based on ETH 149, these selectivity factors were only slightly improved by using other carriers. The 14-crown-4 derivative 7 [Kitazawa et al. 85] and the diamide 8 [Metzger et al. 84] exhibit the best rejection of Na$^+$ and K$^+$ ions. Microelectrodes containing these carriers have not yet been developed.

Figure 9.13 shows the Li$^+$, Na$^+$, and Ca^{2+} electrode functions of a solvent polymeric membrane based on the carrier ETH 149 (Güggi et al. 75). The selectivities of the corresponding microelectrodes are compiled in Table 9.6. [Thomas et al. 75;

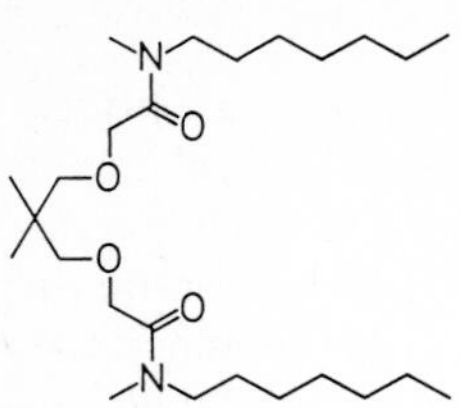

1 (ETH 149) 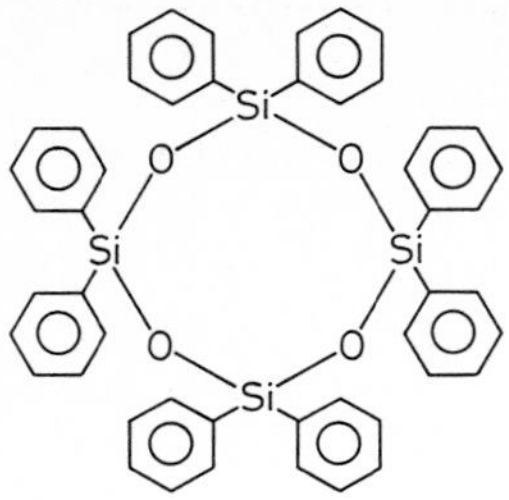   2 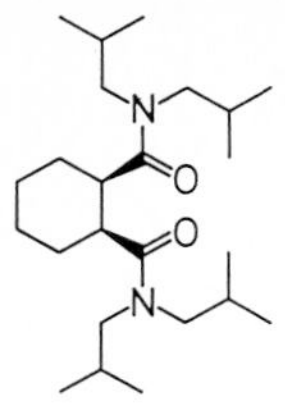   3 (ETH 1644)

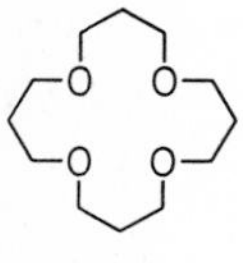

4 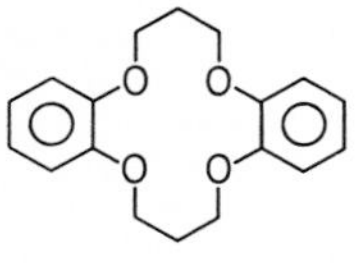   5 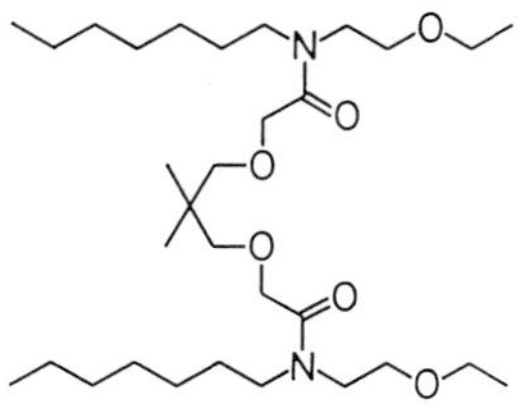   6

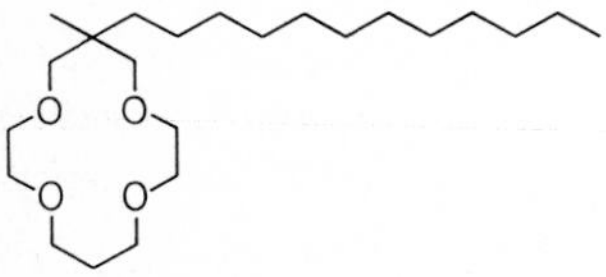

7 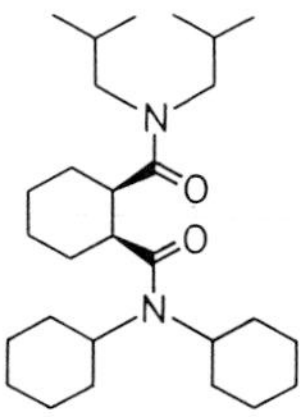   8 (ETH 1810)

Fig. 9.12. Constitutions of Li$^+$-selective carriers (1 [Güggi et al. 75; Thomas et al. 75]; 2 [Schindler et al. 78]; 3 [Zhukov et al. 81; Erne et al. 82]; 4 [Aalmo and Krane 82]; 5 [Olsher 82]; 6 [Shanzer et al. 83; Gadzekpo et al. 85]; 7 [Kitazawa et al. 84; Kitazawa et al. 85]; 8 [Metzger et al. 84])

Oehme 77]. The selectivity of Li$^+$ over Na$^+$ (factor of 25) and over K$^+$ (factor of 200) allows the measurement of Li$^+$ activities in the therapeutic millimolar range. However, adequate calibrations are necessary [Thomas et al. 75; Grafe et al. 82b]. Figure 9.14 illustrates the considerable interference caused by Na$^+$ and K$^+$ ions under typical intracellular conditions and a therapeutic level of Li$^+$ ions. The slope of the Li$^+$ electrode function is reduced to about 35 mV. The central point of the curves (see Fig. 9.14) is only 0.2 log units above the detection limit. As a result, the sensitivity is reduced: i.e. 1 mV is equivalent to a 6.8% change in Li$^+$ activity.

In spite of the poor selectivity, the microelectrodes have already been used in several studies [Thomas et al. 75; Ullrich et al. 80; ten Bruggencate et al. 81; Grafe et

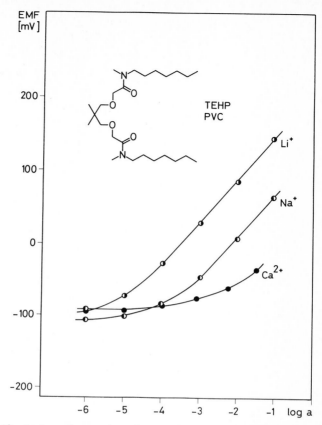

Fig. 9.13. EMF response of a Li⁺-selective solvent polymeric membrane electrode based on the neutral carrier ETH 149 to aqueous solutions of Li⁺, Na⁺ and Ca²⁺ (chloride salts) [Güggi et al. 75]. TEHP: tris(2-ethyl-hexyl) phosphate

Table 9.6. Selectivity factors, log K_{LiM}^{Pot}, of a liquid membrane microelectrode based on the neutral carrier ETH 149[a] [Thomas et al. 75; Oehme 77]

M^{z+}	log $K_{LiM}^{Pot b}$	M^{z+}	log $K_{LiM}^{Pot b}$
H^+	-0.5	Cs^+	-0.9
Li^+	0	Mg^{2+}	-2.7
Na^+	-1.4	Ca^{2+}	-0.6
K^+	-2.3	Sr^{2+}	-1.4
Rb^+	-2.2	Ba^{2+}	-1.7

[a] composition of the membrane solution: 9.7 wt.-% Li⁺ carrier ETH 149, 4.8 wt.-% sodium tetraphenylborate, 85.5 wt.-% tris (2-ethyl-hexyl) phosphate
[b] separate solution method, 0.1 M metal chloride solutions (see Sect. 5.5)

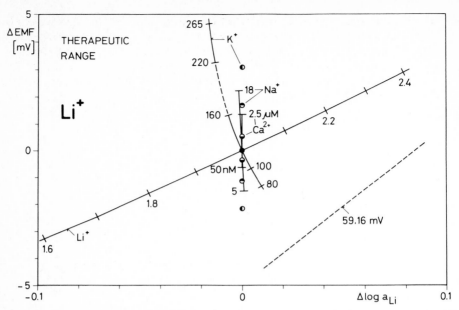

Fig. 9.14. The influence of Na$^+$, K$^+$ and Ca^{2+} on the Li$^+$ electrode function in the millimolar concentration range of Li$^+$. Details of the evaluation procedure are given in Sect. 7.8. The curves were calculated using the selectivity factors given in Table 9.6. The symbols ◑, ◐, ◒ indicate the shifts of the central point due to a change in log K^{Pot}_{LiM} of ± 0.2 for M = Na$^+$, K$^+$ and Ca^{2+}, respectively. Central point: typical therapeutic concentration of Li$^+$ (2.0 mM) and typical intracellular concentrations of the interfering ions (see Tab. 7.3)

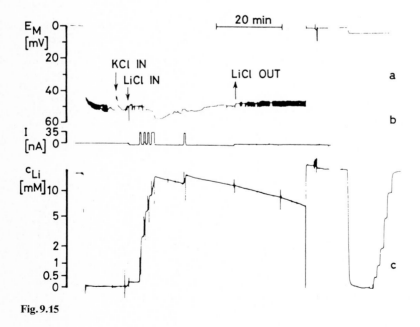

Fig. 9.15

al. 82b]. Typical applications are illustrated in Figs. 9.15. and 9.16. [Thomas et al. 75]. Both figures contain external calibrations at typical intracellular backgrounds (see also [Grafe et al. 82b]). The response of the Li$^+$ microelectrode towards changes in the intracellular activity of Li$^+$ produced by ionophoretic injections is given in Fig. 9.15. The experiment guarantees a response of the intracellular microelectrode towards Li$^+$ ions. Figure 9.16 also shows the uptake of Li$^+$ ions into the cell due to elevated extracellular concentrations of Li$^+$. Using Li$^+$-loaded cells, it was demonstrated that Li$^+$ ions are actively transported out of nerve cells.

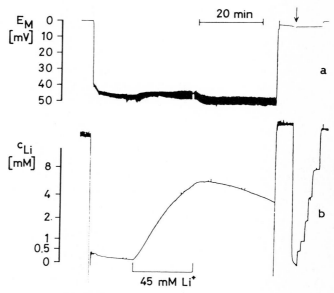

Fig. 9.16a, b. Li$^+$ uptake by a neuron exposed to 45 mM Li$^+$ [Thomas et al. 75]. **a:** membrane potential. **b:** potential of the Li$^+$ microelectrode. Where indicated, the normal Ringer solution was changed to one in which half of the Na$^+$ was replaced by Li$^+$. After the withdrawal of the microelectrodes the Li$^+$ microelectrode was calibrated. The six calibration solutions contained the same concentration of NaCl (2 mM), KCl (90 mM) and TRIS/maleate (10 mM), but different concentrations of LiCl (0, 0.5, 1, 2, 4 and 8 mM). Redrawn from Thomas, Simon and Oehme, Nature, 1975 [Thomas et al. 75]

Fig. 9.15a–c. Comparison of the response of a Li$^+$-selective microelectrode to increasing intracellular concentrations of Li$^+$ in a neuron with that of calibration solutions [Thomas et al. 75]. **a:** potential difference between the intracellular reference microelectrode and a reference electrode in the bath. **b:** ionophoretic injection current. **c:** potential difference between the reference microelectrode and the Li$^+$ microelectrode. The Li$^+$ microelectrode was inserted first, then the reference microelectrode and finally two current-passing microelectrodes, one filled with KCl and the other with LiCl. At the end of the experiment the microelectrodes were withdrawn from the cell and the Li$^+$ microelectrode was calibrated by superfusing the preparation with six calibration solutions. All of them contained the same levels of NaCl (4 mM), KCl (90 mM) and TRIS/maleate (10 mM, pH 7.5), but different levels of LiCl (0, 0.5, 1, 2, 5 and 10 mM). Redrawn from Thomas, Simon and Oehme, Nature, 1975 [Thomas et al. 75]

9.3 Intracellular Measurements of Na$^+$

9.3.1 Techniques Used in Studies of Intracellular Na$^+$

Early attempts to measure Na$^+$ in collected biological fluids relied on the use of flame photometry [Bureš and Křivánek 60]. At the same time, ^{24}Na was used as a radioactive label for studies of ion fluxes in brain tissues [Bureš and Křivánek 60; Brinley 63]. The two techniques lacked both spatial and temporal resolution. On the other hand, electron-probe X-ray microanalysis has become a very useful technique for the determination of the elemental composition within single cells or subcellular compartments. In this technique the cell sample is subjected to a beam of electrons, which generate X-rays of a specific energy (depending on the atom present). As in NMR spectroscopy, the element itself is the probe. A spatial resolution of 1 μm or less is achievable. This allows a quantitative mapping (errors of about 10%) of the intracellular distribution of a certain element [Gupta and Hall 79; Galvan et al. 84]. The major difficulties of the method are the sample preparation (for example, the preservation of the cellular distribution of the elements in solution) and the quantification of the results [Dörge et al.78]. Cryo-ultramicrotomic procedures allow a tissue preparation under conditions where the temperature never rises above $-80\,^\circ$C. This probably maintains the initial distribution of the diffusible ions during the preparation.

The gelatin reference-phase analysis method was developed for the quantitative determination of free and bound Na$^+$ in the cytoplasm [Horowitz et al. 79]. The method is based on the introduction of a gelatine reference-phase (about 2% of the cell volume) into the cytoplasm. The gelatine forms droplets that are separated by trapped amounts of interstitial cytoplasm. After an adequate equilibration time, which allows the dissolved small species to diffuse between the cytoplasm and reference phase, the cells are quickly frozen in liquid nitrogen. Individual cells are then dissected at low temperatures (about $-40\,^\circ$C). Samples are removed from the reference phase and the cytoplasm and are analyzed by atomic absorption spectrometry. Solutes were shown to distribute themselves uniformly between an electrolyte solution and the water of the gelatine gel. The distribution of ions is not influenced by the presence of ion-exchange sites or Donnan potentials. Diffusion equilibrium is usually attained after about 15 minutes. However, the technique is restricted to very large cells and is not suited for the study of dynamic processes.

By 1965 ^{23}Na NMR had already been suggested as a noninvasive technique for the study of sodium in tissue samples [Cope 65]. A direct observation of ^{23}Na signals from free and bound Na$^+$ within the multiple extra- and intracellular compartments is extremely difficult [Yeh et al. 73]. However, refined ^{23}Na NMR analyses have yielded physiologically relevant results [Civan 83]. In recent approaches, anionic complexes of dysprosium (III) have been added as paramagnetic shift reagents to the extracellular fluid. The splitting of the extracellular signals then allows a differentiation from the intracellular ^{23}Na signals [Gupta and Gupta 82; Balschi et al. 82].

Many difficulties have been overcome by the local and continuous activity measurements of Na$^+$ with microelectrodes. Indeed, potentiometry with microelectrodes is now the most widely used technique for the study of the physiological role of Na$^+$ ions.

9.3.2 Survey of Na$^+$-Selective Microelectrodes

The chronological survey of the course of the development of Na$^+$-selective microelectrodes shown in Table 9.7 clearly demonstrates that the interest became focussed on liquid membrane microelectrodes after a period of intensive work on Na$^+$ glass membrane microelectrodes. The development of Na$^+$-selective glass membrane microelectrodes began in 1959 and reached an optimal stage with the introduction of single- and double-barrelled recessed-tip microelectrodes. Although

Table 9.7. Course of the development of Na$^+$-selective microelectrodes

Membrane	Development (for components see Fig. 9.17)	Year, Authors, Reference
glass	protruding-tip	1959, Hinke, [Hinke 59]
glass	reversed-tip	1969, Thomas, [Thomas 69]
glass	recessed-tip	1970, Thomas, [Thomas 70]
glass	double-barrelled recessed-tip	1976, Zeuthen, [Zeuthen 76 b]
liquid	monensin (free acid) as Na$^+$ carrier	1976, Kraig, Nicholson, [Kraig and Nicholson 76]
liquid	potassium tetrakis(p-chloro-phenyl) borate as ion-exchanger salt in tris(2-ethyl-hexyl) phosphate	1977, Palmer, Civan, [Palmer and Civan 77]
liquid	monensin (free acid) as Na$^+$ carrier, dissolved in K$^+$ ion-exchanger solution	1979, Kotera, Satake, Honda, Fujimoto, [Kotera et al. 79]
liquid	synthetic Na$^+$ carrier (ETH 227)	1979, Steiner, Oehme, Ammann, Simon, [Steiner et al. 79]
liquid	synthetic Na$^+$ carrier (ETH 227), 3-nitro-o-xylene as solvent	1979, O'Doherty, Garcia-Diaz, Armstrong, [O'Doherty et al. 79]
liquid	synthetic Na$^+$ carrier (ETH 227), addition of K$^+$ ion-exchanger solution	1980, Wills, Lewis, [Wills and Lewis 80]
liquid	synthetic Na$^+$ carrier (ETH 157)	1985, Ammann, Anker, [Ammann and Anker 85]

some of the technical difficulties encountered in the preparation of glass microelec-trodes have been overcome by subtle construction techniques [Thomas 76; Thomas 78], problems involving the volume of the recess, the risk of tip blockage and the rel-atively slow electrode response have not yet been solved. The construction of very small microelectrodes (tip diameters well below 1 µm) and of double- and multibar-relled microelectrodes is extremely difficult. Liquid membrane microelectrodes are less affected by these drawbacks and, in addition, their preparation is much easier and less limited. Although none of the liquid membrane Na^+ microelectrodes de-scribed to date exhibit Na^+ selectivities as good as the NAS_{11-18} glass membrane microelectrodes, the neutral carrier-based microelectrodes have now almost com-pletely replaced the Na^+ glass membrane microelectrodes.

9.3.3 Available Liquid Membrane Microelectrodes

To date, Na^+-selective microelectrodes based on either a classical ion-exchanger salt (1, Fig. 9.17), an electrically charged carrier (4) or neutral carriers (5, 6) have been proposed. Only recently have Na^+-selective solvent polymeric membranes, containing a bis-crown compound (7) [Shono et al. 82] or a hemispherand (8) [Toner et al. 84], been presented.

9.3.3.1 Microelectrodes Based on Potassium Tetrakis-(p-chlorophenyl) borate (1)

The potassium salt of the lipophilic anion tetrakis(p-chlorophenyl) borate is nor-mally used as a component in classical cation-exchanger membrane microelec-trodes for the determination of K^+ (see Sect. 9.4.2). These microelectrodes show a preference for K^+ over Na^+ by a factor of about 70. It is known, however, that the selectivity of ion-exchanger membranes can be influenced to a certain extent by the membrane solvent [Baum and Lynn 73; Palmer and Civan 77]. PVC membranes plasticized with tris(2-ethyl-hexyl) phosphate (2 in Fig. 9.17) extract alkali ions with a selectivity sequence, $Li^+ > Na^+ > K^+$. Indeed, a mixture of 1.5% 1 in 2 yields liq-uid membrane microelectrodes with a preference of Na^+ over K^+ by a factor of about 4 [Palmer and Civan 77; Palmer et al. 78]. A mixture which also contains tri-n-octyl-phosphine oxide (3 in Fig. 9.17) shows a similar preference of Na^+ over K^+ [Cohen and Fozzard 79; Cohen et al. 82a]. For most applications these ion-ex-changer Na^+ microelectrodes exhibit an insufficient Na^+/K^+ selectivity.

9.3.3.2 Microelectrodes Based on the Antibiotic Monensin (4)

In 1970 it was shown that liquid membranes containing monensin (4 in Fig. 9.17) ex-hibit a preference for Na^+ over all of the other alkali ions [Lutz et al. 70]. Six years later, corresponding microelectrodes with a Na^+/K^+ selectivity of about 15 were described [Kraig and Nicholson 76]. Ca^{2+} and Mg^{2+} are rejected by factors of 140 and 7, respectively (see Table 9.8). The Na^+ response is insensitive to changes in pH

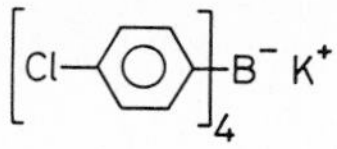

1

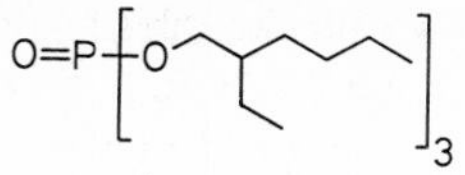

2

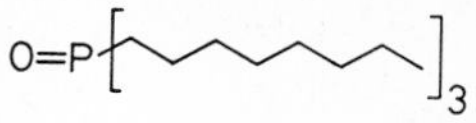

3

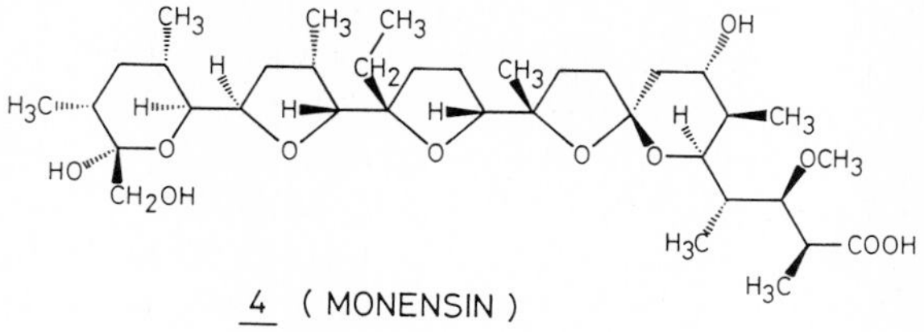

4 (MONENSIN)

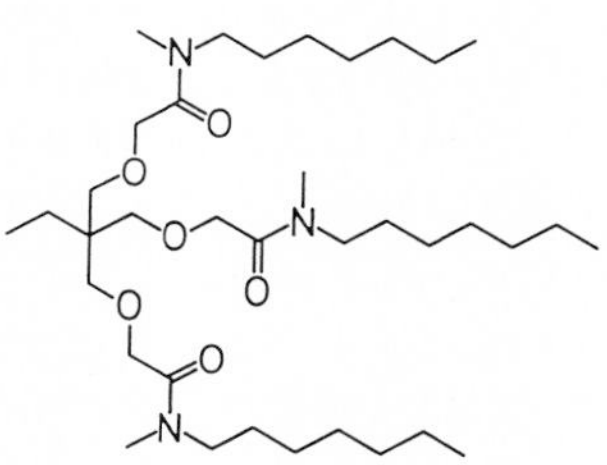

5 (ETH 227)

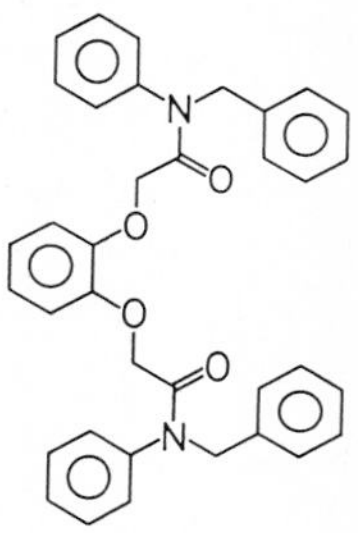

6 (ETH 157)

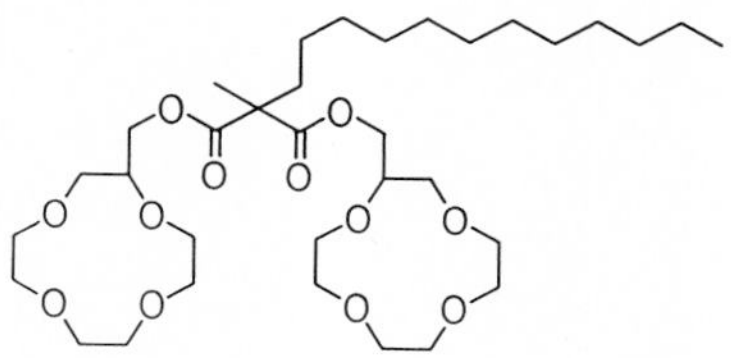

7

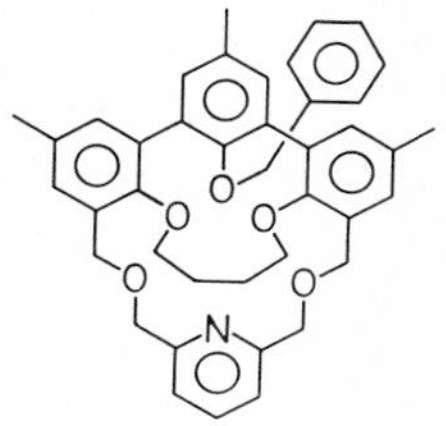

8

Fig. 9.17. Components for Na$^+$-selective liquid membrane electrodes

in the pH range 5 to 9. Response times obtained from concentration steps in aqueous solutions are around 1 s [Kraig and Nicholson 76]. Although the Na$^+$/K$^+$ selectivity is relatively poor, monensin-based microelectrodes have nevertheless found an application in intracellular measurements. The microelectrodes possess interesting properties for extracellular Na$^+$ studies (see Sect. 9.3.3.4). Modifications of the original composition of the membrane solution, for example 10% monensin dissolved in a Corning K$^+$-exchanger solution, show similar characteristics (Na$^+$/K$^+$ selectivity of 12) [Fujimoto and Honda 80].

9.3.3.3 Microelectrodes Based on the Neutral Carrier ETH 227 (5)

The synthetic Na$^+$ carrier ETH 227 (5 in Fig. 9.17) induces a Na$^+$/K$^+$ selectivity as high as 250 in solvent polymeric membranes [Güggi et al. 76]. A membrane solution has been developed which exhibits a Na$^+$/K$^+$ selectivity of about 200 [Steiner et al. 79]. It has been modified by using another membrane solvent [O'Doherty et al. 79] or by adding Corning K$^+$-exchanger solution [Wills and Lewis 80; Hansen and Zeuthen 81]. The three slightly different membrane solutions yield microelectrodes with almost the same performance characteristics. The modified membrane solutions show a reduced electrical resistance and, therefore, a slightly shorter electrode response time [Hansen and Zeuthen 81]. Although the Na$^+$/K$^+$ selectivity is still not as high as that of glass membrane microelectrodes, the neutral carrier microelectrodes can be used for most intracellular studies. Initially, two membrane solutions based on ETH 227 were proposed: with and without sodium tetraphenylborate. Membrane solutions containing the salt show a lower resistance but poorer Na$^+$/Ca^{2+} selectivity than those without the salt. Most of the physiological studies published to date have been carried out with membrane solutions containing this additive. The characteristics given in the following sections therefore refer mainly to this type of membrane solution.

Electrode Function and Selectivity Factors. Figure 9.18 shows the electrode functions obtained in pure aqueous NaCl and KCl solutions. A linear regression in the NaCl concentration range from 10^{-3} to 10^{-1}M yields (after correction for liquid-junction potentials) a slope of 53.0 ± 2.5 mV (std. dev., n=3). Calibration curves obtained in the presence of a typical intracellular background exhibit reduced slopes in the

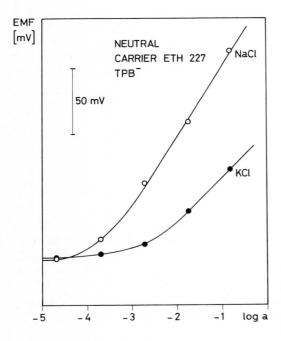

Fig. 9.18. EMF response of a Na$^+$ microelectrode (ETH 227 with NaTPB) to aqueous NaCl and KCl solutions

physiological activity range of Na$^+$. A direct comparison of reported slopes and detection limits is difficult since different ionic backgrounds have been employed. However, the data appear to be very consistent [Sheu and Fozzard 82; Cohen et al. 82a; Schümperli et al. 82; Dagostino and Lee 82; Wang et al. 83; Oberleithner et al. 83a; Deitmer and Schlue 83; Lang et al. 84; Giraldez 84; Greger and Schlatter 84]. The measured and calculated Na$^+$ electrode functions in the presence of interfering ions are shown in Fig. 9.19. The monensin-based microelectrode is compared with microelectrodes containing ETH 227 (with or without NaTPB). Using the Nicolsky-Eisenman formalism and the selectivity factors of Tab. 9.8 the electrode functions of Fig. 9.19 were calculated. Assuming a representative intracellular fluid containing 200 mM K$^+$, 2.0 mM Mg^{2+} and 0.01 mM Ca^{2+} and varying amounts of Na$^+$, it can be shown that the neutral carrier electrode possesses a detection limit (indicated by

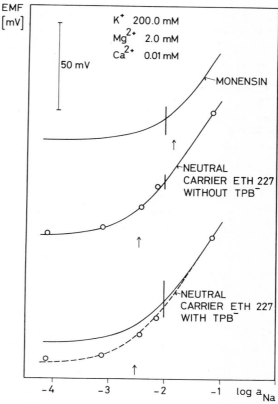

Fig. 9.19. EMF response of microelectrodes to various Na$^+$ activities at a constant ionic background. *Solid curves:* calculated response using the selectivity data for K$^+$, Mg^{2+} and Ca^{2+} given in Tab. 9.8 and the measured slopes of the electrode functions (ETH 227: see Fig. 9.18; monensin: see [Kraig and Nicholson 76]). *Dashed curve:* calculated response using the selectivity data for Mg^{2+} and Ca^{2+} given in Tab. 9.8 and a Na$^+$/K$^+$ selectivity of 200 (found in a best fit analysis of the experimental values (dots)). Arrows indicate the detection limit and vertical bars represent a typical intracellular Na$^+$ activity

Table 9.8. Selectivity factors, log K_{NaM}^{Pot}, for different Na^+ microelectrodes

Membrane solution[a] [wt.-%]	Method[b]	log K_{NaM}^{Pot}; M^{z+} =								
		Li^+	K^+	Rb^+	Cs^+	Mg^{2+}	Ca^{2+}	Sr^{2+}	Ba^{2+}	H^+
10% ETH 227 90% o-NPOE	SSM 0.1 M	0.4	−1.7	−2.0	−2.2	−2.7	−0.7	−1.1	−1.3	−0.9
10% ETH 227 0.5% NaTPB 89.5% o-NPOE	SSM 0.1 M	0.4	−1.7 −2.3[c]	−1.6	−1.8	−2.4	0.2	−0.2	−0.4	−0.8
10% monensin 90% nitrobenzene	FIM 1 M		−1.1			−0.8	−2.2			−1.1
10% ETH 157 0.5% NaTPB 89.5% o-NPOE	SSM 0.1 M	−1.7	−0.4 −0.5[c]			−3.4	−1.3			−1.2

[a] ETH 227, ETH 157, monensin: see Fig. 9.17; o-NPOE: o-nitrophenyl-n-octyl ether; NaTPB: sodium tetraphenylborate
[b] SSM: separate solution method; FIM: fixed interference method (see Sect. 5.5)
[c] values adjusted to a Na^+ electrode function in the presence of a typical intracellular (ETH 227, Fig. 9.19) or extracellular (ETH 157, Fig. 9.21) ion background

arrows) which is about an order of magnitude lower than that of the monensin-based microelectrode. An analysis of the Nicolsky-Eisenman equation indicates that the detection limit of the monensin-based sensor is mainly governed by the interference due to K^+ and Mg^{2+}, whereas the neutral carrier electrode suffers almost exclusively from an interference by K^+. In contrast to the monensin-based electrode, where the detection limit is slightly above the physiological activity of Na^+ (10^{-2} M, vertical bar in Fig. 9.19), the neutral carrier electrode makes the measurement of intracellular Na^+ activities possible. This is especially true if the intracellular concentration of K^+ becomes smaller, as is usually observed in living cells.

Figure 9.20 is helpful in a more detailed discussion of the interference due to Li^+, K^+ and Ca^{2+}. The Na^+ calibration curve was calculated using the selectivity factors of Tab. 9.8 (log K_{NaK}^{Pot} = −2.3). The computation is presented in detail in Sect. 7.8. The central point of the curve (10 mM Na^+) is about 0.8 log units above the detection limit. The slope near the central point is 51.3 mV. Hence, 1 mV is approximately equivalent to a 4.6% change in the Na^+ activity. A considerable interference due to Ca^{2+} (above about 10^{-6} M Ca^{2+}) and a strong interference due to Li^+ (at therapeutic Li^+ levels) can be determined from Fig. 9.20.

The Na^+/K^+ and Na^+/Ca^{2+} selectivities have been discussed in several contributions. Selectivity factors, K_{NaK}^{Pot}, of 0.01–0.02 [Dagostino and Lee 82], 0.025 [Grafe et al. 82a], 0.02 [Cohen et al. 82a], 0.024 [Sheu and Fozzard 82], 0.01 [Oberleithner et al. 83a], 0.015 [Lang et al. 84], 0.02 [Harvey and Kernan 84a], 0.02–0.03 [Giraldez 84], 0.009 [Donaldson and Leader 84] and 0.036 [Greger and Schlatter 84] have been measured. The reported K_{NaCa}^{Pot} values are also in good agreement: 2–4 [Grafe et al. 82a], 2.5 [Sheu and Fozzard 82], 5.6 [Bers and Ellis 82], 12 [Lang et al. 84], 2 [Dagostino and Lee 82] and 0.2 [Harvey and Kernan 84].

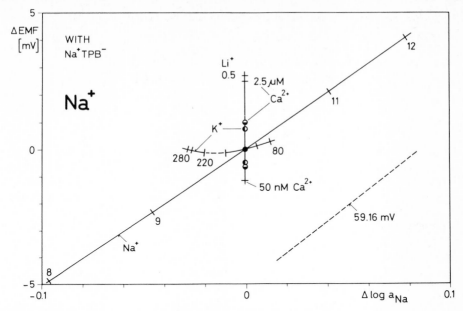

Fig. 9.20. Influence of Li⁺, K⁺ and Ca²⁺ on the electrode function of a Na⁺-selective microelectrode based on ETH 227 and NaTPB. Details of the calculations are given in Sect. 7.8. The curves were computed using the selectivity factors of Tab. 9.8. The symbols ◓ and ◑ indicate shifts in the central point for an assumed change in log K_{NaCa}^{Pot} and log K_{NaK}^{Pot} of ± 0.2

For most intracellular studies of Na⁺, the interference due to K⁺ becomes negligible if Na⁺ calibration solutions with a representative background of K⁺ are used. The interference of intracellular Ca²⁺ has been discussed [Bers and Ellis 82; Dagostino and Lee 82] (see also [Grafe et al. 82 a; Chapman et al. 83; Wang et al. 83]). Since the Na⁺ microelectrode exhibits a slight preference of Ca²⁺ over Na⁺, Ca²⁺ activities over 10^{-6} M contribute significantly to the measured EMF. The disturbances caused by the following concentrations of Ca²⁺ have been studied: 2.5 mM (5–11 mV) [Aickin 84 b]; 1.8 mM (<15 mV) [Glitsch et al. 82]; 1.8 mM (6.6 mV) [Glitsch and Pusch 84 b]; 2 mM (5–10 mV) [Giraldez 84]; 1.3 mM (0.7 mV) [Greger and Schlatter 84]; and 1.8 mM (8 mV) [Dagostino and Lee 82]. If the selectivity factors do not change during an experiment, corrections using the Nicolsky-Eisenman equation can be made [Lee and Dagostino 82; Deitmer and Schlue 83]. Constant selectivity factors have been observed during experiments of 10–15 hours duration [Lee and Dagostino 82].

Whereas Na⁺-selective glass membrane microelectrodes exhibit a preference of H⁺ over Na⁺ by a factor of about 1000, the neutral carrier microelectrode rejects H⁺ by a factor of about 10. Therefore, changes in the pH of a sample solution in the pH range 2 to 10 do not influence the EMF if the concentration of Na⁺ is kept constant at either 10^{-1} M (± 1 mV (2 < pH < 10)) or 10^{-2} M (± 5 mV (2 < pH < 10)) [Steiner et al. 79]. Obviously, changes in pH in the physiological range will not interfere with a determination of Na⁺.

Only very little information is available on the interference due to other substances. Inhibitors such as ouabain [Deitmer and Schlue 83] or strophanthidin [Lee and Dagostino 82; Chapman et al. 83] as well as physiologically active concentrations of catecholamines (e.g. $2 \cdot 10^{-7}$ to 10^{-6} M isoproterenol or $2 \cdot 10^{-7}$ to $4 \cdot 10^{-7}$ M norepinephrine) [Wasserstrom et al. 82] do not influence the electrode response. Acetylcholine is rejected by a factor of about 100 [Steiner et al. 79]. Concentrations of tetramethylammonium ions as high as 0.15 M do not disturb the Na^+ response [Chapman et al. 83].

Response Time. A response time (t_{90}) of about 5 s is observed when 1 ml of 1 M NaCl solution is added to 50 ml of 10^{-3} M NaCl solution [Steiner et al. 79]. Since the rate of mixing of the NaCl solutions is relatively slow, the intrinsic response time of the neutral carrier microelectrode is expected to be considerably shorter than 5 s. A comparison with values from the literature is difficult since the response times are defined differently and depend on many parameters (Sect. 5.9). Nevertheless, similar or shorter response times have been claimed: 8 s (tip diameter < 1 µm; resistance $\sim 5 \cdot 10^{10} \,\Omega$ [Lee and Dagostino 82]); < 1 s [Cohen et al. 82a]; 10–20 s [Dagostino and Lee 82]; < 5 s (shank painted with conductive silver) [Dagostino and Lee 82]; 1 s ($\sim 10^{11} \,\Omega$ [Oberleithner et al. 83a]); 0.2–0.5 s (< 1 µm, double-barrelled [Deitmer and Schlue 83]); 13 s (< 1 µm, bevelled [Chapman et al. 83]); 10–15 s ($5 \cdot 10^{11} \,\Omega$ [Greger and Schlatter 84]); and 0.2 s [Harvey and Kernan 84a]. For the modified membranes based on ETH 227 (Tab. 9.7) response times of 10 s ($2 \cdot 10^9 \,\Omega$ [O'Doherty et al. 79]), 1 s (double-barrelled [Hansen and Zeuthen 81]), and < 1 s (0.3 µm, double-barrelled [Zeuthen 82]) have been observed.

Electrical Membrane Resistance. Reported values are: $10^{10} \,\Omega$ (2 µm, double-barrelled [Steiner et al. 79]); $5 \cdot 10^{10} \,\Omega$ (< 1 µm [Lee and Dagostino 82]); $10^{10} \,\Omega$ (0.5–1.0 µm, double-barrelled [Grafe et al. 82a]); $7 \cdot 10^{10} \,\Omega$ (< 1 µm [Dagostino and Lee 82]); $\sim 10^9 \,\Omega$ [Shabunova and Vyskočil 82]; $\sim 10^{11} \,\Omega$ [Oberleithner et al. 83a]; $\sim 10^{10} \,\Omega$ [Harvey and Kernan 84a]; and $5 \cdot 10^{11} \,\Omega$ [Greger and Schlatter 84].

EMF Stability. The drift in the EMF of a typical neutral carrier microelectrode was found to be less than 1 mV/day [Steiner et al. 79]. The stability has been confirmed in similar drifts during periods of at least one week [Sheu and Fozzard 82] and in the measurement of reproducible calibration curves during a period of more than one week [Schümperli et al. 82].

Lifetime. The lifetime of a microelectrode in continuous contact with an aqueous electrolyte is several weeks [Steiner et al. 79]. Hence, it can be assumed that the performance of neutral carrier microelectrodes does not change during physiological studies. Recalibrations after the experiments have confirmed these observations.

Shunt Artifacts. It has been shown that the high-impedance neutral carrier Na^+ microelectrodes may exhibit disturbing electrical shunts across the glass wall at the tip of the micropipette [Lewis and Wills 80]. These shunts may be responsible for the high values of the intracellular Na^+ activities [Lewis and Wills 80]. The elimination of shunts and the simultaneous improvement of the response time can be achieved by shielding the microelectrodes (see Sect. 6.8).

Extracellular Measurements of Na$^+$. Due to their poor Na$^+$/Ca^{2+} selectivity, membrane solutions based on ETH 227 cannot be recommended for extracellular measurements of Na$^+$ activities unless Ca^{2+} activities are simultaneously measured and corrections for Ca^{2+} interference (using the Nicolsky-Eisenman equation) are carried out. However, a recently described membrane solution based on the synthetic Na$^+$ carrier ETH 157 is much more suited for extracellular studies of Na$^+$ (see below).

9.3.3.4 Microelectrodes Based on the Neutral Carrier ETH 157 (6)

Synthetic electrically neutral carriers which exhibit much higher Na$^+$/Ca^{2+} selectivities than ETH 227 and at the same time sufficient Na$^+$/K$^+$ selectivities for extracellular studies have been available for quite a long time [Ammann et al. 74; Ammann et al. 76; Ammann et al. 83]. Only recently, however, has a microelectrode based on a membrane solution containing the Na$^+$ carrier ETH 157 (6 in Fig. 9.17) been developed [Ammann and Anker 85]. Indeed, microelectrodes based on the carrier ETH 157 exhibit improved Na$^+$/Ca^{2+} selectivities in membranes compared to microelectrodes based on ETH 227 (Tab. 9.8). Since the extracellular Na$^+$ activity

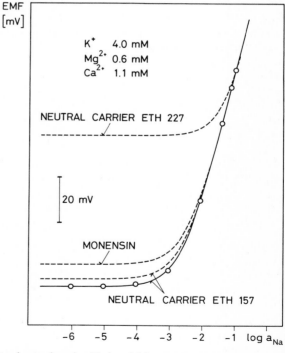

Fig. 9.21. EMF response of microelectrodes to changing Na$^+$ activities at a constant extracellular ionic background. *Dashed curves:* calculated response using the selectivity factors of Tab. 9.8. *Dots:* experimental values. *Solid curve:* calculated response using the selectivity data of Table 9.8 and a value of log K_{NaK}^{Pot} equal to -0.5 to best fit the experimental results

is considerably higher than the K^+ activity, the relatively poor Na^+/K^+ selectivity is sufficient for extracellular studies. On the other hand, intracellular measurements with microelectrodes containing ETH 157 are not feasible due to interference by K^+. The charged antibiotic monensin yields membrane solutions with better Na^+/K^+ and Na^+/Ca^{2+} selectivities than those containing ETH 157 (Tab. 9.8). Indeed, monensin-based microelectrodes have been used in extracellular fluids [Fujimoto et al. 80; Nicholson and Kraig 81]. However, as shown in Fig. 9.21, the microelectrodes based on ETH 157 show a lower detection limit in Na^+ solutions containing a typical extracellular background of K^+, Mg^{2+} and Ca^{2+} activities. Analyses of the response curves have shown that the detection limit of the monensin-based microelectrode is mainly governed by the interference due to Mg^{2+}, whereas K^+ and Ca^{2+} are the major interfering ions for microelectrodes based on ETH 157.

The 90% response time of the microelectrode, obtained by adding 1.5 ml of a 10^{-1} M NaCl solution with a typical extracellular ionic background (see Fig. 9.21) to 5 ml of a 10^{-2} M NaCl solution with the same ionic background, is about 3 s. Since the rate of mixing of the sample solution after the stepwise addition of the NaCl solution is rather slow, the intrinsic response time of the microelectrode could be considerably less.

Microelectrodes with tip diameters of about 0.7 μm exhibit electrical membrane resistances of $3 \cdot 10^{10}$ Ω. The drift of the microelectrode in 140 mM NaCl solutions with an extracellular ionic background is less than 0.2 mV/3 h [Ammann and Anker 85].

The first successful applications of this microelectrode to the diluting segment of frog kidney [Oberleithner et al. 84a] and to the extracellular space of slices of drone retina [Orkand et al. 84] confirm the reliability of this microelectrode. The latter study was performed with triple-barrelled microelectrodes (Na^+, Ca^{2+}, resting potential).

9.3.4 Concluding Remarks

Due to the performance of neutral carrier-based Na^+ microelectrodes and the simplicity of the construction of liquid membrane microelectrodes, neutral carrier-based Na^+ microelectrodes are now predominantly used in physiological studies of Na^+. Different membrane solutions have to be considered for intra- or extracellular studies. A membrane solution based on the Na^+ carrier ETH 227 enables intracellular determinations to be made, while a membrane solution containing ETH 157 is suitable for extracellular measurements of Na^+ activity. If adequate calibration procedures are applied, both types of microelectrodes allow a direct potentiometric measurement of Na^+ activity without significant interference from other ions. Na^+-selective carriers with high Na^+/K^+ and Na^+/Ca^{2+} selectivities, which would be suitable for both extra- and intracellular measurements, have yet to be developed.

9.4 Intracellular Measurements of K$^+$

The same analysis techniques are usually employed for the determination of Na$^+$ and K$^+$. The methods described in Sect. 9.3 (flame photometry, radioactive labels, gelatine reference phase, electron-probe X-ray microanalysis, and nuclear magnetic resonance) for physiological measurements of Na$^+$ have also been used for measurements of K$^+$ at the cellular level. The introduction of K$^+$-selective microelectrodes was an important step forward in electrophysiology, and at present, most of the intracellular measurements of K$^+$ activity are performed potentiometrically.

9.4.1 Survey of K$^+$-Selective Microelectrodes

A survey of the important developments in the field of K$^+$-selective microelectrodes is given in Tab. 9.9. In contrast to the H$^+$- and Na$^+$-selective types of glass, K$^+$-selective glass has never played an important role in the construction of K$^+$ microelectrodes. The first liquid membrane ion-exchanger K$^+$ microelectrodes

Table 9.9. Course of the development of K$^+$-selective microelectrodes

Membrane	Development	Year, Authors, Reference
glass	first K$^+$-selective glass membrane microelectrode, protruding-tip type	1959, Hinke, [Hinke 59]
liquid	first ion-exchanger liquid membrane microelectrode (Corning 477317)	1971, Walker, [Walker 71]
glass	paraffin wax between an outer borosilicate pipette and an inner K$^+$-selective glass membrane pipette	1974, Lee, Armstrong, [Lee and Armstrong 74]
liquid	first valinomycin-based liquid membrane microelectrode	1976, Oehme, Simon, [Oehme and Simon 76]
liquid	valinomycin-based double-barrelled microelectrode	1977, Coles, Tsacopoulos, [Coles and Tsacopoulos 77]
liquid	modified valinomycin-based liquid membrane microelectrode (lower electrical resistance)	1979, Wuhrmann, Ineichen, Riesen-Willi, Lezzi, [Wuhrmann et al. 79]
liquid	low-impedance coaxial ion-exchanger microelectrode (Corning 477317)	1979, Ujec, Keller, Machek, Pavlik, [Ujec et al. 79]
liquid	ion-selective field-effect transistor microelectrode (Corning 477317)	1980, Hämmerli, Janata, Brown [Hämmerli et al. 80]

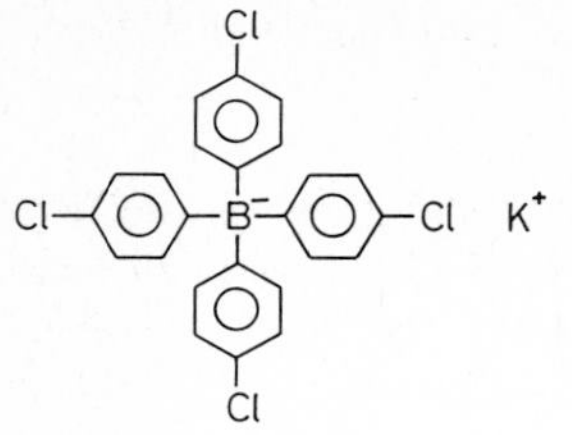

Fig. 9.22. Constitutions of the components of K^+-selective electrodes

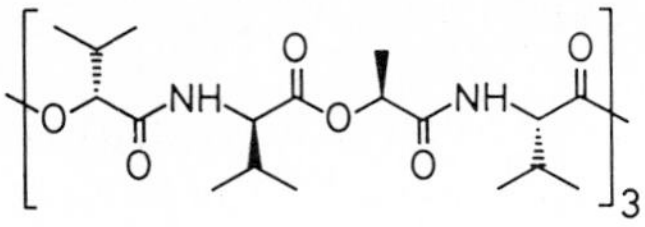

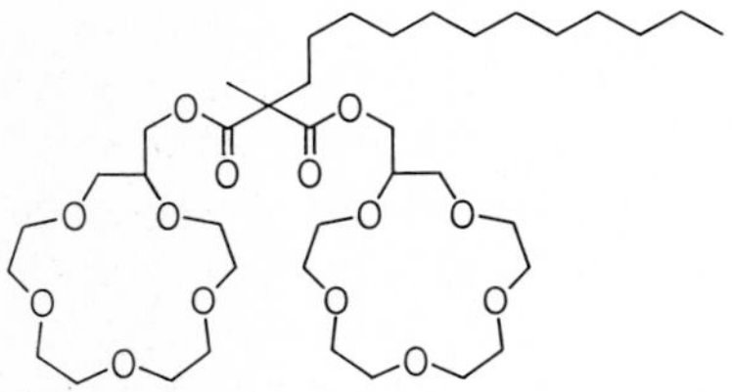

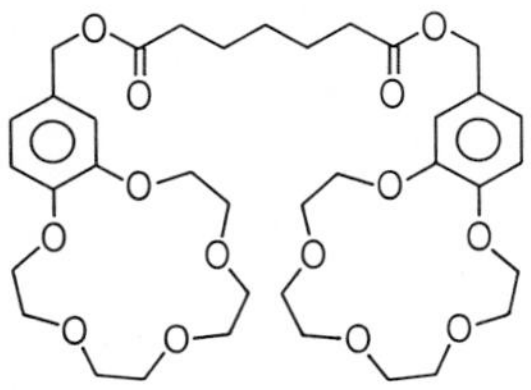

(Corning 477317) were developed about 15 years ago [Walker 71] and are still the most widely used. Using these ion-exchanger membrane solutions, various types of microelectrodes have been prepared: double-barrelled [Vyskočil and Křiž 72; Lux and Neher 73; Fujimoto and Kubota 76; Coles and Tsacopoulos 77; Hansen and Zeuthen 81], triple-barrelled [Fujimoto and Honda 80], coaxial [Ujec et al. 78; Ujec et al. 79] and ISFET [Hämmerli et al. 80].

An alternative to the ion-exchanger microelectrode of relative poor selectivity is the highly selective valinomycin-based microelectrode (Sect. 9.4.2). Unfortunately, its rather high resistance limits its use in physiological studies. Nevertheless, neutral carrier-based microelectrodes have been successfully applied in intracellular studies of K^+ (Tab. 10.1). Double- [Oehme 77; Coles and Orkand 83], triple- [Dufau et al. 82] and four-barrelled [Kessler et al. 76b] microelectrodes have been prepared. In certain studies, it was shown that the K^+ responses of ion-exchanger microelectrodes are prone to interferences and that the K^+ activities are only measurable using the neutral carrier-based microelectrodes.

Despite this limitation, classical ion-exchanger microelectrodes are preferentially applied in physiological studies because of their low resistances and short response times. Valinomycin-based microelectrodes will only find increasing applications if their electrical resistances can be further reduced. An alternative to either ion-exchanger or neutral carrier-based microelectrodes could result from the development of microelectrodes based on the highly selective crown ethers [Kimura et al. 79; Fung and Wong 80; Yamauchi et al. 82; Kimura et al. 83] (e.g. 3 and 4 in Fig. 9.22).

9.4.2 Comparison of Properties of Classical Ion-Exchanger and Neutral Carrier-Based Microelectrodes

The membrane solutions of classical ion-exchanger microelectrodes contain the lipophilic salt potassium tetrakis(p-chlorophenyl) borate (Fig. 9.22) dissolved in an organic solvent (usually 2,3-dimethyl-nitrobenzene). Membrane solutions have been made commercially available by two companies (Corning and Orion). However, there are almost no reports on measurements using the Orion ion-exchanger. Recently, a direct comparison of the two membrane solutions showed that the Orion ion-exchanger exhibits a much lower K$^+$/Na$^+$ selectivity (Corning: $K_{KNa}^{Pot} = 0.03$; Orion: $K_{KNa}^{Pot} = 0.3$ [Laming and Djamgoz 83]).

To date, two membrane solutions based on valinomycin (Fig. 9.22) have been described in the literature. Improvements in the originally proposed microelectrode [Oehme and Simon 76] which lowered the membrane resistance were suggested by Wuhrmann [Wuhrmann et al. 79]. Thus, three types of K$^+$-selective membrane solutions are available. For the sake of clarity, they are termed ion-exchanger, original neutral-carrier and modified neutral-carrier:

- ion-exchanger membrane solution: Corning 477317 (KTpClPB in solvent of high dielectric constant)
- original neutral-carrier membrane solution: 5 wt.-% valinomycin, 2 wt.-% KTpClPB, 93 wt.-% dioctyl phthalate
- modified neutral-carrier membrane solution: 5 wt.-% valinomycin, 2 wt.-% KTpClPB, 93 wt.-% of a mixture of 2,3-dimethyl-nitrobenzene (25%) and dibutyl sebacate.

A comparison of the performance of the microelectrodes based on these membrane solutions is given in the following sections.

Electrode Functions. Typical response curves obtained with a classical ion-exchanger microelectrode and an original neutral-carrier microelectrode in aqueous metal chloride solutions are shown in Fig. 9.23. The valinomycin-based microelectrode exhibits a linear response from 10^{-1} to $5 \cdot 10^{-6}$ M with a slope of 58.4 ± 0.3 mV (std. dev., n = 5). The linear range of the ion-exchanger microelectrode only extends to about 10^{-4} M (slope: 58.1 ± 0.5 mV, std. dev., n = 4). Typical extra- or intracellular concentrations of Na$^+$ (Tab. 7.3) do not influence the K$^+$ electrode function at physiological activities of K$^+$. Intracellular measurements with classical ion-exchanger microelectrodes are only slightly influenced by the presence of typical intracellular concentrations of Na$^+$ (Fig. 9.24). Under exceptional intracellular conditions (low K$^+$, high Na$^+$) and extracellular conditions in general, the contribution of Na$^+$ to the measured EMF can no longer be neglected (see below).

Selectivities. The selectivities of the three types of K$^+$ microelectrodes are compared in Tab. 9.10. Valinomycin-based microelectrodes exhibit much higher selectivities with respect to alkali and alkaline-earth cations and lipophilic ammonium ions are rejected 10^5 times more strongly than in classical ion-exchanger microelectrodes.

Figure 9.24 illustrates the influence of Na$^+$ (5–18 mM) on the K$^+$ electrode function of the ion-exchanger microelectrode within the K$^+$ concentration range of

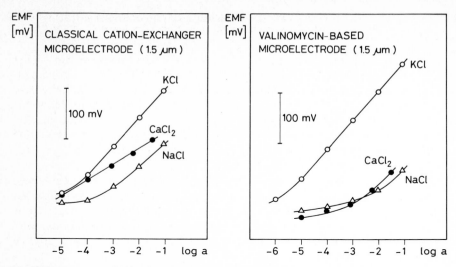

Fig. 9.23. Response of K^+ microelectrode in aqueous solutions of KCl, NaCl and CaCl$_2$

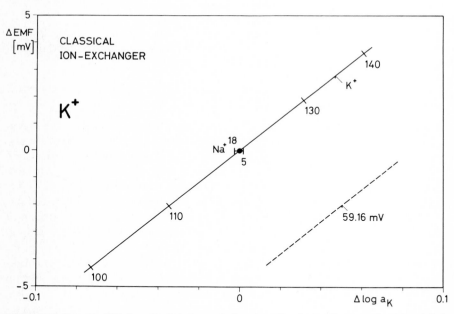

Fig. 9.24. Influence of Na$^+$ on the K$^+$ electrode function of an ion-exchanger microelectrode in an intracellular environment. The selectivity factor, log K_{KNa}^{Pot}, was fixed at -1.3 [Meier et al. 82]. Details of the evaluation are given in Sect. 7.8

Table 9.10. Selectivity factors, log K_{KM}^{Pot}, of ion-exchanger (Corning 477317), original neutral-carrier [Oehme and Simon 76] and modified neutral-carrier microelectrodes [Wuhrmann et al. 79]

Interfering species	Selectivities (K_{KM}^{Pot}), remarks, references		
	classical ion-exchanger	valinomycin (original)	valinomycin (modified)
Li^+	−2.5[a]	−4.0[a]	−4.0[b]
Na^+	−1.85 to −2.0[c] −1.7 to −2.0[d] −2.0[e] −1.96 to −2.15[f] −1.9[a] −1.96[g] −1.18 to −1.70[h] −0.4 to −1.8[i] −1.6[k] −1.7 to −2.0[l] −2.0[m] −1.49[n] −1.57[o]	−3.5[a]	−3.2[b]
Rb^+	0.6[a]	0.5[a]	
Cs^+	1.1[a]	−0.4[a]	
H^+	−0.5[a] no effect pH 5.6 to 7.8[g]; no effect pH 4 to 10[l]; no effect pH 6 to 8.5[e]; 0.2–0.3[c]	−4.4[a]	
Mg^{2+}	−2.7[a] < −3.0[g]	−5.1[a]	−5.0[b] −4.2[p]
Ca^{2+}	−2.1[a] < −3.0[g] −1.5 to −2.7[c]	−4.4[a]	−4.5[b] −4.7[p]
Sr^{2+}	−2.3[a]	−5.0[a]	
Ba^{2+}	−2.0[a]	−5.0[a]	
acetylcholine$^+$	2.7[a]	−2.4[a]	−2.5[b]
$(CH_3)_4N^+$	2.7[a]	−3.5[a]	
$(C_4H_9)_4N^+$	7.0[a]	0.8[a]	
NH_4	−0.7[g]		
glucose	no effect[g]		
urea	no effect[g]		
bovine serum albumin	little effect of 0.1 mM protein[g]		
anesthetics	interference of 29 µM procaine or 73 µM lidocaine[r]		

[a] [Oehme and Simon 76]; [b] [Wuhrmann et al. 79]; [c] [Walker 71]; [d] [Vyskočil and Kříž 72]; [e] [Lux and Neher 73]; [f] [Lux 74]; [g] [Fujimoto and Kubota 76]; [h] [Edelman et al. 78]; [i] [Fujimoto and Honda 80]; [k] [Hämmerli et al.80]; [l] [Hansen and Zeuthen 81]; [m] [Zeuthen 82]; [n] [Laming and Djamgoz 83]; [o] [Greger et al. 84a]; [p] [Dufau et al. 82]; [q] [Shimazaki 83]; [r] [Greenwood et al. 79].

100 to 140 mM [Meier et al. 82]. The calculated curve was obtained for a selectivity factor log K_{KNa}^{Pot} equal to -1.3. In spite of the presence of Na^+, the slope of the calibration curve is almost linear (1 mV is approximately equal to a 4% change in activity). The central point of the curve is more than 2.5 logarithmic activity units above the detection limit. However, if the microelectrode is used for extracellular measurements, the contributions from sodium (61%) and calcium (4.5% for log K_{KCa}^{Pot} $= -1.7$) have to be considered. In contrast, the valinomycin-based microelectrodes only show a 2% contribution from Na^+ to the measured EMF of the intracellular environment.

The superiority of the neutral carrier-based microelectrodes is further demonstrated by the selectivity of K^+ over lipophilic ammonium ions (Tab. 9.10). The risk of interference from such lipophilic cations has been analyzed for ion-exchanger microelectrodes [Lux 74]. These studies showed that an interference due to acetylcholine [Lux 74; Shimazaki 83] or other substances [Wuttke and Schlue 82] cannot be excluded. Indeed, classical ion-exchanger microelectrodes can be used for the direct measurement of lipophilic cations. Tetramethylammonium or choline ions are measured to monitor changes in the extracellular volume [Dietzel et al. 80] or to study the diffusion of cations in the extracellular space [Nicholson and Phillips 79; Nicholson et al. 79]. Certain anesthetics interfere at concentrations of about 10^{-5} M (see Tab. 9.10. [Greenwood et al. 79]). This is not surprising since the anesthetics that were studied (procaine (pK = 8.98 [Forth et al. 77]) and lidocaine (pK = 8.0 [Forth et al. 77])) are almost entirely protonated at the physiological pH's under consideration and may therefore interfere as lipophilic ammonium ions. Pentylenetetrazol, a drug used to elicit seizure activity, was shown to have a considerable influence on the K^+ response [Walden et al. 82]. Indeed, ion-exchanger microelectrodes for the direct measurement of pentylenetetrazol have been developed [Walden et al. 84]. Albumin, glucose and urea do not influence the measurements of K^+ using ion-exchanger microelectrodes [Fujimoto and Kubota 76].

Response Time and Electrical Membrane Resistance. Tables 9.11 and 9.12 clearly show an additional marked difference between ion-exchanger and neutral carrier-based microelectrodes. While the use of valinomycin leads to rather slow, high-impedance microelectrodes, the classical ion-exchanger membrane solution yield sensors of low resistances and short response times. As a result, physiologists tend to use ion-exchanger microelectrodes.

It was recently shown that a valinomycin-based membrane solution that does not contain any solvent of low dielectric constant (5 wt.-% valinomycin, 2 wt.-% KTpClPB, 93 wt.-%, 2,3-dimethyl nitrobenzene) still exhibits a sufficient selectivity for K^+ over divalent interfering ions (for a discussion of the influence of the dielectric constant of the membrane solvent on the monovalent/divalent selectivity see Sect. 4.1.2). Because of the polar membrane phase the microelectrodes exhibit a slightly decreased resistance and an improved response time (<0.6 s) [Shimazaki 83].

Stability of EMF and Lifetime. The measured EMF of the valinomycin-based microelectrode shows a stability of ± 0.3 mV (std. dev., 16 h, 10^{-1} M KCl) [Oehme and Simon 76]. The drift during the intracellular experiments is less than 2 mV [Wuhrmann et al. 79]. Using triple-barrelled neutral carrier-based microelectrodes, it has

Table 9.11. Response times of the ion-exchanger (Corning 477317), the original neutral-carrier [Oehme and Simon 76] and the modified neutral-carrier microelectrodes [Wuhrmann et al. 79]

Microelectrode	Response time		
	classical ion-exchanger	valinomycin (original)	valinomycin (modified)
single-barrelled	<0.5 s (<0.2 μm [Edelman et al. 78])		
double-barrelled	2 to 10 ms (2–3 μm [Lux and Neher 73]), 2 to 10 ms (1–3 μm [Lux 74]), <1 s [Fujimoto and Kubota 76], <0.2 s [Hansen and Zeuthen 81], <1 s [Zeuthen 82]	~30 s (1.5 μm [Oehme and Simon 76])	~30 s (1.5 μm [Wuhrmann et al. 79])
triple-barrelled			<0.2 s (1–2 μm [Dufau et al. 82])
coaxial	<2 ms [Ujec et al. 76]		
ISFET	0.1 to 0.5 s (<1 μm [Hämmerli et al. 80])		

Table 9.12. Electrical membrane resistances of the ion-exchanger (Corning 477317), the original neutral-carrier [Oehme and Simon 76] and the modified neutral-carrier microelectrodes [Wuhrmann et al. 79]

Microelectrode	Electrical membrane resistance		
	classical ion-exchanger	valinomycin (original)	valinomycin (modified)
single-barrelled	$1.5 \cdot 10^{10}\,\Omega$ (<0.2 μm [Edelman et al. 78]), $7 \cdot 10^9\,\Omega$ (<0.1 μm [Greger et al. 84a])		
double-barrelled	10^8–$5 \cdot 10^9\,\Omega$ (2–3 μm [Lux and Neher 73]), $2 \cdot 10^9\,\Omega$ (1.5 μm [Oehme and Simon 76])	$3 \cdot 10^{11}\,\Omega$ (1.5 μm [Oehme and Simon 76]), $8 \cdot 10^{11}\,\Omega$ (0.8 μm [Coles and Orkand 83])	$\sim 10^{11}\,\Omega$ (1.5 μm [Wuhrmann et al. 79])
triple-barrelled	$1.5 \cdot 10^{10}\,\Omega$ (0.5 μm [Fujimoto and Honda 80])		$7 \cdot 10^9\,\Omega$ (1–2 μm [Dufau et al. 82])
coaxial	1–$5 \cdot 10^7\,\Omega$ (2–3 μm [Ujec et al. 79])		

been observed that the slope of the electrode function decreases from 55.8 mV after 1 h of use to 50.0 mV after 5 days [Dufau et al. 82]. The stability of the EMF measured by an ion-exchanger microelectrode is claimed to be ±0.1 mV (std. dev., 5 h, 10^{-1} M KCl) [Oehme and Simon 76].

As is to be expected from the composition of the membrane solutions, the lifetimes of the ion-exchanger and of the neutral carrier-based microelectrode differ significantly. Ion-exchanger microelectrodes have been reported to exhibit lifetimes of 8 h [Oehme and Simon 76] or 10–15 days [Hämmerli et al. 80]. On the other hand, the lifetimes of the valinomycin-based microelectrodes are expected to be considerably longer because of the considerably higher lipophilicity of the membrane components. Studies have shown that this is indeed the case. For example, a lifetime of about 6 weeks was measured in aqueous solutions [Oehme and Simon 76]. However, as has already been extensively discussed in Sect. 5.10, the lifetimes of microelectrodes are usually determined by other parameters that are not related to the loss of membrane components.

9.4.3 Concluding Remarks

At present, both classical ion-exchanger and valinomycin-based microelectrodes are used for intracellular measurements of K^+. The former are widely used because of their low resistance, fast response and simple composition of the membrane solution. From the selectivity point of view, the neutral carrier-based microelectrode is far superior to the ion-exchanger microelectrode. K^+ measurements performed with ion-exchanger microelectrodes are prone to interference caused by lipophilic cations. Except for measurements of fast transients in K^+ activity, the use of the neutral carrier-based K^+ microelectrode is recommended.

9.5 Intracellular Measurements of Mg^{2+}

Results on the important physiological role of magnesium have accumulated rapidly during the past few years. Magnesium is a major regulator of the functions of enzymes and a number of cellular properties and processes. Its intracellular homeostasis, however, is rather poorly understood, a fact which reflects the analytical difficulties involved in a quantitative measurement of magnesium at the cellular level.

The total amount of magnesium in cells can be evaluated in tracer studies (radioisotope ^{28}Mg), gravimetric procedures and atomic absorption spectrometry. Several techniques are available for the quantitative determination of the concentration of free Mg^{2+}: spectrophotometry, null point for permeabilized cells, nuclear magnetic resonance spectroscopy and potentiometry. Each method has its own temporal and spatial resolution as well as clear-cut limitations which make intracellular studies of Mg^{2+} activity more difficult than the corresponding studies of other cations. The following paragraphs contain a description of the techniques available for the determination of Mg^{2+}, including a more detailed discussion of the potential of currently available Mg^{2+} microelectrodes.

9.5.1 Mg^{2+} Null Point for Permeabilized Cell Membranes

The null point is defined as the concentration of the ion to be measured at which the addition of cells, previously made permeable for the ion under study, does not pro-

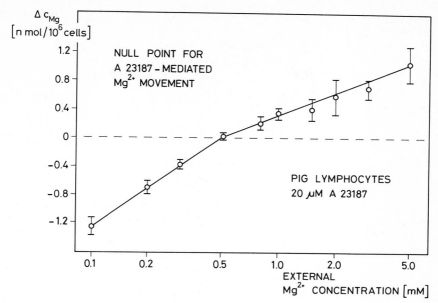

Fig. 9.25. Determination of the null point of magnesium for Mg^{2+} movements induced by the iono-phore A23187 [Rink et al. 82]. The vertical axis indicates the gain or loss in cell magnesium with re-spect to untreated cells. The measurements were carried out after an incubation period of 45 mi-nutes at various external concentrations of magnesium ($[Mg^{2+}]_o$ (abscissa) at pH 7.05 and 20 µM A23187). The total cellular concentration of magnesium at equilibrium was measured by atomic ab-sorption spectrometry after centrifugation and treatment with perchloric acid. Redrawn from Rink, Tsien and Pozzan, J. Cell. Biol., 1982 [Rink et al. 82]

duce any change in concentration in the incubation medium. Obviously, this tech-nique yields an average concentration for the cell population. The method can also be applied to H^+ and Ca^{2+}. The cell membranes are permeabilized by various ap-proaches: H^+: digitonin [Rink et al. 82]; Ca^{2+}: digitonin [Murphy et al. 80], saponin [Wakasugi et al. 82], very low external Ca^{2+} [Streb and Schulz 83]; and Mg^{2+}: charged carrier A23187 [Flatman and Lew 77].

For intracellular measurements of magnesium, the null point technique was first described for red cells using atomic absorption spectrometry [Flatman and Lew 77]. The cell membrane was made permeable to Mg^{2+} ions by adding a relative high concentration of the carboxylic carrier antibiotic A23187. Since this charged carrier preferably binds Ca^{2+}, the external solution has to be free of calcium. Under appro-priate conditions, the antibiotic transports a Mg^{2+} ion in one direction and two H^+ ions in the opposite direction through the cell membrane. Thus, if the equilibri-um extra- and intracellular H^+ activities are known, the intracellular concentration of Mg^{2+} can be calculated from the null point concentration of Mg^{2+} in the exter-nal incubation medium [Rink et al. 82]. As a result of the coupled Mg^{2+}/H^+ coun-tertransport, the intracellular concentration of Mg^{2+} can be determined by observ-ing the null points caused by the movement of either Mg^{2+} (Fig. 9.25) or H^+ [Rink et al. 82].

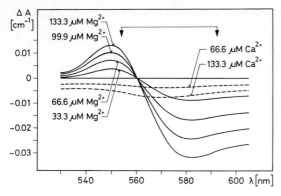
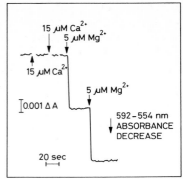

Fig. 9.26. Difference spectra of Eriochrome Blue SE versus Eriochrome Blue SE in the presence of different amounts of MgCl₂ or CaCl₂ in aqueous solutions [Scarpa and Brinley 81]. Both cuvettes contained 100 mM KCl, 30 mM MOPS buffer (pH 7.1) and 100 µM Eriochrome Blue SE. The arrows (left side) indicate the spectral regions in which suitable pairs of wavelength can be selected (e. g. 554 and 592 nm) for differential measurements that minimize the interference due to Ca^{2+}. The right side of the figure shows dual wavelength (554/592 nm) differential measurements of the changes in the absorption of Eriochrome Blue SE due to the addition of MgCl₂ and CaCl₂ as a function of time (1 ml cuvette). Redrawn from Scarpa and Brinley, Fed. Proc., 1981 [Scarpa and Brinley 81]

9.5.2 Metallochromic Indicators

Metallochromic indicators are mainly used for the determination of intracellular concentrations of Ca^{2+} (Sect. 9.6.2). The technique relies on the introduction (usually microinjection) of a dye into the cytosol followed by the continuous in situ measurement of changes in absorbance. Differential measurements are often made using multi-wavelength microspectrophotometry.

In 1974 Eriochrome Blue SE was introduced as an indicator dye for the optical measurement of intracellular concentrations of Mg^{2+} [Scarpa 74]. Its dissociation constant for Mg^{2+} is about 20 mM, which guarantees that it will only cause slight disturbances in the intracellular concentrations of free Mg^{2+} (millimolar range). The time constant for the dissociation of the magnesium complex is about 90 ms which allows the measurement of relative fast transients [Scarpa and Brinley 81]. The indicator exhibits large absorbance differences between the free dye and its Mg^{2+} complex (Fig. 9.26). Interference due to H^+ and Ca^{2+} ($K_d > 3.5$ mM) is minimized by measuring the differential signal of the changes in absorbance at two wavelengths (e. g. 554 and 592 nm (Fig. 9.26) [Scarpa 74]). Disturbances due to a loss of dye or other factors (for example, scattering of light or geometric factors) can be minimized by measuring the changes in absorbance at the isobestic point (Eriochrome Blue SE: 566 nm) [Scarpa and Brinley 81].

The best results have been obtained using the null point approach, in which an internally perfused cell is observed. The null point corresponds to the concentration of the internal dialysis solution, which does not raise or lower the cytosolic concentration of free Mg^{2+} during the dialysis. The approach does not require an accurate knowledge of the characteristics of the dye, but is limited to a determination of resting concentrations of Mg^{2+} in large internally perfused cells (axon of squid [Brinley and Scarpa 75], muscle of barnacle [Brinley et al. 77]).

On the other hand, a direct optical measurement of the concentration of Mg^{2+} requires a full knowledge of the spectral data and the properties of the complex formation of the dye under physiological conditions. There have not yet been any direct quantitative measurements using Eriochrome Blue SE mainly because of a lack of selectivity and the difficulties involved in making calibrations for absolute determinations of Mg^{2+}. During the last ten years, no improved dyes for the measurement of Mg^{2+} have been developed.

9.5.3 ^{31}P Nuclear Magnetic Resonance Spectroscopy

Nuclear magnetic resonance techniques, such as NMR imaging and in vivo measurements are increasingly playing important roles in biomedicine. They are now well established as noninvasive and presumedly nondestructive approaches for the in vivo study of metabolism [Shulman 79; Gadian 83] and H$^+$ activities (Sect. 9.1).

An early study on the influence of metal binding on the NMR spectrum of adenosine triphosphate (ATP) [Cohn and Hughes 62] initiated the use of ^{31}P NMR for intracellular measurements of the concentrations of free Mg^{2+}. The α-, β- and γ-resonances of ATP are usually clearly observable in cells. Their chemical shifts depend on the degree of complexation between ATP and Mg^{2+} (Fig. 9.27). A comparison of the spectrum of ATP in a cell with the spectra of calibration solutions (physiological electrolytes with and without magnesium) yields the fraction f of the total amount of ATP (ATP$_t$) that is not complexed (ATP) [Gupta and Yushok 80]:

$$f = \frac{ATP}{ATP + MgATP} = \frac{ATP}{ATP_t}. \tag{9.16}$$

Using the dissociation constant K$_d$ of the ATP-magnesium complex under physiological conditions, the concentration of free Mg^{2+} (Mg) within the cell can be calculated [Gupta and Moore 80; Gupta and Yushok 80]:

$$K_d = \frac{Mg \cdot ATP}{MgATP}, \tag{9.17}$$

and therefore

$$Mg = \frac{K_d(ATP_t - ATP)}{ATP}$$
$$= K_d\left(\frac{ATP_t}{ATP} - 1\right). \tag{9.18}$$

Using Eq. (9.16) one obtains:

$$Mg = K_d(1/f - 1). \tag{9.19}$$

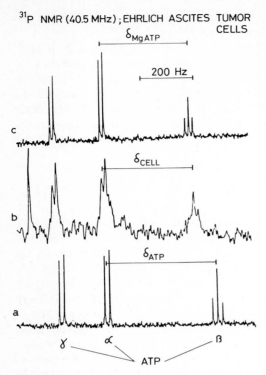

Fig. 9.27 a–c. ^{31}P NMR spectra (40.5 MHz) showing α-, β- and γ-resonances of intracellular ATP in ascites tumor cells **(b)** and in aqueous calibration solutions **(a, c)** [Gupta and Yushok 80]. The quantity δ is equal to the difference in the chemical shifts of the central component of the β-triplet and the center of the α-doublet. **a:** 4 mM ATP (pH 7.2), **c:** the same as a) plus 10 mM Mg^{2+}. For details see [Gupta and Yushok 80]. Redrawn from Gupta and Yushok, Proc. Natl. Acad. Sci., 1980 [Gupta and Yushok 80]

In order to avoid the problem of defining an internal reference in the cell sample, it has been suggested that the differences in chemical shift between the resonance signals of the α- and β-phosphorus in the ^{31}P NMR spectrum of ATP should be used to evaluate the fraction f [Gupta and Yushok 80]. Therefore, as indicated in Fig. 9.27, the δ values given below correspond to the distance between the chemical shifts of the α- and β-resonances. It can be shown that the fraction f can be directly determined from the ^{31}P NMR spectra:

$$\begin{aligned} ATP_t\, \delta_{cell} &= ATP\, \delta_{cell} + MgATP\, \delta_{cell} \\ &= ATP\, \delta_{ATP} + MgATP\, \delta_{MgATP} \end{aligned} \tag{9.20}$$

and

$$ATP = MgATP \frac{\delta_{MgATP} - \delta_{cell}}{\delta_{cell} - \delta_{ATP}}. \tag{9.21}$$

Substitution into Eq. (9.16) yields

$$f = \frac{MgATP\dfrac{\delta_{MgATP} - \delta_{cell}}{\delta_{cell} - \delta_{ATP}}}{MgATP\dfrac{\delta_{MgATP} - \delta_{cell}}{\delta_{cell} - \delta_{ATP}} + MgATP\dfrac{\delta_{cell} - \delta_{ATP}}{\delta_{cell} - \delta_{ATP}}},$$ (9.22)

and finally [Gupta and Yushok 80]:

$$f = \frac{\delta_{cell} - \delta_{MgATP}}{\delta_{ATP} - \delta_{MgATP}}.$$ (9.23)

A combination of Eqs. (9.19) and (9.23) leads to an expression for the intracellular concentration of the free Mg^{2+} as a function of the differences in the chemical shifts in cell suspensions (δ_{cell}) and in calibration solutions (without Mg^{2+}: δ_{ATP}; with Mg^{2+}: δ_{MgATP}) (Fig. 9.27) and of the dissociation constant of the magnesium complex of ATP.

In calibration solutions without Mg^{2+} the separation of the signals of the α- and β-phosphorus is found to be 438 ± 1 Hz, while in the spectrum of the MgATP complex the separation is 349 ± 0.5 Hz (at 40.5 MHz) [Gupta and Yushok 80]. In cell samples the differences in chemical shift can be measured with a precision of $\leqslant 2$ Hz, and therefore, the determination of the fraction f is quite accurate [Gupta and Yushok 80]. However, there is some disagreement concerning the dissociation constant K_d under physiological conditions. Depending on the value of K_d that is used, considerable lower [Gupta and Moore 80; Gupta and Yushok 80] or similar [Wu et al. 81] concentrations of Mg^{2+} are obtained compared to the results of other techniques.

9.5.4 Potentiometry

9.5.4.1 Mg^{2+}-Selective Carriers

The development of molecules exhibiting high selectivities for Mg^{2+} has been extremely difficult. Although several antibiotics with an affinity for Mg^{2+} are known and several classes of synthetic ion carriers have been intensively investigated, a highly selective Mg^{2+} microelectrode is not yet available.

Some naturally occurring ionophores such as the monocarboxylic polyether antibiotic-6016 (1 in Fig. 9.28 [Otake and Mitani 79]), the N-methyl derivative of calcimycin (2 in Fig. 9.28 [Prudhomme and Jeminet 83]), or other pyrrol ether antibiotics (e.g. X-14885A, 3 in Fig. 9.28 [Westley et al. 83; Liu et al. 83]) have been reported to show Mg^{2+} selectivity in extraction studies. However, it has not yet been possible to make use of these selectivities in solvent polymeric membranes or membrane solutions [Erne et al. 80b; Ammann, Selle, Läubli, Simon, unpublished]. The larger alkaline-earth cations are clearly preferred over Mg^{2+}.

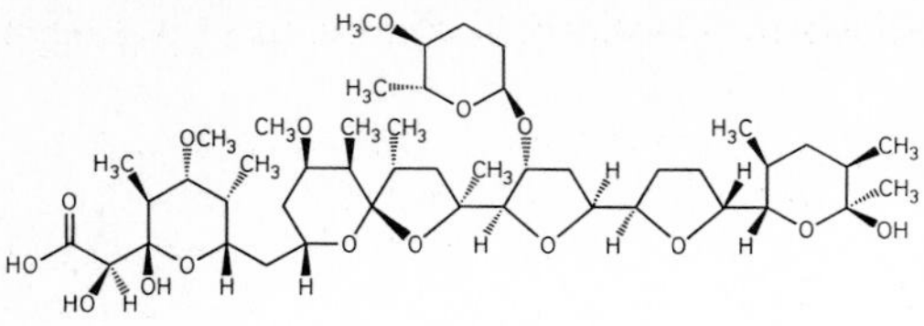

1 (ANTIBIOTIC - 6016)

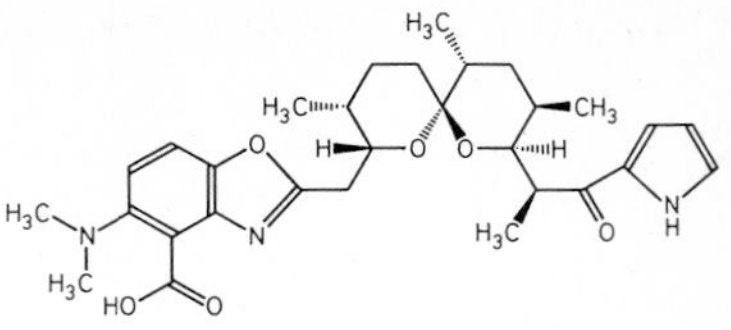

2

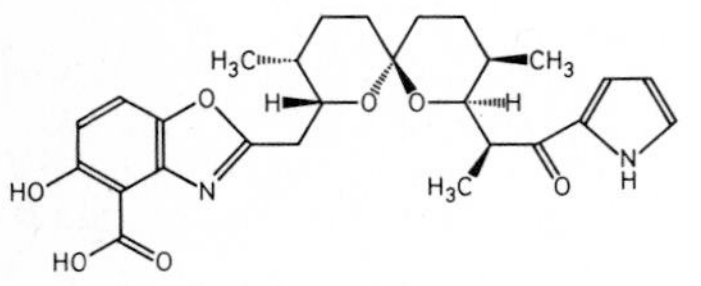

3 (ANTIBIOTIC X - 14885 A)

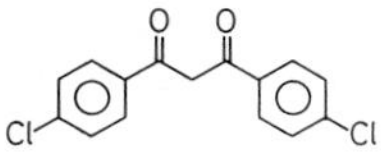

4 (ETH 248)

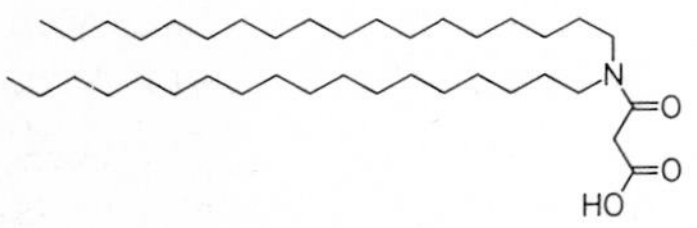

5 (ETH 1645)

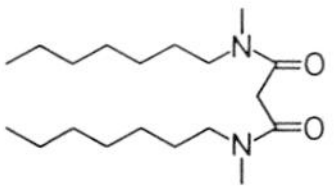

6 (ETH 1611)

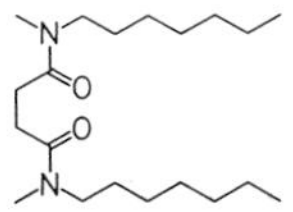

7 (ETH 1117)

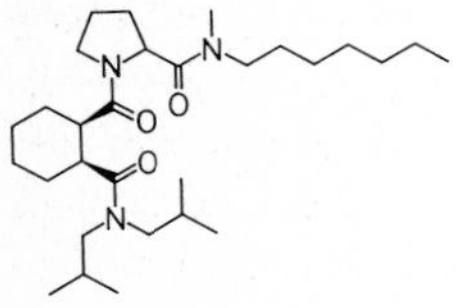

8 (ETH 1759)

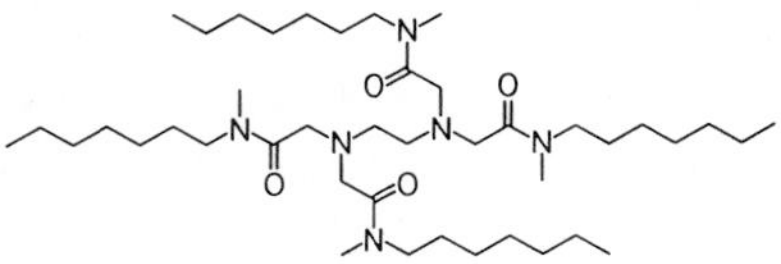

9 (ETH 605)

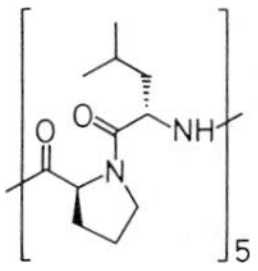

10

Fig. 9.28. Constitutions of the components of Mg^{2+}-selective electrodes

Lipophilic β-diketones ($\underline{4}$ in Fig. 9.28 [Erne et al. 79b]) and N,N-dioctadecyl monoamides of dicarboxylic acids ($\underline{5}$ in Fig. 9.28 [Maj-Żurawska et al. 82]) have been synthesized as possible Mg^{2+} carriers for membranes. Membranes based on such carriers have been shown to transport Mg^{2+} ions when a pH gradient is applied across the membrane [Erne et al. 79b; Maj-Żurawska et al. 82]. Mg^{2+} selectivity is only observed if the carriers are predominantly in the deprotonated form within the membrane. In order to exploit the selectivity of β-diketones the internal filling solution of the membrane electrode must have a pH above 8. Monoamides of dicarboxylic acids are in the deprotonated form within a membrane if the external electrolyte solutions have physiological values of the pH. Attempts to utilize the ion selectivity of these synthetic charged carriers in membrane solutions have not been successful.

A large series of potential neutral carriers which exhibit a selectivity for Mg^{2+} have been prepared. The strategy used was to design an octahedral coordination site surrounded by oxygen atoms from functional groups with high dipole moments (amides) and/or nitrogen atoms from functional groups with high polarizabilities (amines) (see Chap. 3). Lipophilic diamides and triamides (6–8 in Fig. 9.28 [Erne et al. 80b; Erne et al. 82; Zhukov et al. 81]) and tetraamides containing tertiary amino groups (9 in Fig. 9.28 [Erne et al. 80a]) seemed to be especially promising. A liquid membrane based on the neutral carrier 7 (ETH 1117) exhibits sufficient selectivity for Mg^{2+} in order to be used in studies of intracellular Mg^{2+} activity [Lanter et al. 80]. Recently, oligopeptides have been investigated as a possible class of neutral carriers exhibiting selectivity for Mg^{2+} [Behm et al. 85]. Cyclo(L-Pro-L-Leu)$_5$ (10 in Fig. 9.28) exhibits a Mg^{2+}/Na^+ selectivity of about 200. Ca^{2+} is rejected only slightly. Membranes containing cyclo(L-Pro-D-Leu)$_5$ show a Mg^{2+}/Ca^{2+} selectivity of more than 100 [Behm et al. 85].

9.5.4.2 Microelectrodes Based on the Neutral Carrier ETH 1117

Microelectrodes based on membrane solutions containing the lipophilic neutral carrier N,N'-diheptyl-N,N'-dimethylsuccinic acid diamide (ETH 1117, 7 in Fig. 9.28) have been used in measurements of intracellular Mg^{2+} activity [Lanter et al. 80]. The response of such a microelectrode to different Mg^{2+} activities in a sample solution is given in Fig. 9.29. In the absence of interfering ions, the slope of the electrode function is 28.0 ± 0.7 mV (std. dev., n = 4). On the basis of the selectivity factors given in Table 9.13, the response (dashed line) presented in Fig. 9.29 was calculated for solutions with a typical intracellular ionic background using the Nicolsky-Eisenman formalism. The actual behaviour of the microelectrode (middle curve with open circles in Fig. 9.29) indicates slightly higher selectivities and a lower detection limit than log $a_{Mg} = -3.5$. A useful response is obtained for the intracellular concentration range of 1 to $6 \cdot 10^{-3}$ M of free Mg^{2+} (log $a_{Mg} = -3.4$ to -2.7) (Fig. 9.30). In this range of Mg^{2+} activities an appropriate calibration procedure has to be chosen. In preparing the calibration solutions it is advantageous to make use of the known or estimated concentrations of all the ions present. The ion that interferes the most is K^+. Hence, care should be taken to know the concentration of K^+ as accurately as possible. The interference due to K^+ is discussed in more detail in Fig. 9.31. With a typical intracellular background of ions the slope of the Mg^{2+} electrode function is reduced to about 22.3 mV. The central point of the curve is 0.5 logarithmic activity units above the detection limit. Thus, 1 mV is about equivalent to a 10.9% change in activity.

Disturbance by K^+ has also been observed during other studies of intracellular Mg^{2+} activities [Hess and Weingart 81; Hess et al. 82; Alvarez-Leefmans et al. 84]. Consideration of K^+ in the calibration solutions has been shown to be mandatory. Although sodium is a minor interfering ion, it should also be considered [Hess et al. 82] (see also [Alvarez-Leefmans et al. 84]). A detectable interference by Ca^{2+} or H^+ cannot be found in Ca^{2+} solutions as concentrated as 10^{-5} M or in the pH range 6.7

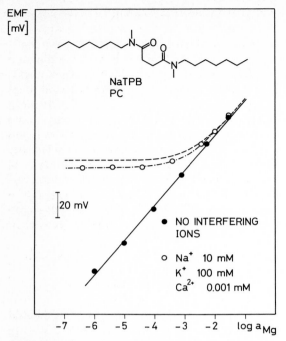

Fig. 9.29. EMF response of a microelectrode based on the neutral carrier ETH 1117 to different Mg^{2+} activities in a sample solution. Dots: experimental values. *Lower trace* (black dots): response in pure solutions of $MgCl_2$. *Middle trace:* Mg^{2+} response at a constant ionic background. *Upper trace:* calculated Mg^{2+} response using selectivity data of Table 9.13 and actual slope of the electrode function at the constant ionic background indicated

Table 9.13. Selectivity factors, log K_{MgM}^{Pot}, of microelectrodes based on the neutral carrier ETH 1117[a] [Lanter et al. 80]

Interfering ion	log K_{MgM}^{Pot}[b]	Interfering ion	log K_{MgM}^{Pot}[b]
H^+	2.8	Ca^{2+}	1.1
Li^+	0.1	Sr^{2+}	0.6
Na^+	-1.1	Ba^{2+}	0.7
K^+	-1.4	NH_4^+	-0.1
Rb^+	-1.3	acetylcholine$^+$	-0.1
Cs^+	-0.9		

[a] composition of the membrane solution: 20 wt.-% Mg^{2+} carrier ETH 1117, 1 wt.-% sodium tetraphenylborate, 79 wt.-% propylene carbonate [Lanter et al. 80]
[b] separate solution method, 0.1 M metal chloride solutions (see Sect. 5.5)

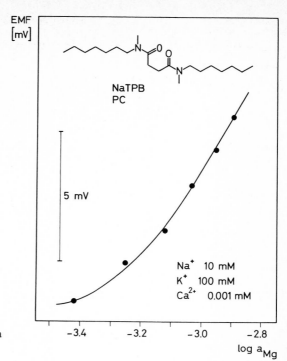

Fig. 9.30. EMF response of a microelectrode based on the neutral carrier ETH 1117 to different Mg^{2+} activities in the physiological range at a constant ionic background

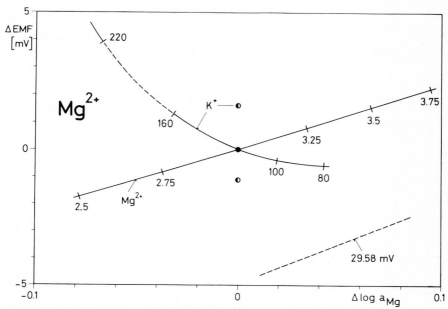

Fig. 9.31. The influence of K^+ on the Mg^{2+} electrode function of a neutral carrier-based microelectrode in the millimolar activity range. Details of the evaluation of the curves are given in Sect. 7.8. The curves were calculated using a selectivity factor $\log K_{MgK}^{Pot} = -1.4$ (Tab. 9.13). The shift of the central point (3 mM Mg^{2+}, 120 mM K^+) for a change in $\log K_{MgK}^{Pot}$ of ± 0.2 is indicated by the points (◑)

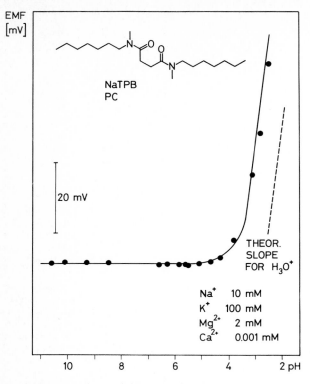

Fig. 9.32. Dependence of the EMF of a microelectrode based on the neutral carrier ETH 1117 on the pH of a sample solution at a $MgCl_2$ concentration of 2 mM and a constant ionic background. The pH's were adjusted by adding diluted KOH or HCl solutions to the sample

to 7.7 [Hess et al. 82]. Figure 9.32 shows that pH's as low as 5 can be tolerated at typical concentrations of free Mg^{2+} (2 mM).

Due to the relative low resistance of these neutral carrier-based microelectrodes ($2-4 \cdot 10^{10} \, \Omega$ ($1-2.5 \, \mu m$ [Lanter et al. 80]), $1-5 \cdot 10^{10} \, \Omega$ ($< 1 \, \mu m$ [Hess and Weingart 81]), $1.5-4 \cdot 10^{10} \, \Omega$ ($< 1 \, \mu m$ [Gamiño and Alvarez-Leefmans 83]), $2 \cdot 10^{10} \, \Omega$ ($< 1 \, \mu m$ [Hess et al. 82]), $1.5-4 \cdot 10^{10} \, \Omega$ [Alvarez-Leefmans et al. 84]), the electrode response to changes in the Mg^{2+} activity is rather fast: $< 5 \, s$ ($\sim 1 \, \mu m$, Fig. 9.33. [Lanter et al. 80]), $< 0.5 \, s$ ($< 1 \, \mu m$ [Hess et al. 82]).

Typical cells show a drift in EMF of less than 1 mV/h over at least 4 days in aqueous Mg^{2+} solutions [Lanter et al. 80]. Stable calibrations have been observed over several days [Hess et al. 82]. Using 14 microelectrodes and 27 calibrations in 1 mM and 10 mM solutions of Mg^{2+}, a reproducibility of $16.1 \pm 0.4 \, mV$ was obtained [Alvarez-Leefmans et al. 84].

Figure 9.34 illustrates the first application of the microelectrode to the determination of Mg^{2+} activities within a cardiac Purkinje fibre of a sheep [Lanter et al. 80]. E_M is the potential of the membrane measured with a reference microelectrode filled with 3 M KCl relative to a reference electrode in an extracellular Tyrode solution (top). $E_{Mg} + E_M$ is the potential difference between the Mg^{2+}-selective micro-

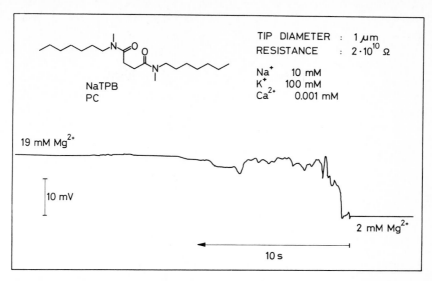

Fig. 9.33. Response of a microelectrode based on the neutral carrier ETH 1117 to the addition of about 0.4 ml of a 10^{-1} M MgCl₂ solution to 2 ml of a $2 \cdot 10^{-3}$ M MgCl₂ solution with the indicated ionic background. Before the experiment the microelectrode was conditioned for at least 4 h in $2 \cdot 10^{-3}$ M MgCl₂ with the same ionic background. The irregularities in the response can be traced to an inadequate mixing of the electrolytes. A 90% response time (t_{90}) of < 5 s can be determined [Lanter et al. 80]. More sophisticated experiments show response times below 0.5 s [Hess et al. 82]

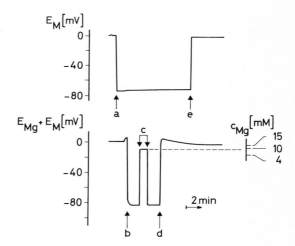

Fig. 9.34. Measurement of the resting of concentration of free Mg²⁺ in a cardiac Purkinje fibre of a sheep using a bevelled neutral carrier-based microelectrode with a tip diameter of about 1 μm [Lanter et al. 80]

electrode and the reference electrode in the Tyrode solution (bottom). At (a) the cell was impaled by the reference microelectrode and at (b) by the ion-selective microelectrode; (d) and (e) mark the withdrawal of the electrodes. In the range indicated by (c), the membrane potential E_M was subtracted from the sum $E_{Mg} + E_M$ to determine E_{Mg}. The right-hand ordinate was obtained from the calibration of the Mg^{2+} microelectrode in 4, 10 and 15 mM solutions of Mg^{2+} and a background of 7 mM Na^+, 120 mM K^+ and 0.1 mM EGTA ($Ca^{2+} < 10^{-7}$ M) at a pH of 7.2. Through the external calibration of the cell assembly, the intracellular concentration of the free magnesium was found to be $8.4 \cdot 10^{-3}$ M ($a_{Mg} = 2.9 \cdot 10^{-3}$ M).

Similar measurements with more stringent impalement criteria yielded somewhat lower concentrations of free Mg^{2+} in cardiac Purkinje fibres of sheep [Hess et al. 82]. Concentrations as low as 0.66 mM have been reliably measured in snail neurons [Alvarez-Leefmans et al. 84].

9.5.5 Comparison of the Different Methods

The total amount of cellular magnesium can be determined by gravimetric methods, a tracer studies or atomic absorption spectrometry. The average amount of magnesium in the cytosol or the entire intracellular space is obtained. These values will be considerably higher than the cytosolic concentrations of free Mg^{2+} because of the binding of Mg^{2+} to various cellular components. The compartmentation of the cell causes further deviations from average and local determinations. It is well accepted that the concentration of free Mg^{2+}, or more importantly the activity of Mg^{2+}, is the physiological relevant magnesium fraction. Techniques for the measurement of the concentrations of free Mg^{2+} have been briefly described in Sects. 9.5.1–9.5.4. A comparison of the typical features of each method is made in Table 9.14.

Table 9.14. Comparison of methods used to determine intracellular concentrations of free Mg^{2+}

Null point of permeabilized cells	Spectrophotometry (Eriochrome Blue SE)	^{31}P NMR spectroscopy	Potentiometry with microelectrodes
– considerably dense cell suspensions – possible errors due to contamination of the cytosol by simultaneously permeabilized organelles – requires independent measurement of intracellular pH – cell population, average over cytoplasm	– large cells – microinjection and internal perfusion – limited selectivity, strong interference due to H^+ – nonspecific changes in absorbance – no absolute measurements based on calibrations in external solutions – relative fast response – single cell, average over cytoplasm	– high-density suspensions – noninvasive and possibly nondestructive – no direct assessment of Mg^{2+}; assumption of MgATP dissociation constant in the cytosol – expensive instrumentation – cell population, average over cytoplasm	– fairly small cells – insertion of microelectrode(s) – limited selectivity; intracellular activity of K^+ has to be considered in the calibration – quantitative determination of the concentration or activity of free Mg^{2+} – limited temporal resolution; certain dynamic processes are accessible – single cell, local

9.5.6 Concluding Remarks

Microelectrodes based on the neutral carrier ETH 1117 enable the direct potentiometric measurement of intracellular Mg^{2+} activities. The microelectrodes are prone to considerable K$^+$ interference at intracellular concentrations. Simultaneous measurements of intracellular K$^+$ and careful calibration procedures are necessary in order to obtain physiologically relevant intracellular activities of Mg^{2+} in the millimolar range.

9.6 Intracellular Measurements of Ca^{2+}

Indirect methods and studies of the relationship between the concentration of Ca^{2+} and physiological processes have led to a considerable understanding of the role of calcium in cells (for a discussion of the indirect techniques see [Thomas 82]). The introduction of direct techniques for the measurement of intracellular concentrations of free Ca^{2+} has been an important step forward in Ca^{2+} research. To date, three such methods are routinely used: luminescence studies using photoproteins (Sect. 9.6.1), spectrophotometric studies using metallochromic indicators (Sect. 9.6.2) and potentiometric studies using ion-selective microelectrodes (Sect. 9.6.3). Calcium ions are involved in several equilibria having fast exchange rates and the concentration of free Ca^{2+} is at a very low, submicromolar level. Hence, the techniques have to satisfy a series of severe criteria with respect to selectivity, detection limit and response time. The merits and disadvantages of each of the three techniques are discussed in Sect. 9.6.4.

9.6.1 Photoproteins

9.6.1.1 Origin of Bioluminescence

Classical bioluminescent systems found in bacteria, fungi and protozoa are based on enzymatic oxidation-reduction reactions in which the change in free energy serves to excite the substrate. Subsequently, the molecules relax to their ground states by emitting visible light. For example, the luciferin found in fireflies reacts with ATP to form luciferyl adenylate which is tightly bound to the enzyme luciferase. When this enzyme-bound substrate is exposed to molecular oxygen, oxyluciferin is formed which emits light on returning to the ground state (Fig. 9.35). One photon is emitted for each molecule of luciferin that is oxidized. This reaction is the basis of a widely used assay for the highly selective detection of ATP [Lundin and Baltscheffsky 78].

    Another interesting class of compounds which show luminescence are the photoproteins. These have been isolated from marine coelenterates (for a review see [Morin 74]). The term photoprotein has been suggested for the association of an

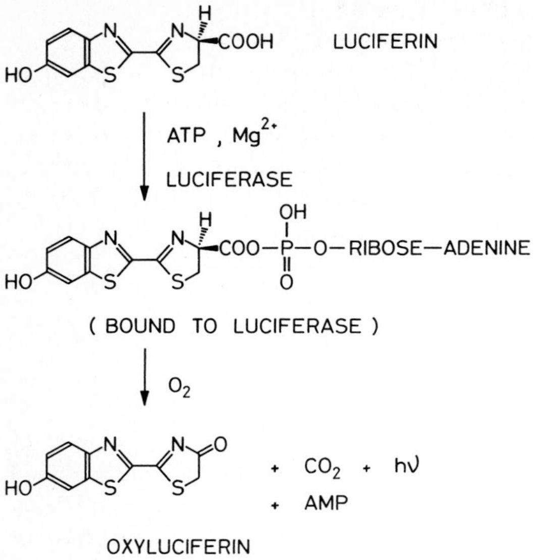

LUCIFERIN

ATP , Mg^{2+}

LUCIFERASE

(BOUND TO LUCIFERASE)

O_2

+ CO_2 + $h\nu$

+ AMP

OXYLUCIFERIN

Fig. 9.35. Mechanism for the bioluminescence of luciferin in the firefly

apoprotein with a chromophore which acts as the light-emitting moiety. The photoprotein is therefore the undischarged bioluminescent species [Shimomura and Johnson 66]. In contrast to the classical compounds, the photoproteins are highly sensitive to the presence of free Ca^{2+} and do not require oxygen or other cofactors such as ATP in order to show luminescence (Fig. 9.36).

It was observed that the binding of Ca^{2+} does not supply the required energy for luminescence, but that Ca^{2+} seems to influence the conformation of the binding sites of the prosthetic group. The Ca^{2+}-activated change in conformation allows the bound oxygen to oxidize the chromophore, which in turn induces the emission of light (Ca^{2+}-triggered luminescence). Certain aspects of the processes involved in aequorin luminescence are not yet understood and agreement on a model for the molecular mechanisms has not yet been reached. Impressive work has been carried out to elucidate the reaction mechanisms of photoproteins [Campbell et al. 79; Shimomura and Johnson 79; Prendergast 82; Blinks et al. 82].

9.6.1.2 Availability of Photoproteins

The acquisition of photoproteins involves extreme efforts compared to the synthesis of neutral carriers or indicator dyes. The specimens (e.g. *Aequorea*) have to be collected during a favourable season at a limited number of locations throughout the world. In order to isolate 125 mg of aequorin, which was required for structural studies, two metric tons of the parent luminous organisms had to be collected. Furthermore, aequorin cannot be used repeatedly, since the light-emitting reaction of an aequorin molecule is irreversible. In order to utilize aequorin in Ca^{2+} measurements all of the procedures carried out before using the photoprotein must be per-

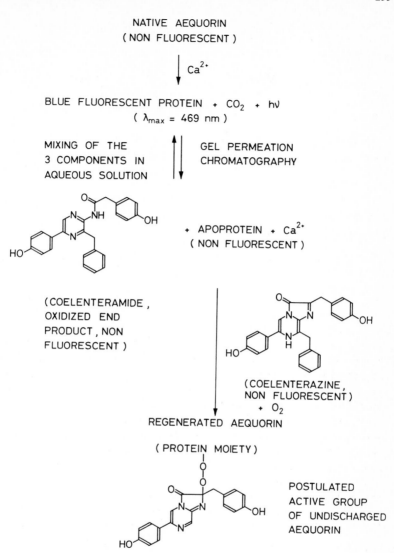

NATIVE AEQUORIN
(NON FLUORESCENT)

Ca^{2+}

BLUE FLUORESCENT PROTEIN + CO_2 + hν
(λ_{max} = 469 nm)

MIXING OF THE
3 COMPONENTS IN
AQUEOUS SOLUTION

GEL PERMEATION
CHROMATOGRAPHY

+ APOPROTEIN + Ca^{2+}
(NON FLUORESCENT)

(COELENTERAMIDE,
OXIDIZED END
PRODUCT, NON
FLUORESCENT)

(COELENTERAZINE,
NON FLUORESCENT)
+ O_2

REGENERATED AEQUORIN

(PROTEIN MOIETY)

POSTULATED
ACTIVE GROUP
OF UNDISCHARGED
AEQUORIN

Fig. 9.36. Reactions of the bioluminescent photoprotein aequorin

formed under Ca^{2+}-free conditions. Details of the sources, extraction, purification, activity, storage and handling have been described in the literature (see for example [Blinks et al. 82]).

9.6.1.3 Properties and Use of Photoproteins

Shimomura and Johnson first showed that the photoprotein aequorin obtained from the jelly-fish *Aequorea forskålea* could be stimulated to emit light by the addition of Ca^{2+} [Shimomura et al. 62]. Five years later, in 1967, and nine years before the first intracellular study with Ca^{2+}-selective microelectrodes [Owen et al. 76], aequorin was injected into giant muscle fibres in order to study of Ca^{2+} transients [Ridgway and Ashley 67]. Since that time, much information on the properties of photoproteins and on the bioluminescence technique has been published (for reviews see [Morin 74; Ashley and Campbell 79; Thomas 82; Blinks et al. 82]). A brief discussion of the most important parameters of the technique is given below.

Structural Properties. The constitution of the light-emitting moiety (coelenterazine) is known and a coelenterazine-oxygen-protein complex has been postulated (Fig. 9.36). The oxidized product (coelenteramide), isolated from discharged aequorin, has also been characterized [Shimomura and Johnson 72]. However, the structure of the protein moiety has not been fully elucidated. The protein has a molecular weight of about 20000 [Blinks et al. 82], and possesses several negatively charged centres at physiological values of the pH [Campbell et al. 79].

Luminescence as a Function of the Concentration of Free Ca^{2+}. The measurement of the intensity of the emitted light allows a quantification of the luminescence. A common expression for the relationship between the intensity of light and the concentration of Ca^{2+} is based on the intensity of light (L) actually measured as a fraction of the intensity emitted at a saturating concentration of Ca^{2+} (L_{max}). Figure 9.37 shows a typical double-logarithmic representation of the fractional emission of light (L/L_{max}) as a function of the concentration of free Ca^{2+}. The detection limit at low concentrations of Ca^{2+} is given by an EGTA-irreducible luminescence, while the limit at high concentrations corresponds to the maximal luminescence of the saturated photoprotein (completely occupied Ca^{2+} binding sites). The maximum slope of the linear part of the curve is about 2.5 [Allen et al. 77]. Often, a quadratic relationship between the concentration of free Ca^{2+} and the emission of light is accepted to give valuable results [Thomas 82]. A detailed quantification of the method is still under discussion. A description of the process by a two-state model assuming three calcium binding sites seems to be satisfactory [Allen et al. 77].

Selectivity. Certain interfering compounds are also capable of triggering luminescence. The lanthanides are more effective triggers than Ca^{2+}. Sr^{2+} ions are less effective by a factor of about 100, while Ba^{2+} ions are even less effective. Mg^{2+} ions decrease the luminescence in the absence of Ca^{2+} and therefore act as an antagonist to Ca^{2+} [Blinks et al. 78]. The opposite effect of Mg^{2+} was considered in the two-state model mentioned above [Allen et al 77]. Under intracellular conditions, the influence of Mg^{2+} ions cannot be neglected. A Mg^{2+} concentration of 3 mM reduces the luminescence at 10 μM Ca^{2+} by about one order of magnitude. Another group of interfering ions decreases the sensitivity or the detection limit of the Ca^{2+} response. As an antagonist, Mg^{2+} belongs to this class of interfering ions. At typical physiological levels, Na^+ and K^+ do not significantly disturb the luminescence/Ca^{2+} relationship. However, the calibrations have to be performed in an ionic environment as similar as possible to the intracellular fluid. At high levels (several hundred millimolar) potassium exhibits an inhibitory effect comparable with that of magnesium. Ag^+ ions inactivate aequorin even at 10^{-8} M. Hg^{2+} ions show a similar poisoning effect. On the other hand, there are some substances, including some

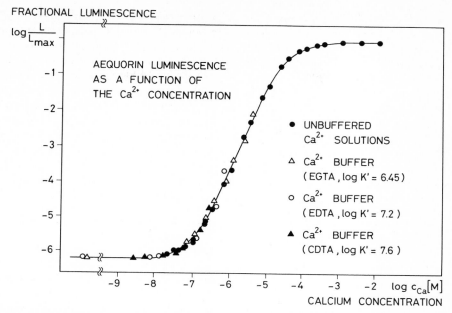

Fig. 9.37. Relationship between luminescence of aequorin and concentration of free Ca^{2+} [Allen et al. 77]. A double-logarithmic plot of the results obtained in $CaCl_2$ and Ca^{2+}-buffered solutions is shown. All of the solutions contained 150 mM KCl and were buffered to pH 7 with 5 mM PIPES. K′ is the apparent association constant for the complexones under the conditions of the experiment [Allen et al. 77]. Redrawn from Allen, Blinks and Prendergast, Science, 1977 [Allen et al. 77]

anesthetics, which enhance the emission of light from aequorin [Thomas 82]. The influence of changes in pH on the aequorin reaction is rather weak, so that severe disturbances are not expected to occur between pH 5.5 and 8.5.

Detection Limit. The lower limit of detection is governed by a residual, Ca^{2+}-independent level of luminescence which is influenced by the presence of physiologically relevant cations. For example, a change in the concentration of K^+ from 10 to 500 mM produces a shift in the luminescence curve (Fig. 9.37) by 1.5 log units to higher values. Detection limits that have been reported for typical intracellular conditions are $3 \cdot 10^{-8}$ M [Allen et al. 77], $< 10^{-8}$ M [Shimomura and Johnson 79] and $< 10^{-7}$ M [Blinks et al. 82]. The detection limit depends heavily on the temperature. The Ca^{2+}-independent luminescence is about 70 times greater at 40 °C than at 0 °C [Allen et al. 77].

Response Time. The kinetics of the Ca^{2+}-triggered bioluminescence reaction of aequorin has been carefully studied with regards to understanding the molecular processes involved and the evaluation of practical response times under various conditions [Hastings et al. 69; Ashley and Campbell 79; Thomas 82; Blinks et al. 82]. It was found that the emission of light takes place at a rate of about $100 \, s^{-1}$ (rate constant for the decomposition of an activated intermediate state to a discharged molecule of aequorin [Hastings et al. 69]). The response time of the luminescence of aequorin to a stepwise change in the concentration of Ca^{2+} is usually given as 10 ms.

Introduction of Photoproteins into Cells. A sufficient amount of photoprotein has to be incorporated into the cytoplasm without causing any damage to the cell membrane. Microinjection has proved to be the most suitable method. Ionophoretic application of the photoprotein is not possible because of the low mobility and the high negative charge of the protein. As a result, the electrical charge is carried by positively charged sample ions moving into the injection microelectrode rather than by an outflow of aequorin. Pressure injection works well and is only limited by the viscosity of the solution to be injected. The contamination of the photoprotein by Ca^{2+} from the glass of the micropipette is a serious problem. For small cells microinjection is very demanding. Hence, a series of other techniques have been introduced [Ashley and Campbell 79]. Liposome fusion has been widely proposed. However, the amount of aequorin within one liposome is often not sufficient to obtain an adequate signal in the cytosol [Prendergast 82].

Distribution Within Cells by Diffusion. Quantitative measurements of luminescence require a uniform distribution of the injected photoprotein within a cell. Erroneous results are obtained with nonuniform distributions because of the nonlinear relationship between luminescence and concentration of free Ca^{2+}. The degree of error has been illustrated in a consideration of hypothetical nonuniform distributions [Blinks et al. 82]. Several values for the diffusion coefficient of aequorin have been cited in the literature, and a value of $7 \cdot 10^{-7} \, cm^2 \, s^{-1}$ is widely accepted [Campbell et al. 79]. However, it has been observed that the diffusion of aequorin in muscle fibres is considerably slower (diffusion coefficient of $5 \cdot 10^{-8} \, cm^2 \, s^{-1}$) than predicted [Blinks et al. 82]. There is good evidence that photoprotein do not penetrate internal membranes of cells, since it has been observed that aequorin is excluded from organelles such as the sarcoplasmic reticulum or mitochondria. The penetration of photoprotein molecules into the nuclei of cells from insect salivary glands has been observed (see [Blinks et al. 82]). A theoretical consideration has shown that it takes 8–10 minutes to load myofibrils to about 90% with photoprotein [Campbell et al. 79].

Calibration. Several calibration and evaluation procedures have been discussed. Each assumes a uniform distribution of the concentration of free Ca^{2+} within the cell. For the determination of Ca^{2+} concentrations under resting conditions, the first approach mentioned below seems to be very suitable. A series of calcium-EGTA buffers are injected into a cell loaded with aequorin until a buffer solution is found which just alters the luminescence of the cell (before treatment with Ca^{2+} buffer; resting glow). Obviously, for such an internal calibration method the luminescence relationship need not be known. The procedure is however limited to large cells in which the compositions of the cytoplasmic electrolytes are well known. It is also important that the Ca^{2+} buffers exhibit almost the same pH, Mg^{2+} concentration and ionic strength as the cytosol. If this is not the case, the assumed dissociation constant for the Ca^{2+} complex of EGTA in an intracellular environment is not reliable.

Another approach relies on the double-logarithmic plot of the luminescence relationship (see Fig. 9.37). The curve is evaluated by performing an external calibration in solutions whose compositions, ionic strengths and temperatures are comparable to the sample. Photoprotein originating from the same batch must be employed

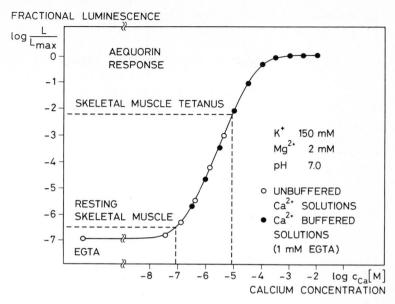

Fig. 9.38. Luminescence of aequorin as a function of the concentration of Ca^{2+} determined in vitro under conditions simulating an intracellular environment [Blinks et al. 82]. The point designated EGTA *(lower left)* was determined in a 1 mM solution of EGTA (no calcium). Dashed horizontal lines indicate the fractional luminescence of a muscle cell injected with aequorin (at rest and during tetanic contraction). Redrawn from Blinks, Wier, Hess and Prendergast, Progr. Biophys. Molec. Biol., 1982 [Blinks et al. 82]

for both the calibration and the intracellular experiment. It is advantageous to plot the relative (fractional) luminescence (log L/L$_{max}$) as a function of the logarithm of the concentration of the free Ca^{2+} (Fig. 9.38). L$_{max}$ can be determined by instantaneously exposing the intracellular aequorin to a saturated solution of Ca^{2+}. This is achieved by adding a detergent such as Triton X-100 to the cell which rapidly destroys the integrity of the membrane. Figure 9.38 illustrates such an intracellular measurement of Ca^{2+} [Blinks et al. 82].

Further Parameters. A series of other properties of photoproteins have been studied: Toxicity, physiology, reactivation [Ashley and Campbell 79], binding of calcium [Ashley and Campbell 79; Thomas 82; Blinks et al. 82], contamination of calcium and contamination of chelating agent [Blinks et al. 82].

9.6.1.4 Measurement of the Luminescence of Photoproteins

A series of commercially available photometers have been designed for luminescent reactions. They have been developed for the assay of ATP using the luciferin/luciferase system of the firefly (see Sect. 9.6.1.1). However, users of photoproteins usually construct their own measuring equipment in order to enhance the sensitivity. In brief, the apparatus consists of a cuvette in a light-tight chamber in front of a

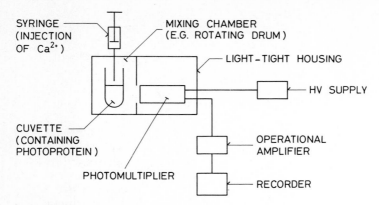

Fig. 9.39. Schematic representation of an equipment for the measurement of bioluminescence

photomultiplier connected to a recording system (see Fig. 9.39). The most important demands put on such a device are a fast, reproducible mixing and a high sensitivity (low detection limits). Mixing times of the order of 100 ms are feasible using typical chambers. More sophisticated procedures allow the mixing of reactants in about 1 ms [Campbell et al. 79]. Photomultiplier tubes are used as highly sensitive detectors of photons (for a discussion see [Campbell et al. 79; Thomas 82]). The light of the bioluminescence is measured against a dark background which makes the detection of the signal easy. A combination of the photoprotein approach with image-intensifier techniques [Reynolds 79; Thomas 82] is very attractive; for example, for the spatial resolution of Ca^{2+} gradients within the cytoplasm of a cell.

9.6.2 Metallochromic Indicators

9.6.2.1 Currently Used Indicators

Metallochromic dyes are compounds of relatively low molecular mass which show changes in their absorbance or fluorescence upon binding to metal ions. The difference between the spectra of the free and complexed indicator usually allows quantitative measurements of the concentrations of free metals in solution to be made. The first application of such indicators for the determination of Ca^{2+} in biological systems was reported in 1963 [Ohnishi and Ebashi 63]. Since that time, a series of metallochromic indicators have been proposed. They are now classified according to their structure and their time of appearance, i.e. first (e.g. murexide), second (e.g. arsenazo III, antipyrylazo III) or third generation (e.g. quin 2) of dyes (Fig. 9.40).

9.6.2.2 Properties of the Indicators

The spectrophotometric method is based on the reversible formation of a complex between the indicator molecule and Ca^{2+} ions. A knowledge of the dissociation

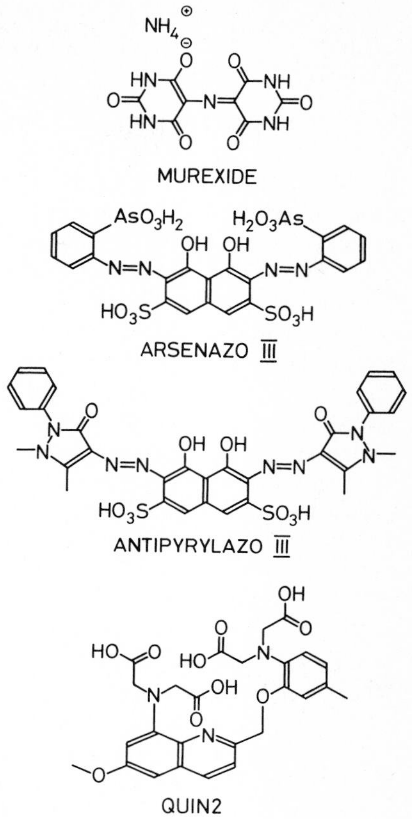

MUREXIDE

ARSENAZO $\overline{\underline{\text{III}}}$

ANTIPYRYLAZO $\overline{\underline{\text{III}}}$

Fig. 9.40. Constitution of metallochromic indicators

QUIN2

constant K_d of this equilibrium enables the concentration of free Ca^{2+} to be evaluated. The optical detection allows a fast and sensitive measurement. The technique has been extensively discussed in the literature [Ashley and Campbell 79; Scarpa 82; Blinks et al. 82; Thomas 82]. Some aspects of the method are summarized in the following sections.

Dissociation Constant. The dissociation constant K_d characterizes the formation of a complex between the indicator and Ca^{2+}. To minimize disturbances the indicator should be present predominantly in the free form during the measurements of Ca^{2+}. Indicators can therefore only be used in limited concentration ranges of Ca^{2+}. Different Ca^{2+} indicators have to be used if the concentration of Ca^{2+} varies widely during an experiment. An ideal indicator for the measurement of resting intracellular concentrations of free Ca^{2+} should exhibit a relative high affinity, i.e. a K_d of about 100 nM. Indicators of lower affinity (K_d around 10 μM) are more suited for the measurement of changes in Ca^{2+} at higher concentrations. Murexide (K_d ~2 mM) has an optimal working range from 10^{-4} to 10^{-3} M Ca^{2+}, arsenazo III (K_d ~30 μM) from 10^{-6} to 10^{-7} M Ca^{2+} and antipyrylazo III (K_d ~150 μM) from 10^{-5} to 10^{-6} M Ca^{2+} [Scarpa 82]. Indicators with much lower K_d's are needed for measurements at very low resting concentrations (10–100 nM). The fluorescent dye quin 2 (Fig. 9.40) has a value of K_d (0.1 μM) that is suitable for use in this range [Tsien 80].

Differential Measurement. The spectral characteristics of the currently used indicators have been summarized in the literature (see e. g. [Scarpa 79; Scarpa 82; Thomas 82]). Maximal differences in molar extinction coefficients ε (free indicator versus indicator in saturated Ca^{2+} solution) have been reported [Thomas 82]: 6000 (murexide), 25000 (arsenazo III), 7000 (antipyrylazo III) and 30000, or a five-fold fluorescence (quin 2). These differences make it possible to use the indicators at concentrations of 10–100 μM. It is advantageous to carry out the measurements at long wavelengths because of the normal background absorbance of the cell. The changes in absorbance to be measured are usually very small (e. g. $\Delta A = 0.0005$). Thus, in order to minimize any side effects (scattering of light, instability of light source, volume changes), the absorbance is measured at two wavelengths (see discussion in Sect. 9.1.2). Furthermore, the differential measurement helps to eliminate contributions to the observed absorbances from interfering ions (differential measurement denotes the recording of the difference of the two signals measured simultaneously from the same sample at two different wavelengths). Suitable pairs of wavelengths have been proposed for various indicators [Scarpa 82]: murexide 507/540 nm, arsenazo III 675/685 nm or 650/690 nm, antipyrylazo III 670/690 nm or 720/790 nm. The pairs of wavelengths are chosen close together, so that the corrections for nonspecific changes in absorbance are optimal. Using the Lambert-Beer law, calibration curves for a selected pair of wavelengths as a function of the concentration of free Ca^{2+} can be obtained [Blinks et al. 82]. The use of such calibration curves implies that the stoichiometry of the complex, the dissociation constant K_d and the differences in the extinction coefficients are the same in both the calibration solutions and the cell. The situation is more complicated for dyes which form complexes of higher stoichiometry. The calibration curves will then vary considerably if the concentration of the indicator is changed [Blinks et al. 82; Thomas 82].

Selectivity. The selectivity of a dye can be defined as the ratio of the dissociation constants of the complexes formed with different cations. This description of the selectivity assumes that the changes in the absorbance due to binding with different cations are the same. This has not been confirmed in practice. The selectivity depends on the choice of the two wavelengths and on the concentration of the indicator if different stoichiometries are involved [Thomas 82] (see also equation for selectivity in [Blinks et al. 82]).

The most strongly interfering ions are usually H^+ and Mg^{2+}. Both yield absorption spectra similar to that of the Ca^{2+} complex, but compete with Ca^{2+} for binding to the indicator. The latter effect may cause a reduction in the concentration of dye available for complexing to Ca^{2+}. The Ca^{2+}/H^+ and Ca^{2+}/Mg^{2+} selectivities of the indicators mentioned above have been characterized in the following manner:

Murexide: a $2 \cdot 10^{-5}$ M solution of murexide and $5 \cdot 10^{-3}$ M Ca^{2+} does not show a significant shift in its spectrum if Mg^{2+} is added up to 10^{-2} M [Ogawa et al. 80]; a Ca^{2+}/Mg^{2+} selectivity of about 2000 was calculated (500/544 nm) [Ohnishi and Ebashi 78]; only a slight interference due to H^+ ions is observed at a pH of about 7.

Arsenazo III: different Ca^{2+}/Mg^{2+} selectivities have been reported: 39, 87 and 4000 (see literature given in [Blinks et al. 82]); significant interferences are to be expected if the concentration of Mg^{2+} [Scarpa 79] and/or the pH [Scarpa 79; Blinks et al. 82] change during the measurement; the value of K$_d$ stronlgy depends on the H$^+$ activity [Blinks et al. 82].

Antipyrylazo III: optimal wavelengths have to be selected because of interference by Mg^{2+}; the interference due to H$^+$ is comparable to that encountered with arsenazo III.

Quin 2: This fluorescent dye binds Ca^{2+} better than Mg^{2+} by a factor of $> 2 \cdot 10^4$; quin 2 seems to be less sensitive to Mg^{2+} and H$^+$ ions than the other dyes [Blinks et al. 82].

Absorbing cellular chromophores (cytochromes, carotenoids) can cause some interference, but can be avoided by caring out the measurements in the 600–700 nm range.

Detection Limit. The detection limit strongly depends on the K$_d$ of the indicator. The sensitivity is further influenced by the optical detection system, the concentration of the indicator, the molar extinction coefficient of the complex and the path length in the sample [Blinks et al. 82]. The dyes currently used exhibit sufficiently low detection limits to allow the measurement of intracellular concentrations of free Ca^{2+}.

Response Time. A major advantage of metallochromic indicators is the very fast rate of complex formation, which allows the observation of short Ca^{2+} transients within cells. Temperature-jump experiments yielded relaxation times of 2 µs (murexide), < 2.8 ms (arsenazo III) and 180 µs (antipyrylazo III) [Scarpa 79]. Accordingly, the response time can be expected to lie in the millisecond range for arsenazo III and in the submillisecond range for the other indicators. These extremely short response times allow the observation of most changes in Ca^{2+} concentration within cells.

Other Properties. Other important properties of various indicators including the stoichiometry, buffer capacity, binding to cellular constituents, diffusion into the cell membrane and toxicity have been discussed in review articles [Ashley and Campbell 79; Scarpa 82; Thomas 82; Blinks et al. 82].

9.6.2.3 Measurements with Metallochromic Indicators

Most of the metallochromic indicators are available commercially, but they often have to be purified before use. The dyes are incorporated into the cells by ionophoresis, pressure injection or internal dialysis. An interesting approach has been used for the introduction of fluorescent dyes (quin 2) into cells. The compounds are made temporarily membrane-permeable by masking their carboxylic acid groups as ester groups. After penetration into the cell, enzymatic hydrolysis regenerates the free acid [Tsien 81]. For example, quin 2 (Fig. 9.40) diffuses through the cell membrane as acetoxymethyl tetraester. The subsequent intracellular cleavage of the esters by an esterase releases the nonpermeating and optically active tetracarboxylic indicator into the cytoplasm. It is uncertain whether the by-product of the ester cleavage (formaldehyde) has a toxic effect on the cells. Accumulation of indicator

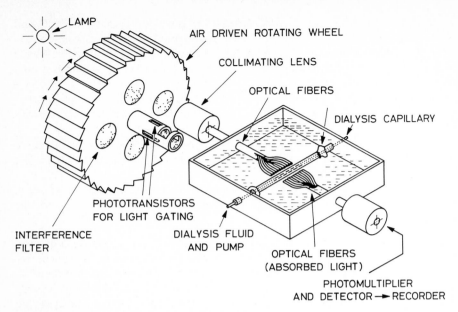

Fig. 9.41. Schematic representation of a chopped beam, multi-wavelength spectrophotometer for measurements of Ca^{2+} in a large single cell (e.g. axon of giant squid or fibre of barnacle) [Scarpa 82]. Redrawn from Scarpa, Techn. Cell. Physiol., 1982 [Scarpa 82]

has not been observed in organelles [Pozzan et al. 81]. However, the technique has not proven to be successful in all types of cells (see [Blinks et al. 82]).

As shown above, differential measurements are advantageous. Multiwavelength spectrometers are therefore necessary. A schematic view of a chopped beam, multi-wavelength, differential spectrophotometer is shown in Fig. 9.41 [Scarpa 82]. Light emitted from a tungsten iodine lamp is passed through a rotating disk (up to 1000 Hz) containing four interference filters.

For a gating of the light (encoder), stationary axial rings containing phototransistors are attached. The light which passes through the filters is focussed with a collimating lens, transmitted through the sample, and detected with a photomultiplier tube [Scarpa 82]. A detailed description of other arrangements for the determination of changes in absorbance of metallochromic indicators in physiological systems is described by [Thomas 82].

Conventional fluorometers are inappropriate for most applications using fluorescent indicators such as quin 2. Recordings from single cells require highly sensitive microfluorometers (see Sect. 9.1 and [Blinks et al. 82]). Microscope image intensifier studies on single cells loaded with quin 2 have been carried out [Ashley et al. 85].

In principle, determinations of absolute absorption can serve as calibrations and lead to correct results if K_d, the extinction coefficients, the concentration of the indicator and the path length of the light are known. However, uncertainties in the concentration of the injected indicator and the dependence of K_d and the extinction coefficients on the pH and the ionic strength usually prevent an application of this simple calibration and evaluation procedure.

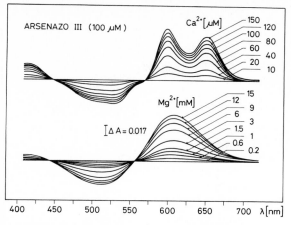
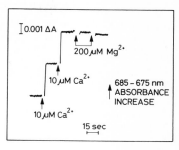

Fig. 9.42. Difference spectra of arsenazo III and arsenazo III in the presence of various concentrations of Ca^{2+} and Mg^{2+} [Scarpa 82]. Both cuvettes contained 500 mM KCl, 10 mM MOPS (pH 6.8) and 100 μM arsenazo III. Ca^{2+} and Mg^{2+} were added to the measuring cuvette before each scan at the concentrations indicated. The traces on the right illustrate the influence of Ca^{2+} and Mg^{2+} on the differential measurement of arsenazo III as a function of the time. Recordings were made at 675/685 nm in a single cuvette using a dual wavelength spectrophotometer. Redrawn from Scarpa, Techn. Cell. Physiol., 1982 [Scarpa 82]

Changes in absorbance can be calibrated in vitro using aqueous solutions whose ionic strength, pH and Mg^{2+} concentration are as close as possible to the sample. For this purpose, the concentration of the indicator and the optical path length must be known. In order to compare the optical, geometric and ionic conditions of the calibration solutions and the sample, capillaries of a shape similar to that of the cell are employed [Brinley 78].

A more satisfying calibration is obtained by adding a known concentration of Ca^{2+} or EGTA to the sample (standard addition or standard subtraction). Such in vivo calibrations can be achieved by internal dialysis (Fig. 9.41) [Scarpa 82].

An illustration of typical difference spectra, the differential technique and the influence of the interfering ion Mg^{2+} is given in Fig. 9.42 [Scarpa 82].

Discussions of other calibration procedures, the multi-wavelength technique, the stoichiometry problem and the buffering of Ca^{2+} by the injected indicators are given in the literature [Scarpa 82; Thomas 82; Blinks et al. 82].

9.6.3 Potentiometry

9.6.3.1 Survey of Ca^{2+}-Selective Microelectrodes

Two types of Ca^{2+}-selective liquid membrane microelectrodes have been developed. One type, using membrane solutions based on electrically charged lipophilic esters of phosphoric acid, was introduced in 1976 [Christoffersen and Johansen 76; Brown et al. 76] (see Tabl. 9.15).

Table 9.15. Course of the development of Ca^{2+}-selective microelectrodes

Membrane solution, Reference (see Fig. 9.43)	Development	Year, Authors, Reference
liquid (PVC), ion-exchanger 1 dissolved in 3	first ion-exchanger liquid membrane microelectrode, tip diameter 20 μm	1976, Christoffersen, Johansen, [Christoffersen and Johansen 76]
liquid (PVC), ion-exchanger 2 dissolved in 3	improved ion-exchanger liquid membrane microelectrode, tip diameter 1–2.5 μm	1976, Brown, Pemberton, Owen, [Brown et al. 76]
liquid, neutral carrier 5 dissolved in 4, 6 as the additive	first neutral carrier-based microelectrode, tip diameter 2 μm	1976, Oehme, Kessler, Simon, [Oehme et al. 76]
liquid, neutral carrier 5 dissolved in 4, 9 as the additive	new additive (9), improved stability, reduced hysteresis, tip diameter 0.4 μm	1980, Tsien, Rink, [Tsien and Rink 80]
liquid, neutral carrier 5 dissolved in 3-nitro-o-xylene, 8 as the additive	new additive (8), improved Ca^{2+} electrode function	1980, O'Doherty, Youmans, Armstrong, [O'Doherty et al. 80]
liquid (PVC), neutral carrier 5 dissolved in 4, 7 as the additive	use of PVC, new additive (7), elimination of shunts, very small tips	1980, Marban, Rink, Tsien, Tsien, [Marban et al. 80]
liquid (see [Oehme et al. 76])	coaxial-type of microelectrode, reduced resistance, fast response	1980, Ujec, Keller, Kriz, Pavlik, Machek, [Ujec et al. 80]
liquid (PVC), membrane solution as in [Oehme et al. 76]	original membrane solution with PVC for microelectrodes with very small tip diameters	1982, Lanter, Steiner, Ammann, Simon, [Lanter et al. 82]

In the same years, Oehme, Kessler and Simon [Oehme et al. 76] developed a membrane solution containing a synthetic neutral carrier (ETH 1001, [Ammann et al. 74]) as the ion-selective component (see Fig. 9.43 and Tab. 9.15).

Both types of microelectrodes exhibit a high preference of Ca^{2+} over Na^+ and K^+. However, the neutral carrier-based microelectrode exhibits much better Ca^{2+}/Mg^{2+} and Ca^{2+}/H^+ selectivities. Indeed, the use of an organo-phosphate-based microelectrode is severely limited by the low rejection of Mg^{2+} ions. A direct comparison of the two liquid membranes has been carried out for physiological concentrations of electrolytes and has confirmed the advantages of the neutral carrier-based microelectrode (Fig. 9.44) [Tsien and Rink 81].

The composition of membrane solutions containing ETH 1001 has been modified several times with regard to improving the performance of the microelectrode

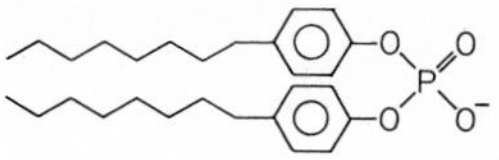

1 (DOPP$^-$)

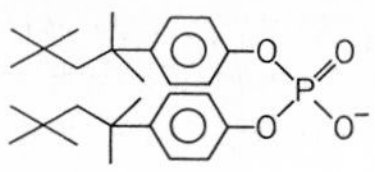

2 (DTMBPP$^-$)

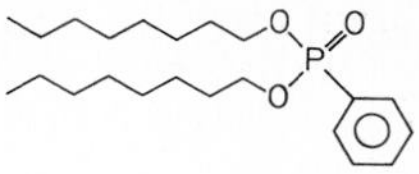

3 (DOPP)

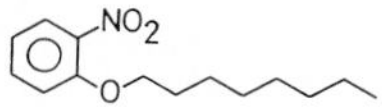

4 (o-NPOE)

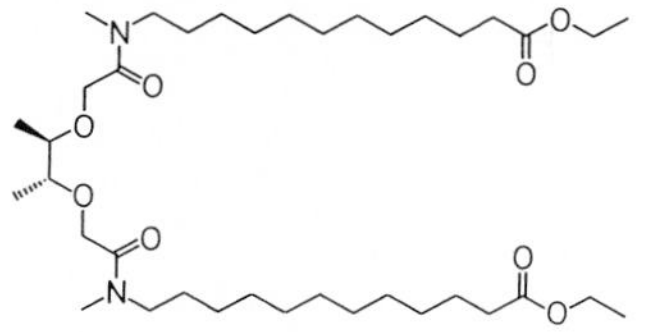

5 (ETH 1001)

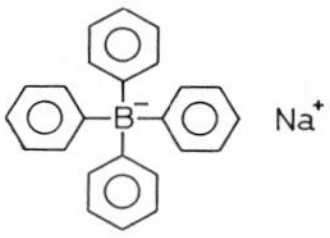

6 (NaTPB)

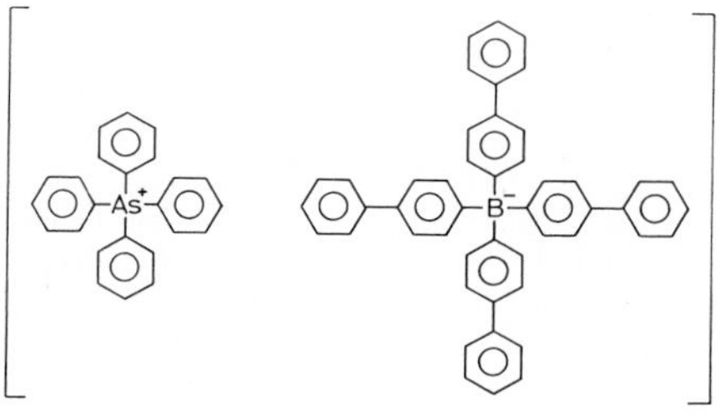

7 (TPAs-TBPB)

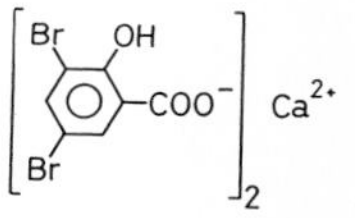

8 (Ca-3,5 DBS)

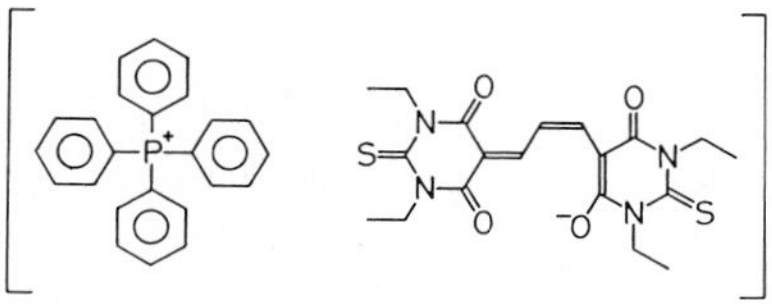

9 (TPP-BDTBTO)

Fig. 9.43. Components for Ca^{2+}-selective liquid membrane microelectrodes

(Tab. 9.15). Tsien and Rink [Tsien and Rink 81] gave a better explanation of the reasons the deviations from ideal electromotive behaviour which are occasionally observed and proposed a procedure for improving the membrane solutions to be used in very small microelectrodes (see Tab. 9.15).

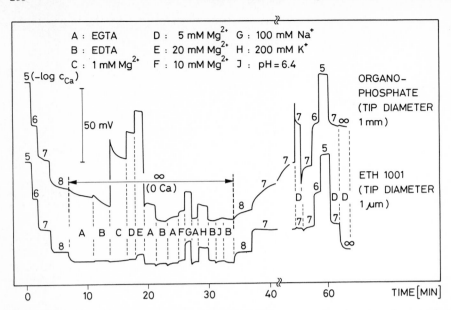

Fig. 9.44. Comparison of the electrode responses of an organophosphate-based macroelectrode [Craggs et al. 79] and a neutral carrier-based (ETH 1001) microelectrode to low concentrations of Ca^{2+} and other intracellular cations [Tsien and Rink 81]. Redrawn from Tsien, Rink, J. Neurosci. Meth., 1981 [Tsien and Rink 81]

9.6.3.2 Course of the Development of Optimized Neutral Carrier-Based Ca^{2+} Microelectrodes

Most of the initial experiments using neutral carrier-based membrane solutions were carried out with microelectrodes that had tip diameters larger than 1 μm. However, disturbances in the electrode function were repeatedly observed with very small microelectrodes (<1 μm). The major disadvantages of the submicron microelectrodes filled with the original membrane solution [Oehme et al. 76] were an increase in the detection limits and/or the larger-than-expected slopes and hysteresis of the electrode function.

Some examples of reported Ca^{2+} electrode functions are shown in Fig. 9.45. Using the original membrane solution [Oehme et al. 76], a larger-than-Nernstian slope in the Ca^{2+} concentration range between 10^{-4} and 10^{-6} M and a relative high detection limit were observed by Coray and coworkers (curve 1) and Tsien and Rink (curve 2). However, these defects were not observed with each microelectrode [Tsien and Rink 80]. Indeed, Lee and coworkers reached extremely low detection limits with the original membrane solution in very small microelectrodes (curve 5).

Initially, it was assumed that the defects were caused by kinetic limitations of the diffusion of Ca^{2+} and other ions between the membrane phase and the sample solution. Such phenomena are well documented for ion-exchanger membrane electrodes [Hulanicki and Lewandowski 74; Seto et al. 75]. The proposed theories have been useful in explaining the hysteresis effect (by assuming the presence of a hydro-

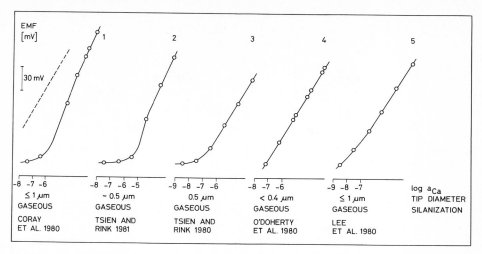

Fig. 9.45. Ca²⁺ electrode functions of various microelectrodes based on the neutral carrier ETH 1001. *Dashed line:* theoretical slope. *Curves: 1* [Coray et al. 80]; *2* [Tsien and Rink 81]; *3* [Tsien and Rink 80]; *4* [O'Doherty et al. 80]; *5* [Lee et al. 80b]

philic, exchangeable cation in the membrane phase (Na⁺ from the additive sodium tetraphenylborate)). Consequently, membrane solutions in which the NaTPB was replaced by other salts have been proposed: tetraphenylphosphonium bis(1,3-diethyl-2-thiobarbiturate) trimethineoxonol (9 in Fig. 9.43. [Tsien and Rink 80], curve 3 in Fig. 9.45), tetraphenylarsonium tetrakis(p-biphenyl) borate (7 in Fig. 9.43 [Marban et al. 80; Tsien and Rink 81]) and calcium 3,5-dibromosalicylate (8 in Fig. 9.43 [O'Doherty et al. 80], curve 4 in Fig. 9.45). Membrane solutions containing the additive 9 yielded Ca²⁺ electrode functions without hysteresis, i.e. the measurements of calibration curves from low to high and from high to low concentrations did not deviate by more than 2 mV [Tsien and Rink 80]. However, the detection limit was reduced when very small micropipettes were used; (an advantage of 9 is its bright red colour which makes the microelectrodes easier to localize during experiments). The performance of microelectrodes containing the salt 8 is not reproducible [Tsien and Rink 81; Lanter et al. 82]. It seems that changes in the relative humidity during the preparation of the microelectrodes have an influence on the characteristics of the microelectrodes [O'Doherty et al. 80].

Diffusion phenomena cannot satisfactorily explain all of the encountered problems. Considerable improvements were achieved by treating the membrane solutions containing the salt 7 with poly(vinyl chloride). The corresponding microelectrodes with small tip diameters showed ideal electrode functions and low detection limits throughout [Tsien and Rink 81]. The improvement was thought to be due to the elimination of electrical shunts through the glass wall at the tip of the microelectrode. Equivalent circuits of microelectrodes containing membrane solutions with and without PVC were proposed (Fig. 9.46).

This model was useful for the simulation of the over-Nernstian behaviour of Ca²⁺ electrode functions [Tsien and Rink 81]. Similar artifacts due to electrical

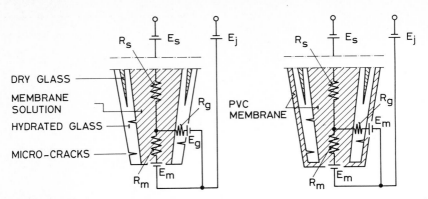

Fig. 9.46. Equivalent circuits for the tips of Ca^{2+}-selective microelectrodes. *Left:* original PVC-free membrane solution [Oehme et al. 76]. *Right:* PVC-containing membrane solution [Tsien and Rink 81]. R_m: electrical resistance at the phase boundary membrane/sample. E_m: potential difference at the membrane/sample boundary phase. R_g: electrical leakage resistance of the thin partly hydrated glass wall near the tip. E_g: potential difference at the glass/sample boundary phase. R_s: electrical resistance in the bulk of the membrane solution (in series to R_m). E_s: potential differences generated inside the microelectrode. E_j: potential differences generated at the reference microelectrode

shunts have been previously discussed [Lewis and Wills 80; Armstrong and Garcia-Diaz 80].

The electrical resistance R_m at the membrane/sample solution boundary phase increases with decreasing concentrations of Ca^{2+} in the sample. Therefore, at very low Ca^{2+} activities, the measured EMF is shunted to a greater extent by the constant electrical resistance of the glass wall at the tip (R_g). Hence, the electrode potential (E_m) approximates the potential generated across the glass wall (E_g). The over-Nernstian response indicates that E_g must be more negative than E_m since the shunt shifts the measured potential to negative values at low concentrations of Ca^{2+}. E_g is virtually independent of the Ca^{2+} concentration since the cation-selectivity of the glass is much larger for H^+ and Na^+ ions. In order to utilize the beneficial effect of PVC-containing membrane solutions, the micropipettes must be filled through the tip, so that the glass wall is coated with membrane material (see Fig. 9.46). The glass at the tip is then protected from the sample solution and the ion-selective membrane material covers both the inside and outside of the micropipette, i.e. E_g can be replaced by E_m.

The usefulness of PVC-containing membrane solutions is illustrated in Fig. 9.47. Microelectrodes containing the original PVC-free membrane solution show response curves that are dependent on the tip diameter (curves 1 and 2). Improved responses are obtained with solutions to which only PVC has been added (curve 3) or to which PVC has been added and the salt has been changed (curve 4). The largest improvement results from the addition of PVC. The lipophilic salt 7 (Fig. 9.43), which is not commercially available, does not seem to be necessary. The original membrane solution is recommended for microelectrodes with tip diameters $> 1\ \mu m$. If microelectrodes with tips $< 1\ \mu m$ are to be used, it is beneficial to treat the original membrane solution with PVC (see Sect. 9.6.3.3).

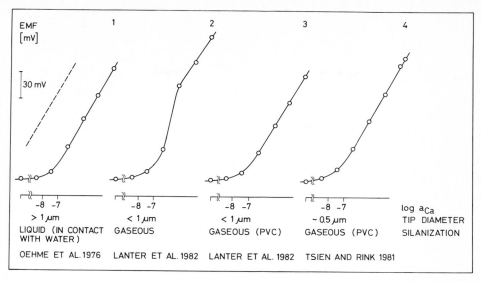

Fig. 9.47. Ca²⁺ electrode functions of various microelectrodes based on the neutral carrier ETH 1001. The responses of microelectrodes containing the original membrane solution *(curves 1 and 2)* are compared with those containing membrane solutions with PVC *(curves 3 and 4)*. Dashed line: theoretical slope. *curves 1,2:* original membrane solution, liquid silanization [Oehme et al. 76]; *3:* original membrane solution with PVC, gaseous silanization [Lanter et al. 82]; *4:* membrane solution with PVC and the salt 7 [Tsien and Rink 81]

Table 9.16. Compositions of the currently used membrane solutions for neutral carrier Ca²⁺ microelectrodes

Membrane composition [wt.-%] (Fig. 9.43)	Remarks	Reference
A) 8.0–12.0% Ca²⁺ carrier ETH 1001 (5) 3.0–8.0% tetraphenyl-arsonium tetrakis(p-biphenyl) borate (7) 12.0–17.0% poly(vinyl chloride) balance: o-nitrophenyl-n-octyl ether	proposed for any tip diameter; to fill the microelectrode the components are dissolved in 1.5–3 times their weight of tetrahydrofuran; the additive 7 is not commercially available	[Tsien and Rink 81]
B) 10.0% Ca²⁺ carrier ETH 1001 (5) 1.0% sodium tetraphenylborate (6) 89.0% o-nitrophenyl-n-octyl ether	proposed for tip diameters > 1 μm; the membrane solution and its components are commercially available (see 3.3)	[Oehme et al. 76; Lanter et al. 82]
C) 86.0% membrane B) 14.0% poly(vinyl chloride)	proposed for tip diameters < 1 μm; to fill the microelectrode the mixture is dissolved in 3 times its weight of tetrahydrofuran	[Lanter et al. 82]

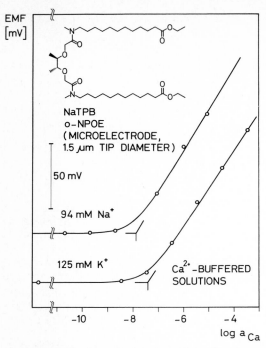

Fig. 9.48. Ca^{2+} electrode function of a microelectrode based on the membrane solution A) (Table 9.16). The Ca^{2+}-buffered solutions contained 94 mM Na$^+$ (according to Růžička et al. 73]) or 125 mM K$^+$ (according to [Tsien and Rink 81]). The detection limit [IUPAC 79] (see Sect. 5.4) is near -7.9 (Na$^+$ background) and -7.4 (K$^+$ background) log a$_{Ca}$ units

Table 9.17. Selectivity factors, log K$_{CaM}^{Pot}$, for a Ca^{2+}-selective microelectrode based on the neutral carrier ETH 1001 (membrane solution C)

Interfering ion	log K$_{CaM}^{Pot}$
Li$^+$	-3.3[a]
Na$^+$	-5.5 ± 0.2[b]
K$^+$	-5.4 ± 0.2[b]
Rb$^+$	-2.7[a]
Cs$^+$	-2.3[a]
Mg^{2+}	< -4.9[c]
Sr^{2+}	-1.5[a]
Ba^{2+}	-2.3[a]

[a] separate solution method (0.1 M solutions)
[b] fixed interference method (Na$^+$: 94 mM, K$^+$: 125 mM) using Ca^{2+}-buffered solutions (Na$^+$: 3 microelectrodes, K$^+$: 7 microelectrodes, standard deviation given)
[c] fixed interference method (Mg^{2+}: 1 M) using unbuffered Ca^{2+} solutions (values as low as -7.0 have been reported [Dagostino and Lee 82])

9.6.3.3 Properties of Currently Used Microelectrodes Based on the Neutral Carrier ETH 1001

The compositions of the currently used Ca^{2+}-selective membrane solutions are given in Tab. 9.16. Microelectrodes based on these membranes all exhibit similar properties provided that the membrane solutions are used as recommended. The following data are representative of the performance of each of the three membrane solutions (exceptions are mentioned).

Electrode Function, Detection Limit. Ca^{2+} electrode functions of Ca^{2+}-buffered solutions are shown in Fig. 9.48 (membrane solution B). Despite the presence of either a high background of Na^+ or a typical intracellular background of K^+, the detection limit is between 10^{-7} and 10^{-8} M Ca^{2+}. Detection limits of pCa 7.3 in the presence of 125 mM K^+ have also been reported (membrane solution A)) [Marban et al. 80].

Selectivity. Very high Ca^{2+} selectivities are difficult to measure. The most useful selectivity factors are obtained from measurements in Ca^{2+}-buffered solutions with a constant background of the interfering ion. So far, such factors are only available for Na^+ and K^+ (Tab. 9.17). Another useful approach for evaluating the selectivity is illustrated in Fig. 9.44 [Tsien and Rink 81] (see also [Alvarez-Leefmans et al. 81]). Figure 9.49 shows the influence of K^+ and Mg^{2+} on the Ca^{2+} electrode function.

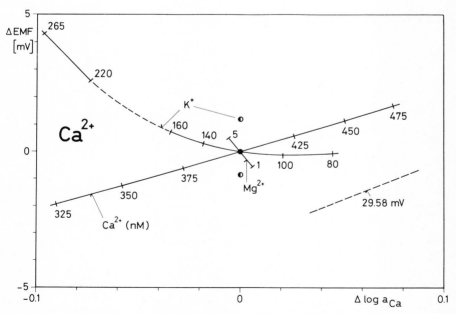

Fig. 9.49. The influence of K^+ and Mg^{2+} on the Ca^{2+} electrode function in an intracellular environment. Details of the evaluation are given in Sect. 7.8. The curves were calculated using the selectivity factors given in Tab. 9.17. Shifts of the central point for an assumed change in log K_{CaK}^{Pot} of ±0.2 are shown (◖) (each ion at its typical intracellular concentrations: Ca^{2+} (400 nM), K^+ (120 mM), Mg^{2+} (2 mM))

INTRACELLULAR Ca^{2+} MEASUREMENTS IN VOLTAGE–
CLAMPED APLYSIA NEURONS WITH NEUTRAL
CARRIER MICROELECTRODES

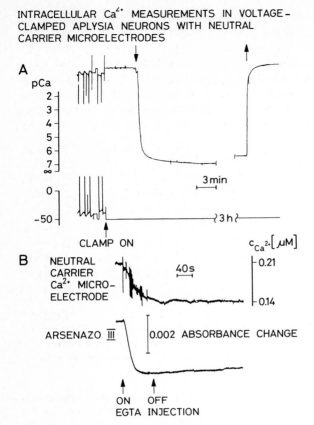

Fig. 9.50a, b. Measurements of intracellular concentrations of free Ca^{2+} during the injection of EGTA [Gorman et al. 84]. **A:** Recording in voltage-clamped neurons (3 h). **B:** Effect of the injection of EGTA on the resting concentration of free Ca^{2+}. EGTA was ionophoretically injected into a neuron. Both the EMF of the neutral carrier-based microelectrode and the absorbance signal of arsenazo III showed that the intracellular concentration of Ca^{2+} decreases with increasing amounts of intracellular EGTA. Redrawn from Gorman, Levy, Nasi, Tillotson, J. Physiol., 1984 [Gorman et al. 84]

The calculated curves (see Sect. 7.8) are based on the selectivity factors given in Tab. 9.17.

The central point of the curve at $4 \cdot 10^{-7}$ M Ca^{2+} is about 0.6 log units above the detection limit. The slope near the central point is 22.2 mV, i.e. a 1 mV change in the EMF is approximately equivalent to a 10.9% change in the Ca^{2+} activity. An analysis of the sum in the Nicolsky-Eisenman equation shows that K^+ is the most strongly interfering ion (see Sect. 7.8). It has been shown experimentally that under resting conditions an electrode response can still be observed at lower Ca^{2+} activities after an intracellular injection of EGTA [Alvarez-Leefmans et al. 81], Fig. 9.50 [Gorman et al. 84].

Response Time. The experimental arrangement may strongly influence the measured response time (e. g. mixing time of solutions). The injection of a Ca^{2+} solution into another Ca^{2+} solution gave a response time of about 10 s [Oehme et al. 76]. When Ca^{2+} was ionophoretically injected close to the tip of the Ca^{2+} microelectrode, response times of 40 to 150 ms were observed [Heinemann et al. 77]. In addition, Ca^{2+} transients of the order of 100 ms [Janus et al. 81] and values of t_{50} of the order of 100 ms have been measured [Gorman et al. 84]. The time constants of the electrode response of coaxial-type microelectrodes are as low as 7 ms [Ujec et al. 80] and 4 ms [Pumain et al. 83]. Response times of several seconds have been reported for microelectrodes with PVC-containing membrane solutions [Tsien and Rink 81].

Electrical Membrane Resistance. Membrane resistances of microelectrodes with the membrane solution B) (Tab. 9.16) are of the order of 10 GΩ: $2 \cdot 10^{10} \, \Omega$ (tip diameter 2 μm [Oehme et al. 76]), $3.5 \cdot 10^{10} \, \Omega$ (0.8 μm [Tsien and Rink 81]); $0.2\text{--}1.5 \cdot 10^{10} \, \Omega$ (2–3 μm [Heinemann et al. 77]), $1.5 \cdot 10^{10} \, \Omega$ (1–2 μm, bevelled [Isenberg and Klockner 82]) and $7 \cdot 10^{10} \, \Omega$ [Dagostino and Lee 82]. The PVC-free membrane solution B) was expelled from microelectrodes with tip diameters of 0.8 μm while the pipettes were refilled with the PVC-containing membrane solution A). The resistance increased slightly from 3.5 to $5.0 \cdot 10^{10} \, \Omega$ [Tsien and Rink 81]. With coaxial-type Ca^{2+} microelectrodes exhibiting an active filling height of only 8 μm, the resistances are reduced by a factor of about 12 [Ujec et al. 80].

Stability in EMF. Stabilities of ± 0.3 mV (std. dev.) over periods of 12 hours have been measured (10^{-1} M $CaCl_2$ solution, membrane solution B)) [Oehme et al. 76]. Improved stabilities were observed for PVC-containing microelectrodes. The EMF's of calibration solutions before and after 5 h of continuous impalement of a ventricular muscle cell agreed within ± 1.5 mV [Tsien and Rink 81].

Lifetime. Most of the reported lifetimes are of the order of days. Such short lifetimes cannot be explained by a loss in components from the membrane solution (see Sect. 3.2.4.3 and Sect. 5.10). In aqueous $CaCl_2$ solutions Oehme measured a lifetime of more than 4 months [Oehme et al. 76]. Thus, in addition to the low lipophilicity of the components of a membrane there must be other factors that are responsible for the increased detection limit frequently observed within a few hours of making a measurement (see e. g. [Lee 81]). Improved lifetimes have been found for microelectrodes with PVC-containing membrane solutions [Marban et al. 80; Tsien and Rink 81]. Calibrations remain stable for several hours and the performance of the microelectrodes is less affected by the impalements of cells (e. g. during 36 hours of physiological studies [Marban et al. 80]).

9.6.3.4 Calibration of Ca^{2+}-Selective Microelectrodes for Intracellular Studies

The general aspects of the calibration of ion-selective microelectrodes are described in Sect. 7.3. However, the calibration of Ca^{2+}-microelectrodes is considered separately because measurements at submicromolar concentrations require the use of Ca^{2+}-buffered solutions. It is not possible to prepare calibrated solutions with Ca^{2+} activities in the submicromolar range. A dilution of more concentrated $CaCl_2$ solutions to these activity levels would produce large errors due to a contamination from the water used and the container walls. Since the equilibrium constants of

Ca^{2+} buffers are defined in terms of concentrations, the calibration solutions allow the determination of the concentrations of free Ca^{2+}.

An excellent proposal for such calibration standards was made by Tsien and Rink [Tsien and Rink 81]. Suitable Ca^{2+} buffers were carefully selected according to their pK and selectivity (especially with respect to Mg^{2+}). The solutions were simultaneously pH-buffered. The slight deviation in the pH's of the different calibration solutions does not interfere with the measurement of Ca^{2+} because of the high Ca^{2+}/H^+ selectivity of the microelectrode. Each calibration solution contains 125 mM K^+, which is the most strongly interfering ion as well as the ion that determines the ionic strength. In order to calculate the concentrations of free Ca^{2+}, H^+ concentrations instead of H^+ activities were inserted into the expression for the equilibrium constant.

Other suitable calibration solutions have been reported (see e.g. [Bers 82; Otto and Thomas 84]). Topics such as the apparent stability constant of the buffers, the influence of ionic strength and temperature, the contribution of Mg^{2+} to the apparent stability constant and the use of activities or free-concentrations have been extensively discussed [Tsien and Rink 80; Lee 81; Thomas 82; Blinks et al. 82; Dagostino and Lee 82; Bers 82; Weingart and Hess 84].

Using such standard solutions the EMF is calibrated as a function of the concentration of free Ca^{2+}. Therefore, an evaluation of the measured intracellular EMF yields the cytosolic free-concentration after correction for the membrane potential. It is not meaningful to set this concentration of free Ca^{2+} equal to the Ca^{2+} activity since the fundamental physico-chemical relationship of electrolyte solutions states that the activity is strictly given by the product of the concentration and the activity coefficient. Thus, the results have to be interpreted in terms of the concentrations of free Ca^{2+} or they have to be converted to Ca^{2+} activities by using an extended Debye-Hückel equation. The latter step has been repeatedly criticized [Thomas 82; Tsien 83; Rink 83]. It seems that either concentrations or activities can be used, but the users have to be aware that the two quantities are defined in a very different way. The application of activity coefficients to intracellular media (the Debye-Hückel coefficients are only strictly valid for single electrolytes) is probably justified since a potentiometric intracellular measurement of Ca^{2+} is inevitably subject to some uncertainties.

9.6.4 Comparison of the Different Methods

The important properties of the three techniques discussed above are summarized in Tab. 9.18. None of the methods is ideal for all types of studies, so that the investigator has to chose a technique by examining the advantages and disadvantages of each approach. Ca^{2+} microelectrodes are especially suitable for determining absolute activities of Ca^{2+} under resting conditions, for monitoring relative slow changes in Ca^{2+} resting conditions, for monitoring relative slow changes in Ca^{2+} activity (in the range of seconds) and for simultaneously assessing other parameters such as the membrane potential. Photoproteins and metallochromic indicators allow the measurement of short Ca^{2+} transients. The three techniques are complementary in many respects, so that the simultaneous use of more than one approach can be very informative. Furthermore, a direct comparison of results obtained by an

Table 9.18. Comparison of methods for the determination of intracellular Ca^{2+}

Property	Photoproteins	Metallochromic indicators	Microelectrodes
availability	supply is limited; aequorin is commercially available; very expensive	commercially available; inexpensive	commercially available; moderate price
detection limit	$\sim 5 \cdot 10^{-8}$ M	$\sim 5 \cdot 10^{-8}$ M	$\sim 5 \cdot 10^{-8}$ M
selectivity	interference from Mg^{2+}; little interference from H^+	interference from Mg^{2+} and H^+	no interference from Mg^{2+} and H^+; K^+ is the most strongly interfering ion
response time	~ 10 ms (aequorin)	> 2.8 ms (arsenazo III); 0.18 ms (antipyrylazo III)	several seconds; for specially constructed microelectrodes ~ 10 ms
calibration, quantitative measurements	non-linear relationship between luminescence and concentration makes accurate quantification difficult; Ca^{2+} gradients within the cytosol should be absent; cell has to be destroyed for a quantitative assay	quantitative measurements are not satisfactory in most situations	easy calibration; useful approach for absolute determinations of Ca^{2+}
method-specific disadvantages	quantification of results is difficult; photoprotein is consumed during assay; precautions are necessary to avoid Ca^{2+} contamination; rather expensive	quantification of results is difficult; cell and its surroundings have to be transparent	relative slow response; measurements of short transients are usually not possible; no information on average cytosolic Ca^{2+}
method-specific advantages	fast response; cell motion does not interfere; easy signal detection	very fast response; no impalement with micropipette in the case of quin 2 and related dyes	quantitative determination of Ca^{2+}; simultaneous supply of membrane potential and other parameters

independent method confirms the reliability of a Ca^{2+} determination. To date, only a few studies using different techniques have been published [Ashley and Campbell 79; Stinnakre 81; Levy 82; Levy et al. 82].

9.6.5 Concluding Remarks

Neutral carrier-based microelectrodes allow the potentiometric measurement of intracellular Ca^{2+} activities. An optimized membrane solution can be used in relatively large micropipettes (> 1 μm), but must be mixed with poly(vinyl chloride) for use

Table 9.19. Other available microelectrodes for extra- and intracellular measurements

Measured species	Remarks	Reference
Cl^-	classical ion-exchanger membrane	[Walker 71; Baumgarten 81]
Cl^-	solid-state membrane, (Ag, 0.3 μm)	[Armstrong et al. 77]
HCO_3^-	tri-n-octyl-propylammonium chloride/ octanol/trifluoracetyl-p-butylbenzene (3:1:6), HCO_3^-/Cl^- selectivity of about 50	[Khuri et al. 74] (see also [Fujimoto et al. 80])
SCN^-, PF_6^-, AsF_6^-, SbF_6^-, salicylate, α-naphthene sulfonate	anion-exchanger membrane based on crystal violet or Aliquat 336	[Nicholson et al. 79; Phillips and Nicholson 79; Nicholson et al. 81]
tetraethyl ammonium	classical cation-exchanger (Corning 477317)	[Nicholson et al. 79]
choline, acetylcholine	ion-exchanger membrane based on acetyl-choline dipicrylaminate, < 50 μm	[Jaramillo et al. 83]
glucose	enzyme microelectrode based on glucose oxidase/Pt, 10–30 μm	[Rehwald et al. 84]
penicillin	classical anion-exchanger membrane, double-barrelled, 3–4 μm	[Speckmann et al. 83]
raffinose (recording of volume fluxes)	enzyme microelectrode based on galactose oxidase/Pt, 15–30 μm	[Geibel et al. 84]
pentylenetetrazol	classical ion-exchanger membrane, double-barrelled, 2–3 μm	[Walden et al. 84]
O_2	amperometry, alloy/gold, 1–2 μm	[Whalen 69]
O_2	amperometry, Pt, double-barrelled, 1 μm	[Tsacopoulos and Lehmenkühler 77]
redox systems (H_2O_2, uric acid, ascorbic acid, cysteine)	Pt microelectrode	[Völkl et al. 84]

Table 9.20. Available membrane systems for the development of microelectrodes

Measuring ion	Remarks	Reference
Cs^+	neutral carrier-based liquid membrane	[Kimura et al. 79b; Fung and Wong 80]
Mg^{2+}	neutral carrier-based liquid membrane	[Behm et al. 85]
Ba^{2+}	neutral carrier-based liquid membrane	[Güggi et al. 77; Läubli et al. 85]
Tl^+	neutral carrier-based liquid membrane	[Tamura et al. 80]
Cd^{2+}	neutral carrier-based liquid membrane	[Schneider et al. 80]
Pb^{2+}	neutral carrier-based liquid membrane	[Lindner et al. 84]
UO_2^{2+}	neutral carrier-based liquid membrane	[Šenkyr et al. 79]
Cl^-	neutral carrier-based liquid membrane	[Wuthier et al. 84]
NO_2^-, SCN^-	charged carrier-based liquid membrane	[Schulthess et al. 84; Schulthess et al. 85]

in very small micropipettes ($< 1 \, \mu m$). Small microelectrodes do not exhibit electrical shunts at the tip when they are filled with the polymeric membrane solution. Response times of the order of seconds only allow the measurement of relative slow changes in Ca^{2+} activity. However, specially constructed microelectrodes (coaxial-type) exhibit time constants for the electrode response of the order of 4–10 ms. The microelectrode technique is the only approach that can simultaneously provide information about the electrical parameters of the cell (membrane potential, membrane resistance). Absolute concentrations of free Ca^{2+} as low as $5 \cdot 10^{-8}$ M can be accurately measured. The technique is especially suited to deliver quantitative information. Reliable calibrations can easily be performed in external solutions before and after an intracellular experiment. The response curves of Ca^{2+} microelectrodes exhibit a large range of sensitivity (10^{-1} to 10^{-8} M Ca^{2+}).

9.7 Other Microelectrodes

Sections 9.1 to 9.6 presented some highly selective neutral carrier-based microelectrodes for the measurement of H^+, Li^+, Na^+, K^+, Mg^{2+} and Ca^{2+}. In certain cases, other types of membranes can also be used to measure these cations (glass membranes, ion-exchanger membranes, solid-state membranes).

There are many other selective membrane systems which can be used in electrodes. A series of membranes has already been successfully employed in microelectrodes (Tab. 9.19). Other membranes have only been characterized in macroelectrodes (Tab. 9.20) and, where useful, the corresponding microelectrodes still have to be developed.

10 Impact of Neutral Carrier Microelectrodes

10.1 Applications of Neutral Carrier-Based Microelectrodes

The widespread use of neutral carrier-based microelectrodes in physiology is illustrated in Tab. 10.1. Only selected contributions are given so that the list is far from complete. The 230 examples disclose several aspects and trends in the applications of neutral carrier-based microelectrodes.

The classification of the experiments according to organs and tissues clearly shows the utility of the microelectrodes in almost every area of electrophysiology. The use of the potentiometric technique is of particular interest for studies in excitable cells (brain, heart, nerve, muscle) and in polar transport cells (kidney, various epithelia). In vivo measurements are frequently performed. Most of the measurements are carried out intracellularly. Double-barrelled neutral carrier-based microelectrodes are now routinely used, whereas triple-barrelled microelectrodes are only used in certain special cases.

The rapidly increasing use of H^+-selective neutral carrier-based microelectrodes indicates that the liquid membrane is progressively replacing the glass membrane.

The Li^+-selective neutral carrier-based microelectrode is the only sensor for potentiometric studies of Li^+ activity. However, the microelectrode has so far only been used in a few applications to determine therapeutic values of Li^+ activities because of its modest selectivity.

The number of studies of intracellular Na^+ activity has increased strikingly since the introduction of neutral carrier-based microelectrodes (see Na^+ studies in epithelia).

To date, the valinomycin-based K^+-selective microelectrode is still used only rarely. Instead, the low-impedance (but much less selective) classical ion-exchanger microelectrode is often applied. However, it is now known that reliable K^+ activities are only obtained when neutral carrier-based microelectrodes are used under certain physiological conditions (high concentration of sodium, presence of lipophilic cations).

Because of the general difficulties encountered in measuring free Mg^{2+} ions, the currently available neutral carrier-based microelectrodes are of importance. The selectivity of the sensor undoubtedly needs to be improved. However, the experiments carried out so far are a valuable contribution to the investigation of the homeostasis of magnesium.

Ca^{2+}-selective microelectrodes are the most widely used neutral carrier-based microelectrodes. Many of the important roles of Ca^{2+} in cells have been elucidated

Table 10.1 Applications of neutral carrier-based microelectrodes

Ion	Extra(e)-, intra(i)-cellular	Single(s)-, double(d)-, triple(t)-, four(f)-barrelled	In vitro in vivo	Preparation	Reference
Brain					
Ca^{2+}	e	–	–	cerebellar cortex	[tenBruggencate and Steinberg 78]
Ca^{2+}	e	–	–	cerebellar cortex	[Steinberg and tenBruggencate 78]
Ca^{2+}	e	d	in vivo	cat, cerebellum	[Nicholson et al. 76]
Ca^{2+}	e	–	in vivo	rat, cerebellar cortex, neocortex	[Gardner-Medwin and Nicholson 77]
Ca^{2+}	e	–	in vivo	cat, cerebellar cortex	[tenBruggencate et al. 77]
Ca^{2+}	e	d	in vivo	cat, cerebral cortex	[Heinemann et al. 77]
Ca^{2+}	e	–	in vivo	rat, cerebellar cortex	[Heuser et al. 77]
Ca^{2+}	e	s, d	in vivo	catfish, rat, cat, cerebellum	[Nicholson et al. 78 a]
Ca^{2+}	e	d	in vivo	rat, cerebellum	[Nicholson et. al. 77]
Ca^{2+}	e	d	in vivo	catfish, cerebellar cortex	[Nicholson 79]
Ca^{2+}	e	d	in vivo	cat, cerebellar cortex	[Nicholson et al. 78 b]
Ca^{2+}	e	d	in vivo	cat, cerebral cortex	[Heinemann et al. 78 a]
Ca^{2+}	e	d	in vivo	rat, cat, cerebellar cortex	[tenBruggencate et al. 78]
Ca^{2+}	e	s	in vivo	cat, cerebral cortex	[Heuser 78]
Ca^{2+}	e	d	in vivo	cat, somatosensory cortex	[Heinemann et al. 78 b]
Ca^{2+}	e	d	in vitro	guinea-pig, hippocampal slices	[Benninger et al. 80]
Ca^{2+}	e	d	in vitro	guinea-pig, slices of olfactory cortex	[Galvan et al. 80]
Ca^{2+}	e	d	in vivo	*Corydoras aneus,* catfish, cerebellar cortex	[Kraig and Nicholson 78]

Table 10.1 (continued)

Ion	Extra(e)-, intra(i)- cellular	Single(s)-, double(d)-, triple(t)-, four(f)- barrelled	In vitro in vivo	Preparation	Reference
Ca^{2+}	e	d	in vivo	cat, cerebral cortex	[Heinemann et al. 77]
Ca^{2+}	e	d	in vivo	cat, cerebral cortex	[Heinemann and Pumain 80]
Ca^{2+}	e	d	in vivo	cat, cerebral cortex	[Heinemann and Pumain 81 a]
Ca^{2+}	e	d	in vivo	cat, cerebral cortex	[Lux and Heinemann 78]
Na^+	e	d	in vivo	cat, cerebral cortex	[Dietzel et al. 80]
Ca^{2+}	e	d	in vivo	cat, cerebellar cortex	[Stöckle and tenBruggencate 78]
Ca^{2+}	e	d	in vivo	cat, somatosensory cortex	[Heinemann and Konnerth 80]
Ca^{2+}	e	d	in vivo	cat, cerebral cortex	[Somjen 80]
Ca^{2+}	e	d	in vivo	rat, cerebellar cortex	[Nicholson 80 a]
Ca^{2+}	e	d	in vivo	rat, hippocampus	[Krnjević et al. 80]
Ca^{2+}	e	d	in vivo	rat, olfactory bulb	[Math and Davrainville 79]
Ca^{2+}	e	–	in vivo	rat, olfactory bulb	[Math and Davrainville 80]
Ca^{2+}	e	s	in vivo	various (CNS)	[Hník et al. 80]
Ca^{2+}	e	d	in vivo	cat, cerebral cortex	[Heinemann and Pumain 81 b]
Na^+, Ca^{2+}	e	s, d	in vivo	rat, cerebral cortex	[Hansen and Zeuthen 81]
Ca^{2+}	e	d	in vivo	cat, sensorimotor cortex	[Heinemann et al. 82]
Na^+, Ca^{2+}	e	d	in vivo	cat, cerebral cortex	[Dietzel et al. 82 a]
Na^+, Ca^{2+}	e	d	in vivo	cat, cerebral cortex	[Dietzel et al. 82 b]
Ca^{2+}	e	d	in vivo	rat, dorsal hippocampus	[Morris and Krnjević 81]
Ca^{2+}	e	d	in vivo	rat, cerebellum, catfish, cerebellum	[Nicholson 80 b]

Table 10.1 (continued)

Ion	Extra(e)-, intra(i)- cellular	Single(s)-, double(d)-, triple(t)-, four(f)- barrelled	In vitro in vivo	Preparation	Reference
Ca^{2+}	e	d	in vivo	cat, cerebellar cortex	[Nicholson et al. 78]
Li^+	e	–	in vivo	rat, cerebellar cortex	[Ullrich et al. 80]
Ca^{2+}	e	d	in vivo	cat, cerebellar cortex	[Stöckle and tenBruggencate 80]
Ca^{2+}	e	d	in vivo	cat, cortex	[Louvel and Heinemann 80]
H^+	e	d	in vivo	rat, cerebellum	[Kraig et al. 83]
Ca^{2+}	e	t	in vivo	baboon, cortex	[Harris et al. 82]
Ca^{2+}	e	d, t	in vivo	baboon, cortex	[Harris et al. 81]
Ca^{2+}	e	s	in vivo	cat, cortical tissue	[Heuser 81]
Ca^{2+}	e	–	–	cerebral cortex	[Somjen 81]
Ca^{2+}	e	d	in vivo	rat, dorsal hippocampus	[Krnjević and Morris 81]
Na^+, Ca^{2+}	e	d	in vivo	rat, cerebellum	[Nicholson and Kraig 81]
Na^+, Ca^{2+}	e	d	in vivo	rat, cortex	[Hansen 81]
Ca^{2+}	e	d	in vivo	rat, cerebral cortex	[Lehmenkühler et al. 81]
Ca^{2+}	e	–	in vivo	cat, sensorimotor cortex	[Heinemann and Pumain 81 a]
K^+, Ca^{2+}	e	d	in vivo	rat, hippocampus	[Morris 81]
Ca^{2+}	e	d	in vitro	hippocampal slices	[Prince et al. 81]
Ca^{2+}	e	d	in vitro	rat, hippocampal tissue	[Somjen et al. 81]
Ca^{2+}	e	d	in vivo	rat, cerebellum	[Nicholson et al. 81]
Li^+	e	–	in vivo	rat, cerebellar cortex	[tenBruggencate et al. 81]
H^+	e	d	in vivo	rat, cerebellar cortex	[Kraig et al. 83]
Ca^{2+}	i	–	in vivo	rat, hippocampal pyramidal neuron	[Morris et al. 83]
Ca^{2+}	e	–	in vivo	cat, sensorimotor cortex	[Heinemann et al. 81]
Ca^{2+}	e	d	in vitro	rat, hippocampal slice	[Dingledine and Somjen 81]

Table 10.1 (continued)

Ion	Extra(e)-, intra(i)-cellular	Single(s)-, double(d)-, triple(t)-, four(f)-barrelled	In vitro in vivo	Preparation	Reference
Na^+, Ca^{2+}	e	d, t	in vivo	rat, cerebellar cortex	[Ullrich et al. 82]
Ca^{2+}	e	d	–	cortical grey matter	[Pumain et al. 83]
H^+	e	d	in vivo	rat, cerebral cortex	[Lehmenkühler et al. 82]
H^+	e	d	in vivo	rat, cortical surface	[Lehmenkühler 82]
Ca^{2+}	e	–	in vitro	hippocampal slice	[Heinemann et al. 84]
H^+	e	d	in vivo	rat, cortex	[Mutch and Hansen 84]
H^+	e	d	in vivo	rat, cortex	[Mutch and Hansen 85]
Ca^{2+}	e	d	in vivo	rat, cortex	[Lehmenkühler et al. 85]
Ca^{2+}	e	d	in vitro	rat, hippocampus	[Wadman and Heinemann 85]
H^+	e	d	in vivo	rat, cerebellum	[Nicholson et al. 85]
K^+	i	s, d	in vitro	guinea-pig, hippocampus	[Sonnhof et al. 85]
K^+	e	d	in vitro	rat, hippocampus	[Haas and Jefferys 84]
Heart					
Ca^{2+}	i	–	in vitro	sheep, Purkinje fibre	[Isenberg and Dahl 78]
Ca^{2+}	i	s	in vitro	sheep, Purkinje fibre	[Sokol et al. 79]
Ca^{2+}	i	s	in vitro	sheep, Purkinje fibre	[Dahl and Isenberg 80]
Ca^{2+}	i	s	in vitro	rabbit, ventricular muscle	[Lee 80]
Mg^{2+}	i	s	in vitro	sheep, Purkinje fibre	[Hess and Weingart 81]
Ca^{2+}	i	s	in vitro	ferret, ventricular muscle	[Marban et al. 80]
Ca^{2+}	i	–	in vitro	sheep, Purkinje fibre	[Bers and Ellis 81]
Na^+	i	–	in vitro	guinea-pig, ventricular muscle	[Daut 82 b]

Table 10.1 (continued)

Ion	Extra(e)-, intra(i)-cellular	Single(s)-, double(d)-, triple(t)-, four(f)-barrelled	In vitro in vivo	Preparation	Reference
Na$^+$	i	–	in vitro	guinea-pig, ventricular muscle	[Daut and Rüdel 82]
Mg^{2+}	i	s	in vitro	sheep, Purkinje fibre, ferret, papillary muscle, frog, skeletal muscle	[Hess et al. 82]
H$^+$	i	d	–	guinea-pig, smooth muscle	[Aickin 82]
Na$^+$, Ca^{2+}	i	s	in vitro	sheep, ventricular muscle, Purkinje fibre	[Sheu and Fozzard 82]
Na$^+$	i	s	in vitro	dog, Purkinje fibre	[Lee and Dagostino 82]
Na$^+$	i	s	in vitro	dog, Purkinje fibre, ventricular muscle	[Wasserstrom et al. 82]
H$^+$	i	–	in vitro	ferret, trabeculae	[Coray and McGuigan 82]
Ca^{2+}	i	s	in vitro	rabbit, papillary muscle	[Lee et al. 80b]
Ca^{2+}	e	d	–	*Rana pipiens,* ventricular muscle	[Dresdner et al. 82]
Ca^{2+}	i	s	in vitro	sheep, Purkinje fibre	[Lado et al. 82]
Na$^+$	i	s	in vitro	sheep, Purkinje fibre	[Cohen et al. 82]
Ca^{2+}	i	s	in vitro	Purkinje fibre	[Ellis et al. 81]
Ca^{2+}	i	s	in vitro	sheep, Purkinje fibre	[Weingart and Hess 81]
Na$^+$	i	s	in vitro	dog, Purkinje fibre	[Vassalle and Lee 84]
Na$^+$	i	s	in vitro	guinea-pig, trabeculae, ferret, trabeculae	[Chapman et al. 83]
Na$^+$, Ca^{2+}	i	s	in vitro	sheep, Purkinje fibre	[Bers and Ellis 82]
Ca^{2+}	i	s	in vitro	sheep, Purkinje fibre, ventricular muscle	[Weingart and Hess 84]
Ca^{2+}	i	s	in vitro	sheep, Purkinje fibre, ventricular muscle	[Coray et al. 80]

Table 10.1 (continued)

Ion	Extra(e)-, intra(i)- cellular	Single(s)-, double(d)-, triple(t)-, four(f)- barrelled	In vitro in vivo	Preparation	Reference
Ca^{2+}	i	–	in vitro	sheep, Purkinje fibre	[Hess and Weingart 80]
Ca^{2+}	i	s	in vitro	sheep, Purkinje fibre	[Dahl and Isenberg 80]
Na^+	i	s	in vitro	sheep, Purkinje fibre	[Glitsch et al. 82]
Na^+	i	–	in vitro	sheep, Purkinje fibre	[Glitsch and Pusch 82]
Na^+	i	–	in vitro	sheep, Purkinje fibre	[Glitsch and Pusch 84a]
Na^+	i	–	in vitro	sheep, Purkinje fibre	[Glitsch et al. 84]
Na^+	i	–	in vitro	sheep, Purkinje fibre	[Glitsch and Pusch 84b]
Na^+	i	–	in vitro	sheep, Purkinje fibre	[Glitsch and Pusch 84c]
Ca^{2+}	i	s	in vitro	guinea-pig, ventricular myocyte	[Isenberg and Klockner 82]
Na^+	i	–	in vitro	sheep, Purkinje fibre	[Boyett and Hart 84]
Ca^{2+}	e	s, t	–	frog, ventricular myocardium	[Dresdner and Kline 85]
Na^+	i	–	in vitro	*Rana esculenta, Rana temporaria,* trabeculae	[Chapman et al. 84]
H^+	i	–	in vitro	sheep, Purkinje fibre	[Ellis and MacLeod 85]
Na^+	i	s	in vitro	Purkinje fibre	[Achenbach 85]
Na^+	i	–	in vitro	sheep, Purkinje fibre	[Glitsch et al. 85]
Nerve					
Ca^{2+}	i	–	in vitro	*Helix pomatia,* neurons	[Hofmeier and Lux 78]
Ca^{2+}	i	–	in vitro	*Helix,* neurons	[Lux and Hofmeier 78]
Ca^{2+}	e	d	–	*Helix pomatia,* neurons	[Heyer and Lux 78]
Ca^{2+}	e	–	–	*Helix pomatia,* neurons	[Lux and Heyer 79]

Table 10.1 (continued)

Ion	Extra(e)-, intra(i)-cellular	Single(s)-, double(d)-, triple(t)-, four(f)-barrelled	In vitro in vivo	Preparation	Reference
Ca^{2+}	e	d	in vivo	cat, spinal cord	[Somjen 80]
Ca^{2+}	i	d	–	*Helix*, neurons	[Lux and Hofmeier 79]
Ca^{2+}	i	d	–	*Helix pomatia*, neurons	[Hofmeier and Lux 81 b]
Ca^{2+}	i	d	in vitro	*Helix pomatia*, bursting pace-maker neuron	[Hofmeier and Lux 82]
Ca^{2+}	i	–	–	*Aplysia*, neurons	[Levy et al. 82]
Ca^{2+}	e	d	in vitro	rat, sympathetic ganglia	[Galvan et al. 79]
Ca^{2+}	i	s	in vitro	*Aplysia califor-nica*, neurons	[Gorman et al. 84]
Mg^{2+}	i	–	–	*Helix aspersa*, neurons	[Gamiño and Alvarez-Leef-mans 84]
Mg^{2+}	i	–	–	*Helix aspersa*, neurons	[Alvarez-Leef-mans et al. 83]
Ca^{2+}	i	s	in vitro	squid, giant axon	[Di Polo et al. 83]
K^+	i	d	–	leech, glial cells	[Schlue and Wuttke 83]
Na^+, Ca^{2+}	i	d	–	leech, neuron	[Deitmer and Schlue 83]
Na^+	i	d	in vitro	*Rana esculenta*, spinal cord	[Grafe et al. 82 a]
Li^+	i	d	–	*Rana esculenta*, spinal cord	[Grafe et al. 82 b]
Ca^{2+}	i	s	–	*Helix aspersa*, neurons	[Alvarez-Leef-mans et al. 80]
Ca^{2+}	i	s	–	*Helix aspersa*, neurons	[Alvarez-Leef-mans et al. 81 b]
Ca^{2+}	e	–	–	spinal cord	[Somjen 81]
Ca^{2+}	e	d	–	frog, spinal cord	[Bührle and Sonnhof 81]
Ca^{2+}	i	s	in vitro	*Helix pomatia*, neurons	[Lux et al. 81]
Ca^{2+}	i	s	in vitro	*Helix aspersa*, giant neurons	[Alvarez-Leef-mans et al. 81 a]
Ca^{2+}	e	d	–	cerebral cortex, spinal dorsal horn	[Somjen et al. 81]

Table 10.1 (continued)

Ion	Extra(e)-, intra(i)- cellular	Single(s)-, double(d)-, triple(t)-, four(f)- barrelled	In vitro in vivo	Preparation	Reference
Ca^{2+}	e	d	in vivo	rat, spinal cord	[Janus et al. 81]
Ca^{2+}	i	s	in vitro	*Helix pomatia,* neuron	[Hofmeier and Lux 81 a]
Li^+	e	–	in vitro	rat, vagus nerve	[tenBruggencate et al. 81]
Ca^{2+}	i	d	in vivo	cat, spinal cord motoneuron	[Krnjević et al. 82]
Li^+	i	s	–	*Helix aspersa,* neuron	[Thomas et al. 75]
Na^+, Ca^{2+}	i	d	–	*Hirudo medicinalis,* leech neuron	[Deitmer and Schlue 84]
Na^+, Ca^{2+}	i	d	–	leech, neuron	[Schlue and Deitmer 82]
K^+	i	d	–	leech, glial cell	[Wuttke and Schlue 82]
Na^+	i	–	in vitro	rat, spinal root	[Bostock and Grafe 84]
Mg^{2+}	i	–	in vitro	*Helix aspersa,* neuron	[Alvarez-Leefmans et al. 84]
H^+	i	d	in vitro	leech, neuron	[Schlue and Thomas 85]
Na^+, K^+, Ca^{2+}	i	d	in vitro	*Hirudo medicinalis,* neuron, glial cell	[Schlue et al. 85]
Na^+	i	d	in vitro	rat, neuron	[Grafe et al. 85]
K^+	i	s, d	in vitro	frog, mouse, spinal cord	[Sonnhof et al. 85]
Na^+	i	d	in vitro	rat, neuron	[Ballanyi et al. 83]
H^+, Ca^{2+}	i	s	in vitro	snail, neuron	[Byerly and Moody 84]
Na^+	i	s	in vitro	mollusc, neuron	[Connor and Hockberger 84]
H^+, Na^+, Ca^{2+}	i	d	in vitro	rat, neuron	[Ballanyi and Grafe 85]
Muscle					
Ca^{2+}	i	d	in vitro	*Balanus nubilus,* muscle fibre	[Ashley et al. 78]
Ca^{2+}	i	s	in vitro	frog, sartorius muscle fibre	[Tsien and Rink 80]
Na^+	i	s	in vitro	*Rana temporaria,* skeletal muscle	[Schümperli et al. 82]

Table 10.1 (continued)

Ion	Extra(e)-, intra(i)- cellular	Single(s)-, double(d)-, triple(t)-, four(f)- barrelled	In vitro in vivo	Preparation	Reference
Mg^{2+}	i	s	in vitro	frog, skeletal muscle	[Hess and Weingart 81]
Mg^{2+}	i	s	–	sartorius muscle	[Lopez et al. 84]
Mg^{2+}	i	–	–	*Rana pipiens,* sartorius muscle	[Lopez et al. 83 a]
Na^+	i	d	–	rat, soleus muscle	[Shabunova and Vyskočil 82]
Na^+	i	s	in vitro	rat, soleus muscle	[Stark and O'Doherty 82]
Ca^{2+}	i	s	in vitro	guinea-pig, papillary muscle	[Coray and McGuigan 81]
Ca^{2+}	i	s	in vitro	rabbit, papil- lary muscle	[Lee and Uhm 81]
Ca^{2+}	i	s	in vitro	rat, soleus muscle	[O'Doherty and Stark 81 b]
Ca^{2+}	i	d	in vitro	frog, tibialis anterior longus	[Delpiano and Acker 81]
Ca^{2+}	i	s	in vitro	*Rana pipiens,* sartorius muscle	[Lopez et al. 83 b]
Ca^{2+}	i	s	in vitro	*Rana temporaria,* skeletal muscle	[Weingart and Hess 84]
Ca^{2+}	i	s	in vitro	frog, skeletal muscle	[Coray et al. 80]
H^+	i	d	in vitro	guinea-pig, vas deferens	[Aickin 84 a]
H^+	i	d	in vitro	guinea-pig, vas deferens	[Aickin and Brading 84]
Na^+	i	d	–	guinea-pig, ureter	[Aickin 84 b]
Na^+	i	s	in vitro	mice, extensor digitorum longus	[Donaldson and Leader 84]
H^+, Na^+	i	s	in vitro	crayfish, stretch receptor	[Moser 85]
Mg^{2+}	i	–	in vitro	frog, skeletal muscle fibre	[Alvarez-Leef- mans et al. 85]
Mg^{2+}	i	s	in vitro	*Rana pipiens,* sartorius muscle	[Lopez et al. 84 b]
Ca^{2+}	e	d	–	rabbit, papillary muscle	[Bers 83]
K^+	i	d	–	*Chilo partellus,* muscle cell	[Dawson and Djamgoz 84]

Table 10.1 (continued)

Ion	Extra(e)-, intra(i)-cellular	Single(s)-, double(d)-, triple(t)-, four(f)-barrelled	In vitro in vivo	Preparation	Reference
H^+	i	s	–	guinea-pig, papillary muscle	[Poole-Wilson and Seabrooke 85]
Kidney					
Na^+	i	d	in vitro	*Necturus*, proximal tubule	[Khuri et al. 78]
Na^+	i	–	–	*Amphiuma*, distal tubule	[Oberleithner et al. 81]
Ca^{2+}	i	s	–	*Necturus*, proximal tubule	[Lee et al. 80a]
Na^+	i	s	in vitro	*Amphiuma*, distal tubule	[Oberleithner et al. 83a]
Na^+	i	s	in vitro	*Amphiuma*, distal tubule	[Oberleithner et al. 82a]
Na^+	–	s, d	in vitro	*Amphiuma*, distal tubule	[Oberleithner et al. 82b]
Na^+	i	d	in vitro	*Necturus*, epithelia	[Giebisch et al. 81]
Na^+	i	d	–	*Necturus, Amphiuma*, distal, proximal tubule	[Khuri and Agulian 81]
H^+	i	–	in vitro	proximal tubule	[Lang et al. 83]
H^+, Na^+, Ca^{2+}	i	s	in vitro	*Rana esculenta*, proximal tubule	[Wang et al. 84]
H^+, Na^+	e	s	in vitro	*Rana esculenta*, diluting segment	[Oberleithner et al. 84a]
Na^+	i	d	in vitro	*Rana esculenta*, proximal tubule	[Lang et al. 84]
Na^+	i	s, d	in vitro	*Rana esculenta*, proximal tubule	[Wang et al. 83]
Na^+	i	s	in vitro	*Amphiuma*, diluting segment	[Oberleithner et al. 83b]
H^+	i	d	in vivo	rat, proximal tubule	[Yoshitomi and Frömter 84a]
H^+	i	s, d	in vivo	rat, proximal tubule	[Yoshitomi and Frömter 84b]
Na^+	e	–	in vitro	frog, diluting segment	[Oberleithner et al. 84b]
Na^+	i	s	in vitro	*Amphiuma*, distal tubule	[Oberleithner et al. 82c]

Table 10.1 (continued)

Ion	Extra(e)-, intra(i)-cellular	Single(s)-, double(d)-, triple(t)-, four(f)-barrelled	In vitro in vivo	Preparation	Reference
Na$^+$	i	s	in vitro	frog, proximal tubule	[Lang et al. 82]
Na$^+$	i	–	–	*Rana esculenta,* proximal tubule	[Wang et al. 82]
Na$^+$	i	s, d	in vitro	rabbit, nephron, *Squalus acanthias,* rectal gland tubule	[Greger and Schlatter 85]
H$^+$, Na$^+$, Ca^{2+} i		s	in vitro	*Rana esculenta,* proximal tubule	[Lang et al. 85]
Liver					
Na$^+$, K$^+$, Ca^{2+} –		–	in vitro	liver tissue, cell damage	[Kessler et al. 77]
Intestine					
Na$^+$	i	–	in vitro	*Necturus,* ephithelial cells of small intestine	[O'Doherty et al. 79]
H$^+$	i	–	–	rabbit, colon	[Duffey and Bebernitz 83]
Na$^+$	i	d	in vitro	*Squalus acanthias,* rectal gland	[Greger and Schlatter 84]
Na$^+$	i	d	in vitro	*Squalus acanthias,* rectal gland	[Greger et al. 84b]
H$^+$	i	s	in vitro	rabbit, colon ephitelia	[Duffey 84]
Bladder					
Na$^+$	i	s	in vitro	New Zealand white rabbits, urinary bladder	[Wills and Lewis 80]
Na$^+$	i	s	in vitro	New Zealand white rabbits, urinary bladder	[Lewis and Wills 80]
Na$^+$	i	d	–	*Necturus,* gallbladder, ephitelium	[Zeuthen 82]
Na$^+$	i	s	in vitro	*Necturus,* gallbladder epithelium	[Armstrong et al. 81]
H$^+$	e, i	s	in vitro	*Necturus,* gallbladder	[Reuss and Costantin 84]
Na$^+$	i	s	in vitro	*Necturus,* gallbladder	[Weinman and Reuss 84]

Table 10.1 (continued)

Ion	Extra(e)-, intra(i)- cellular	Single(s)-, double(d)-, triple(t)-, four(f)- barrelled	In vitro in vivo	Preparation	Reference
Na$^+$	i	s	in vitro	*Necturus,* gall-bladder	[Garcia-Diaz and Armstrong 80]
Na$^+$	i	d	in vitro	*Necturus,* gall-bladder	[Giraldez 84]
Na$^+$	i	d	in vitro	*Necturus maculosus,* gall-bladder	[Giraldez 85]
Na$^+$	i	d	in vitro	*Necturus,* gall-bladder	[Baerentsen et al. 83]
Gland					
Ca^{2+}	i	d	in vitro	*Calliphora,* salivary gland	[Berridge 80]
Ca^{2+}	i	s	in vitro	*Phormia regina,* salivary gland ephitelia	[O'Doherty et al. 80]
Ca^{2+}	i	s	in vitro	*Phormia regina,* salivary gland	[O'Doherty and Stark 81a]
Na$^+$, Ca^{2+}	i	s	in vitro	mice, pancreatic acinar cells	[O'Doherty and Stark 82a]
Na$^+$, K$^+$	i	–	in vitro	*Chironomus tentans,* salivary gland	[Wuhrmann and Lezzi 80]
K$^+$	i	d	in vitro	*Chironomus,* salivary gland	[Wuhrmann et al. 79]
H$^+$	i	s	in vitro	rabbit, gastric gland	[Kafoglis et al. 84]
Eye					
Ca^{2+}	i	s	in vitro	albino rat, retina	[Yoshikami et al. 80a]
Ca^{2+}	i	–	–	*Limulus,* ventral photoreceptors	[Levy 82]
Ca^{2+}	e	–	–	retinal rods	[Yoshikami et al. 80b]
Na$^+$, Ca^{2+}	e	t	in vitro	*Apis mellifera,* retina	[Orkand et al. 84]
H$^+$	e	–	in vivo	cat, vitreous body	[Weingart and Niemeyer 83]
Na$^+$	i	d	–	drone, photo-receptors	[Coles and Orkand 82]
Na$^+$	i	d	in vitro	*Apis mellifera,* photoreceptors	[Tsacopoulos et al. 83]

Table 10.1 (continued)

Ion	Extra(e)-, intra(i)- cellular	Single(s)-, double(d)-, triple(t)-, four(f)- barrelled	In vitro in vivo	Preparation	Reference
K^+	e	d	in vitro	drone, retina	[Coles et al. 81]
K^+	i	d	in vitro	drone, retina	[Coles and Tsacopoulos 77]
K^+	i	d	in vitro	*Apis mellifera*, retina	[Coles and Orkand 83]
Ca^{2+}	i	s	–	*Limulus*, ventral photoreceptors	[Levy and Fein 85]
Na^+, Ca^{2+}	i, e	d, t	in vitro	*Apis mellifera*, photoreceptors, glial cells	[Coles and Orkand 85]
Ear					
Ca^{2+}	e	d	in vivo	rat, cochlear endolymph	[Bosher and Warren 78]
Ca^{2+}	e	d	in vivo	*Xenopus laevis*, cupulae	[McGlone et al. 79]
Ca^{2+}	e	s	–	rat, endolymph	[Bosher 81]
Egg					
Ca^{2+}	i	s	in vitro	*Xenopus laevis*, early embryos	[Rink et al. 80]
Ca^{2+}	i	–	in vitro	*Xenopus laevis*, *Ambystoma mexicanum, Pleurodeles waltlii*, oocytes	[Moreau et al. 80]
H^+	i	s	in vitro	starfish, oocytes	[Picard and Dorée 83 b]
Ca^{2+}	i	s	in vitro	starfish, oocytes	[Picard and Dorée 83 a]
H^+	i	s	in vitro	*Xenopus laevis*, oocytes	[Cicirelli et al. 83]
Na^+	i	s	in vitro	*Xenopus*, oocytes	[Kusano et al. 82]
Ca^{2+}	i	s	in vitro	starfish, oocytes	[Picard and Dorée 83 c]
Plants					
H^+	i	s	in vitro	*Acer pseudoplatanus*, vacuoles	[Kurkdjian and Barbier-Brygoo 83]

Table 10.1 (continued)

Ion	Extra(e)-, intra(i)- cellular	Single(s)-, double(d)-, triple(t)-, four(f)- barrelled	In vitro in vivo	Preparation	Reference
Miscellaneous					
Ca^{2+}	e	d	in vivo	rat, blood plasma	[Math and Davrainville 79]
K^+, Ca^{2+}	e	t	–	cat, carotid body	[Dufau et al. 82]
Ca^{2+}	e	–	–	cat, carotid body	[Acker et al. 81]
Na^+	i	d	in vitro	*Rana temporaria,* skin	[Harvey and Kernan 84a]
Ca^{2+}	e	–	–	cat, carotid body	[Delpiano and Acker 84]
Na^+	i	d	in vitro	*Rana temporaria,* skin	[Harvey and Kernan 84b]
H^+	–	d	in vitro	gingiva, collec- ted human fluid	[Bickel and Cimasoni 85]
Ca^{2+}	i	s	in vitro	frog, skin epithelium	[Civan 84]

using these microelectrodes. Studies using microelectrodes are complementary to the techniques using photoproteins and indicator dyes.

Microelectrodes for the measurement of other species are on hand and promising carrier-based membranes for the development of microelectrodes are under investigation.

Most of the applications given in Table 10.1 are fundamental experiments in physiological research. Many studies are designed so that their results are of potential clinical relevance (experiments under pathological conditions, pharmacological studies) and the knowledge gained is directly applied in clinical approaches. There are only few examples of physiological research that do not have an immediate medical application (see Sect. 10.3). It is unlikely that microelectrodes will become directly applied in clinical laboratories (although, for example, the intracellular activities in erythrocytes are supposed to possess a high diagnostic value).

I hope that the extremely rapidly developing medicine which profits from fundamental research is growing to an institution that is not too intimidated by purely technological issues but meets the role of illness and the rights of patients.

10.2 Animal Experimentation

Microelectrodes are used exclusively for measurements in samples from living organisms. Most of the studies are carried out in cells from laboratory animals although a few experiments have been made in cells from human beings or plants. Some thoughts on animal experimentation are therefore appropriate.

The number of animals used in experiments had been increasing dramatically since 1950. It now seems to have reached a maximum level or may be decreasing slightly again (Fig. 10.1) [Rowan 84] (see also [Griffin and Sechzer 83]).

The number of experiments that have take place in the U.K. is quite accurate. However, it is more difficult to give numbers for world-wide use. A realistic estimate is about 100 million animals per year. The tremendous number is the result of an almost unlimited and uncontrolled use of animals. It is difficult to recognize an ethical limit to animal experiments. The arguments most often put forward in favour of the use of animals are:

- health (human, animal)
- safety (human (e.g. toxicity tests))
- economy
- knowledge.

These arguments disclose that mainly the applied sciences (and to a much lesser extent fundamental scientific research) are responsible for the vast amount of animals used. The dominance of applied sciences in animal use becomes apparent from the following list of important issues in animal experimentation:

- toxicology
- extraction of biological substances
- drug administration
- diagnosis
- welfare
- research
- education.

Several facts including the total number of animals, the large use in education (e.g. 3 million animals annually in the U.S.A. [Rowan 84]), LD_{50} toxicity tests, the cosmetic industry, the war industry and the many painful animal tests have led to a strong opposition to animal experiments. To date, the problems have been treated by several means:

- anti-vivisectionist societies
- animal welfare groups
- legislation
- government (health department, law, right of civil initiative)
- case studies
- conferences
- committees (protection, ethic).

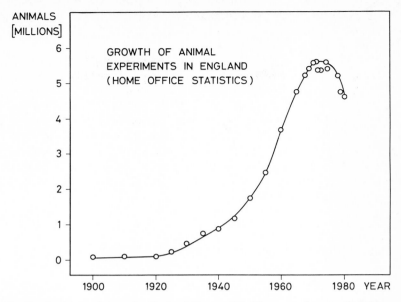

Fig. 10.1. Growth of animal experiments in England [Rowan 84]. The curve is compiled from Home Office statistics (1920–1982). Redrawn from Rowan. Of Mice, Models, and Men, State University of New York Press, 1984 [Rowan 84]

The discussion of ethics in biomedical research continues to increase. The complicated discussion is simultaneously carried out on a very emotional and a very rational basis, and engages small groups (conferences) or entire nations (Swiss civil intiative, "Zur Abschaffung der Vivisektion", 1985). It is obvious that there is increasing public concern. The general society is uncertain about the ethical principles of the scientific community and many demands have been made:

- greater accountability
- adequate caging
- adequate anesthesia
- adequate care
- proper techniques
- no duplications
- no experiments for economic reasons
- no experiments for military reasons
- alternative samples
- no animal experimentation

Many of the public demands as well as the discussions within the scientific community have brought about the development of alternatives to the current practices in animal experiments [Sechzer 83; Rowan 84]. Both a change in the techniques and in the attitudes of the scientists are necessary. Ethical committees have been formed and have proposed standard criteria for the scientific community itself. Some of the alternatives under discussion are:

- limited tests (maximal efficiency, best methods)
- approximate lethal dose tests (no LD_{50} tests)
- statistical evaluations
- computer modelling (structure-activity relationship)
- tissue cultures (in vitro tests)
- microorganisms (probably pain-free organisms)

The guideline for replacement, reduction and refinement (already published in 1959 [Russell and Burch 59]) are now being seriously considered. The scientific community no longer enjoy general confidence of the people since in the applied sciences profit motives often overshadow ethical aspects. Ultimately, the performance of animal experiments is a personal decision. I myself could never kill an animal for a microelectrode study.

10.3 Neurosciences. The Mind-Body Problem

The neurosciences are a rapidly developing and extraordinary field of modern scientific research. They are interested in the structure and physiological performance of the most complex organ, the brain. Hence, the studies are inevitably related to the exploration of the brain-mind relationship. Microelectrodes are irreplaceable tools in many investigations of the brain.

Since its introduction in 1929, electroencephalography has been widely utilized in neurology. However, it seems that the diagnostic value of the method is reaching its limit. Surface recordings suffer from several disadvantages when they are used in the surgical therapy of epilepsy. Hence, there is a great interest in replacing electroencephalography in that area of medicine by systematical depth recordings in the brain. Intracortical measurements of potentials and ion activities can be achieved by using miniaturized electrodes. Depth electrodes are now being increasingly applied in neurology [Wieser 83].

Experience from commissurotomy experiments and other cerebral lesions [Sperry 64; Gazzaniga 70; Penfield 75] first showed that the removal of epileptogenic areas of the brain of patients with focal epilepsy appears to be beneficial. As a result, this practice is being increasingly used [Wieser and Yasargil 82; Wieser 83]. The localization of the foci should be much more precise when depth electrodes rather than electroencephalographs are used. Multielectrodes with shaft diameters of about 0.1 to 0.5 mm have been used for this purpose. The areas of epileptic seizure can consequently be more accurately pin-pointed [Petsche et al. 78; Prohaska 83; Wieser 83]. At present, most of the intracranial recordings are performed presurgically in highly drug-resistant patients with psychomotor seizures. For this purpose, the cortical electrodes are temporarily implanted into the brain under stereotactic conditions over periods of days or several weeks [Wieser 83]. These depth electrodes are obviously considerably larger (100 to 1000 times) than the tips of microelectrodes.

Due to safety and technical reasons, microelectrodes are not applicable in clinical neurology, but they are an important tool for research in the neurosciences:

"Another powerful technique for investigating synapses is the recording from the interior of nerve cells by fine microelectrodes, which has revealed not only the electrical independence of the neurones, but also the mode of operation of synapses." [Popper and Eccles 77, p. 232].

Indeed, microelectrodes have enabled many breakthroughs in the understanding of the mechanisms of various processes in the brain. Different neurophysiological aspects of ionic mechanisms during neuronal activity have been investigated. Phenomena such as spreading depression, generation of epileptic seizures, cell discharges, synaptic transmission and the role of excitatory and inhibitory substances are now partly understood on the ionic level.

Epileptogenesis has been extensively investigated using microelectrodes. Most of the experiments have been performed in the brains of rats and cats. The major aim was the elucidation of the mechanism of the generation of seizure. Typical current animal experiments involve:

- study of stimulus-induced changes in ion activities (repetitive stimulations, chronic foci)
- study of synaptic transmission
- study of the effects of chemicals (excitatory, inhibitory)
- study of the depth-dependence of brain events (cortical layers)
- study of the microenvironment of a brain cell.

So far, most of the experiments have been based on conventional reference microelectrodes. However, it has become apparent that ion-selective microelectrodes are playing an increasing role in the neurosciences. Examples of the wide-spread use of neutral carrier-based microelectrodes in the brain are given in Tab. 10.1. There is strong evidence that Na^+, K^+ and especially Ca^{2+} ions are the major regulators of neural events. Local extracellular fluctuations in Ca^{2+} activity are found in regions of neural activity. The neutral carrier-based Ca^{2+} microelectrode has become an important instrument for the study of events underlying epileptiform activities [Heinemann et al. 77; Krnjević et al. 80; Heinemann et al. 81; Pumain et al. 83]. A spread of hyperexcitability is observed during changes in extracellular Ca^{2+} activities. The changes are very fast and vary with the depth in the cortical grey matter [Pumain et al. 83]. Ca^{2+} microelectrodes based on the neutral carrier ETH 1001 are fast enough to follow these transients (coaxial-type microelectrodes, response time 4 ms [Pumain et al. 83]). The microelectrode technique is especially suited for the measurement of depth profiles (see Fig. 10.2).

However, the organization of animal and human brains differs drastically, so that the information obtained from animals is only partially transferable to human brains. Indeed, it is often claimed that sensory experiments should be performed intracranially on conscious human beings [Penfield 75; Carlson 77; Jung 78; Libet 78; Wieser 83]. The central role of microelectrodes in brain research has been repeatedly stressed:

"Necessarily the crucial experimental investigation of sensory experiences must be carried out on conscious human subjects, but both the design and interpretation of these experiments are dependent on the wonderful successes that have attended in-

RECORDINGS OF DEPTH-PROFILES DURING
SPREADING DEPRESSION WITH DOUBLE-BARRELLED
NEUTRAL CARRIER Ca^{2+} MICROELECTRODES
(RAT CEREBELLUM)

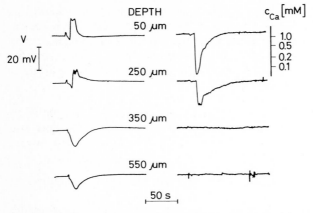

Fig. 10.2. Profiles of slow potentials and concentrations of free Ca^{2+} during spreading depression measured using double-barrelled microelectrodes [Nicholson et al. 81]. The microelectrode was located at different depths in the extracellular space of the benzoate-conditioned cerebellum of a rat. Redrawn from Nicholson, Phillips, Tobias and Kraig, 1981, in Ion-Selective Microelectrodes and Their Use in Excitable Tissues (Syková, Hník, Vyklický (Eds.)) [Nicholson et al. 81]

vestigations on animal, and particularly monkey, sensory systems in the last few decades. The powerful techniques designed for precision and selectivity of stimulation have been matched by microelectrode recording from single neurones." [Popper and Eccles 77, p. 252].

The experimental neurosciences are undoubtedly directly linked to one of the great philosophical issues of our time, the mind-body problem [Spicker and Engelhardt 75; Cheng 75; Penfield 75; Globus et al. 76; Buser and Rougeul-Buser 78; Davidson and Davidson 80; Eccles 80; Sperry 83]. Philosophical statements on the mind-body relationship are now strongly influenced by physiological studies of neural mechanisms including the results obtained with microelectrodes. The rapid experimental progress in neurophysiology and neurobiology has stimulated the interest of many physiologists in the philosophy of the mind. Indeed, after a period of predominant interest in materialism, a further controversy between materialism and idealism has come about from the results of research in the neurosciences. The central issue of the discussion is the extent to which consciousness can be reduced to events in the brain. History has proven that the confrontation between materialistic and idealistic views (which use basically different paradigms) has had important consequences on the development of human society. Dominating ideas lead to guiding paradigms for scientific, moral and social development. Actually, the new plurality of opinions on the mind and body problem is welcome. However a single dominating extreme which could be misused outside of the scientific community should not result.

Today's scientists assume many different positions in the mind-body discussion. Some important philosophical standpoints can be summarized in the following rough classification:

a) reduction of mind to body (consciousness is linked to physiological processes in the brain)
   - radical materialism, behaviourism (Watson, Skinner, Ryle)
   - identity theory, central state theory (Smart, Feigl, Armstrong)
b) reduction of body to mind (physical objects are the result of ideas)
   - phenomenalism (Berkeley)
c) intermediate theories
   - attribute theory (the mind is a form of the body) (Aristotle)
   - panpsychism (everything has material and mental character) (Spinoza)
   - emergent interactionism (mental events are causal emergents of brain processes) (Sperry)
   - dualism (mind and body are independent entities interacting with each other) (Descartes, Popper, Eccles, Penfield).

In this century, the materialist-behaviourist principle was well established until, during the 1960's, mentalistic, holistic, panpsychistic, and dualistic hypotheses reanimated the mind-body problem. Many of these thoughts bring the western scientific concept closer to eastern mystical traditions (see [Capra 75]).

Since the dualist-interactionist theory [Eccles 81] puts forward ideals on the anatomical structure and physiological mechanism of the liaison brain, a brief characterization of the theory with respect to the role of microelectrodes in the mind-body problem has been made here. While the materialistic view does not accept mental qualities, the interactionist model claims that consciousness is emergent and that mental processes cannot be reduced to neurophysiological terms. The dualistic theory requires the existence of areas in the brain (liaison brain) in which the self-conscious mind, i.e. the highest mental experiences, interacts with structures of the brain. These areas, which are the material basis of mental events, are believed to exist in the association cortex (about 90% of the neocortex) [Eccles 81]. In the model by Eccles, the anatomical units [Szentagothái 78] (approximately 3 million modular columns) found in the neocortex create a quasi-unlimited number of spatio temporal patterns [Eccles 81]. Each cognitive memory would have a patterned counterpart of perhaps a few thousand modules and would be stored in and retrieved from these structures.

Experimental facts (including microelectrode studies) exist that indicate actions of the self-conscious mind. This knowledge arises from the following research:

- experiments on the stimulation of the cortex of conscious patients [Penfield 75]
- experiments on the effects of epileptic discharges on various parts of the brain [Penfield 75]
- experiments on commissurotomy patients [Gazzaniga 70; Carlson 77; Sperry 79]
- experiments in which electrical stimulations are given: the conscious experience is delayed about 0.5 s. This indicates that there is no synchrony between physical and mental events [Libet 78 b]

– experiments using electrodes have measured readiness potentials generated
0.5–1.5 s before a willed conscious movement was made. This is another factor
against the identity theory [Kornhuber 78].

Whatever position a neuroscientist takes, his aim is to investigate the relationship
between neuronal activities and conscious experience [Libet 78a; Libet 78b;
Mountcastle 78; Pribram 78; Davidson and Davidson 80]. In his materialistic theory of the mind, Armstrong says:

"It does not attempt to prove the truth of this physicalist thesis about the mind.
The proof must come, if it does come, from science: from neurophysiology in
particular." [Armstrong 68, p. 2],

and the dualist Penfield says:

"Finally, fresh explorers must discover how it is that the movement of potentials
becomes awareness, and how purpose is translated into a patterned neuronal
message. Neurophysiologists will need the help of chemists and physicists in all
this, no doubt." [Penfield 75, p. 84].

The relationship between materialistic interventions and human awareness has
been repeatedly shown. Many examples are known in which chemical (drugs) and
physical (stimulations) actions in the brain change the awareness of an individual.
Furthermore, the dependence of awareness on specific connections in the brain has
been shown by surgical interventions in the human brain. Dissections of the
200 million fibres of the corpus callosum (the connection between the two hemi-
spheres), hemispherectomy and other cerebral lesions as well as electrical stimulations of the cortex dramatically influence the behaviour of animals and human
beings [Sperry 64; Gazzaniga 70; Penfield 75; Gazzaniga and LeDoux 78]. The increasingly important role of microelectrodes in the elucidation of relationships between a brain event and its conscious manifestation has been clearly documented:

"The highly stratified organization and well documented electrophysiology of the
hippocampus, as well as the availability of techniques for localized measurements
with ion-sensitive microelectrodes, have encouraged recent studies of transmembrane ionic shifts which may be associated with the generation of consciousness
and memory formation, and a marked susceptibility to seizure activation and anoxic damage." [Morris 81, p. 241].

"Another point is of course that no one has done any of the detailed microelectrode
searching in the human cerebral cortex in order to discover any special features of
neuronal activity in the modular arrangements. This could come quite soon." [Pop-
per and Eccles 77, p. 241].

The following questions arise: Does electrophysiology have any implications for an
understanding of the mind-body relationship? Does it provide information on consciousness? What is the role of microelectrodes in the discussion of mind and body,
and do they contribute to the view that cognitive processes are totally mechanistic?
A useful scheme for a discussion of the impact of microelectrodes on the mind-

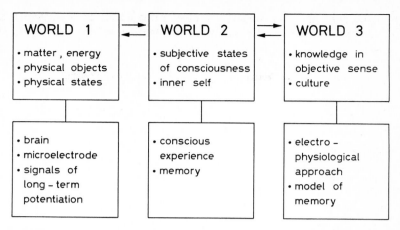

Fig. 10.3. Representation of the three worlds comprising material and mental aspects of life as defined by Popper and Eccles [Popper and Eccles 77]. An electrophysiological approach using microelectrodes for the measurement of long-term potentiation in the brain with regards to the elucidation of the storage of conscious experience in form of memory is incorporated into the tabular representation

body problem is Popper's platonian view of the three worlds [Popper and Eccles 77] (see Fig. 10.3, upper part). Briefly, world 1 includes the physical world, i.e. the world of matter and energy. World 2 includes mental events, i.e. the subjective states of consciousness, and world 3 includes the cultural achievements of mankind.

The various philosophical points of view presented above bear the following relationship to this model:

- radical materialism, behaviourism: world 2 does not exist. Everything is expressed by world 1, i.e. the mind-brain problem does not exist
- identity theory, central state theory: a subworld of physics with mental states is assumed, i.e. world 2 is (because of the identity relationship) attributed to this subworld 1
- dualism, interactionism: world 1 and world 2 are independent entities. The liaison brain, belonging to world 1, is the area of interaction. Mental events of world 2 are of the non-physico-chemical kind.

Memory, a state of consciousness from world 2, has been intensively investigated in electrophysiological studies. Long-term potentiation [Bliss and Lømo 70; Bliss and Lømo 73] has become a neural model that should account for memory and learning. The neurophysiological mechanism for a long-term storage of information is assumed to consist in an increase in the amplitude of the evoked potentials of excitatory synapses after intrahippocampal stimulations. The increased efficacy lasts hours or days and remains confined to the stimulated synapses. The synaptic mechanisms are still unclear [Bliss and Dolphin 82]. Most of the information on long-term potentiation has been gained by using microelectrodes [Bliss and Lømo 73; Eccles 83; Voronin 83; Harris and Teyler 84; Briggs et al. 85]. Ions, especially Ca^{2+}, seem to be crucially involved in the generation of long-term potentiation [Bliss and Dolphin 82; Eccles 83; Briggs et al. 85]. It is therefore likely that neutral carrier-based ion-selective microelectrodes will become increasingly important in the evaluation of a model for memory. Experiments using the neutral carrier-based Ca^{2+} microelectrode have been discussed in the following manner:

> "One can only speculate about the functional significance of such a large Ca^{2+} entry: is this somehow related to "plastic" properties of the hippocampus and the laying down of some form of memory, via changes in RNA and protein synthesis." [Krnjević et al. 80, p. 582].

It has been suggested that there is a relationship between the neurophysiological mechanisms of long-term potentiation and the conscious state of memory [Voronin 83; Eccles 83; Harris and Teyler 84]. Thus, research on long-term potentiation offers an interesting issue for the discussion of the role of microelectrode experiments in the relationship between mind and brain (Fig. 10.3). Following the dualistic-interactionistic theory, all mental experiences are embedded in memories, and long-term potentiation is just the material basis (world 1, Fig. 10.3) of the conscious experience (world 2, Fig. 10.3). The current model of memory, based on the physico-chemical processes of long-term potentiation, is far from being proven, but is a modern scientific attempt to explain the events of world 2 on the basis of electrophysiological studies and physiological models (world 3, Fig. 10.3). The lower part of Fig. 10.3 shows that such a scientific study using microelectrodes cannot lead to a knowledge of the mind. As soon as a subjective, self-conscious experience of world 2 is made objective, it belongs to world 1 or 3. Thus, science can only study phenomena of world 2, but never the self-conscious mind itself.

We have to differentiate on objective measurable material world and a world of perception that is governed by other laws. I believe that the mind, i.e. my subjective consciousness, is not reducible to brain mechanisms. Thus, I align myself with people who feel a spiritual element in man. Mind remains a mystery until humanity learns mutual love.

11 References

K. M. Aalmo and J. Krane, 1,5,9,13-Tetraoxacyclohexadecane and its 3,3,7,7,11,11,15,15-octamethyl derivative as neutral carrier for lithium ion through artificial membranes, Acta Chem. Scand. **A36**, 227 (1982).

C. Achenbach, Effects of thallous ions on the measurement of intracellular ion activities, 1985, in Kessler et al. 85, p. 256.

H. Acker, M. Delpiano, M. Fischer, F. Pietruschka and R. G. O'Regan, Role of calcium in the chemoreceptive process of the carotid body, 1981, in Lübbers et al. 81, p. 122.

S. Addanki, F. D. Cahill and J. F. Sotos, Determination of intramitochondrial pH and intramitochondrial-extramitochondrial pH gradient of isolated heart mitochondria by the use of 5,5-dimethyl-2,4-oxazolidinedione, J. Biol. Chem. **243**, 2337 (1968).

R. H. Adrian, The effect of internal and external potassium concentration on the membrane potential of frog muscle, J. Physiol. **133**, 631 (1956).

D. P. Agin, Electrochemical properties of glass microelectrodes, 1969, in Lavallée et al. 69, p. 62.

J. J. Aguanno and J. H. Ladenson, Influence of fatty acids on the binding of calcium to human albumin, J. Biol. Chem. **257**, 8745 (1982).

C. C. Aickin, Intracellular pH of the smooth muscle cells of the guinea-pig vas deferens, J. Physiol. **334**, 112 P (1982).

C. C. Aickin and A. F. Brading, Measurement of intracellular chloride in guinea-pig vas deferens by ion analysis,[36] chloride efflux and micro-electrodes, J. Physiol. **326**, 139 (1982).

C. C. Aickin and A. F. Brading, The role of chloride-bicarbonate exchange in the regulation of intracellular chloride in guinea-pig vas deferens, J. Physiol. **349**, 587 (1984).

C. C. Aickin, Direct measurement of intracellular pH and buffering power in smooth muscle cells of guinea-pig vas deferens, J. Physiol. **349**, 571 (1984a).

C. C. Aickin, Intracellular sodium activity of the smooth-muscle cells of guinea-pig ureter, J. Physiol. **357**, 48 P (1984b).

B. Alberts, D. Bray, J. Lewis, M. Raff, K. Roberts and J. D. Watson, Molecular Biology of the Cell, Garland Publishing, Inc., New York, London, 1983.

B. J. Alder and T. E. Wainwright, Studies in molecular dynamics. I. General method, J. Chem. Phys. **31**, 459 (1959).

G. Alibert, A. Carrasco and A. M. Boudet, Changes in biochemical composition of vacuoles isolated from *Acer pseudoplatanus* L. during cell culture, Biochim. Biophys. Acta **721**, 22 (1982).

D. G. Allen, J. R. Blinks and F. G. Prendergast, *Aequoria* luminescence: relation of light emission to calcium concentration – a calcium-independent component, Science **195**, 996 (1977).

F. J. Alvarez-Leefmans, T. J. Rink and R. Y. Tsien, Intracellular free calcium in *Helix aspersa* neurones, J. Physiol. **306**, 19 P (1980).

F. J. Alvarez-Leefmans, T. J. Rink and R. Y. Tsien, Measurement of free Ca^{2+} in nerve cell bodies, 1981a, in Syková et al. 81, p. 119.

F. J. Alvarez-Leefmans, T. J. Rink and R. Y. Tsien, Free calcium ions in neurones of *Helix aspersa* measured with ion-selective micro-electrodes, J. Physiol. **315**, 531 (1981b).

F. J. Alvarez-Leefmans, S. M. Gamiño and T. J. Rink, Cytoplasmic free magnesium in neurones of *Helix aspersa* measured with ion-selective micro-electrodes, J. Physiol. **345**, 104 P (1983).

F. J. Alvarez-Leefmans, S. M. Gamiño and T. J. Rink, Intracellular free magnesium in neurones of *Helix aspersa* measured with ion-selective micro-electrodes, J. Physiol. **354**, 303 (1984).

F. J. Alvarez-Leefmans, S. M. Gamiño, F. Giraldez and H. Gonzalez-Serratos, Intracellular free magnesium in frog skeletal muscle fibers measured with ion-selective microelectrodes, Biophys. J. **47**, 458 a (1985).

A. Amidsen, Monitoring of lithium treatment through determination of lithium concentration, Dan. Med. Bull. **22**, 277 (1975).

D. Ammann, E. Pretsch and W. Simon, A synthetic electrically neutral carrier for Ca^{2+}, Tetrahedron Letters **24**, 2473 (1972a).

D. Ammann, E. Pretsch and W. Simon, A calcium ion-selective electrode based on a neutral carrier, Anal. Letters **5**, 843 (1972b).

D. Ammann, E. Pretsch and W. Simon, Darstellung von neutralen, lipophilen Liganden für Membranelektroden mit Selektivität für Erdalkali-Ionen, Helv. Chim. Acta **56**, 1780 (1973).

D. Ammann, E. Pretsch and W. Simon, A sodium ion-selective electrode based on a neutral carrier, Anal. Letters **7**, 23 (1974).

D. Ammann, Darstellung von alkali- und erdalkaliionenselektiven, elektrisch neutralen Trägerliganden und Entwicklung eines hochselektiven Ca^{2+}- Sensors, Thesis, ETH Zürich, No. 5605, Juris Druck + Verlag, Zürich, 1975.

D. Ammann, R. Bissig, M. Güggi, E. Pretsch, W. Simon, I. J. Borowitz and L. Weiss, Preparation of neutral ionophores for alkali and alkaline earth metal cations and their application in ion selective membrane electrodes, Helv. Chim. Acta **58**, 1535 (1975a).

D. Ammann, M. Güggi, E. Pretsch and W. Simon, Improved calcium ion-selective electrode based on a neutral carrier, Anal. Letters **8**, 709 (1975b).

D. Ammann, R. Bissig, Z. Cimerman, U. Fiedler, M. Güggi, W. E. Morf, M. Oehme, H. Osswald, E. Pretsch and W. Simon, Synthetic neutral carriers for cations, 1976, in Kessler et al. 76a, p. 22.

D. Ammann, P. C. Meier and W. Simon, Design and use of calcium-selective microelectrodes, 1979, in Ashley and Campbell 79, p. 117.

D. Ammann, 1981, discussion paper in Lübbers et al. 81, p. 194.

D. Ammann, F. Lanter, R. Steiner, D. Erne and W. Simon, New ion selective liquid membrane microelectrodes, 1981a, in Syková et al. 81, p. 13.

D. Ammann, F. Lanter, R. A. Steiner, P. Schulthess, Y. Shijo and W. Simon, Neutral carrier based hydrogen ion selective microelectrode for extra- and intracellular studies, Anal. Chem. **53**, 2267 (1981b).

D. Ammann, P. Anker, H.-B. Jenny and W. Simon, Valinomycin based silicone rubber membrane electrodes for continuous monitoring of potassium in urine, 1981c, in Ion-Selective Electrodes, E. Pungor and I. Buzás, eds., Akadémiai Kiadó, Budapest, 1981c, p. 179.

D. Ammann, H.-B. Jenny, P. Anker, U. Oesch and W. Simon, Carrier based ion-selective liquid membrane electrodes and their medical applications, in Progress in Enzyme and Ion-Selective Electrodes, D. W. Lübbers, H. Acker, R. P. Buck, G. Eisenman, M. Kessler and W. Simon, eds., Springer Verlag, Berlin, Heidelberg, New York, 1981d, p. 21.

D. Ammann, D. Erne, H.-B. Jenny, F. Lanter and W. Simon, New ion-selective membranes, 1981e, in Lübbers et al. 81, p. 9.

D. Ammann, P. Anker, H.-B. Jenny, P. Schulthess and W. Simon, Use of neutral carrier based electrodes in biomedical systems, in Clinical Chemistry, E. Kaiser, R. Gabl, M. M. Müller and M. Bayer, eds., Walter de Gruyter & Co., Berlin, New York, 1982, p. 1137.

D. Ammann, W. E. Morf, P. Anker, P. C. Meier, E. Pretsch and W. Simon, Neutral carrier based ion-selective electrodes, Ion-Selective Electrode Rev. **5**, 3 (1983).

D. Ammann, E. Pretsch, W. Simon, E. Lindner, A. Bezegh and E. Pungor, Lipophilic salts as membrane additives and their influence on the properties of neutral carrier based macro- and microelectrodes, Anal. Chim. Acta, **171**, 119 (1985a).

D. Ammann, P. Anker, E. Metzger, U. Oesch and W. Simon, Continuous potentiometric measurement of different ion concentrations in whole blood of the extracorporeal circulation, 1985b, in Kessler et al. 85, p. 102.

D. Ammann, F. Lang, M. Paulmichl, U. Oesch and P. Anker, Contamination of cells by neutral carriers from intracellular microelectrodes, 1985c, in preparation.

D. Ammann and P. Anker, Neutral carrier sodium ion-selective microelectrode for extracellular studies, Neurosci. Letters, **57**, 267 (1985).

T. E. Andreoli, M. Tieffenberg and D. C. Tosteson, The effect of valinomycin on the ionic permeability of thin lipid membranes, J. Gen. Physiol. **50**, 2527 (1967).

P. Anker, E. Wieland, D. Ammann, R. E. Dohner, R. Asper and W. Simon, Neutral carrier based ion-selective electrode for the determination of total calcium in blood serum, Anal. Chem. **53**, 1970 (1981).

P. Anker, D. Ammann and W. Simon, Blood pH measurement with a solvent polymeric membrane electrode in comparison with a glass electrode, Mikrochim. Acta 1983 Ia, p. 237.

P. Anker, H.-B. Jenny, U. Wuthier, R. Asper, D. Ammann and W. Simon, Potentiometry of Na^+ in undiluted serum and urine with use of an improved neutral carrier-based solvent polymeric membrane electrode, Clin. Chem. 29, 1508 (1983 b).

P. Anker, H.-B. Jenny, U. Wuthier, R. Asper, D. Ammann and W. Simon, Determination of $[K^+]$ in blood serum with a valinomycin-based silicone rubber membrane of universal applicability to body fluids, Clin. Chem. 29, 1447 (1983 c).

P. Anker, D. Ammann, P. C. Meier and W. Simon, Neutral carrier electrode for continuous measurement of blood Ca^{2+} in the extracorporeal circulation, Clin. Chem. 30, 454 (1984).

F. S. Apple, D. D. Koch, S. Graves and J. H. Ladenson, Relationship between direct-potentiometric and flamephotometric measurement of sodium in blood, Clin. Chem. 28, 1931 (1982).

D. M. Armstrong, A materialist theory of the mind, International Library of Philosophy and Scientific Method, T. Honderich, ed., Routledge & Kegan Paul, New York, 1968.

W. Mc. D. Armstrong, W. Wojtkowski and W. R. Bixenman, A new solid-state microelectrode for measuring intracellular chloride activities, Biochim. Biophys. Acta 465, 165 (1977).

W. McD. Armstrong and J. F. Garcia-Diaz, Ion-selective microelectrodes: theory and technique, Fed. Proc. 39, 2851 (1980).

W. McD. Armstrong, A. Diez de los Rios and N. E. DeRose, Adenosine 3'-5' cyclic monophosphate (cAMP)-induced changes in intracellular ionic activities: relation to epithelial ion transport, 1981, in Lübbers et al. 81, p. 211.

C. C. Ashley, T. J. Rink and R. Y. Tsien, Changes in free Ca during muscle contraction, measured with an intracellular Ca-selective electrode, J. Physiol. 280, 27 P (1978).

C. C. Ashley and A. K. Campbell, eds., Detection and Measurement of Free Ca^{2+} in Cells, Elsevier, North-Holland Biomedical Press, Amsterdam, New York, Oxford, 1979.

C. C. Ashley, J. F. Godber and A. Walton, Microscope-image intensifier studies on single quin-2-loaded muscle fibres from the barnacle *Balanus nubilus* under voltage clamp, J. Physiol. 358, 8 P (1985).

C. R. Bader, D. Bertrand and E. A. Schwartz, Voltage-activated and calcium-activated currents studied in solitary rod inner segments from the salamander retina, J. Physiol. 331, 253 (1982).

H. Baerentsen, F. Giraldez and T. Zeuthen, Influx mechanisms for Na^+ and Cl^- across the brush border membrane of leaky epithelia: a model and microelectrode study, J. Membr. Biol. 75, 205 (1983).

R. R. Baker, Goal orientation by blindfolded humans after long-distance displacement: possible involvement of a magnetic sense, Science 210, 555 (1980).

D. J. Baldwin, Dry beveling of micropipette electrodes, J. Neurosci. Meth. 2, 153 (1980).

K. Ballanyi, P. Grafe and G. ten Bruggencate, Intracellular free sodium and potassium, post-carbachol hyperpolarization, and extracellular potassium-undershoot in rat sympathetic neurones, Neurosci. Letters 38, 275 (1983).

K. Ballanyi and P. Grafe, An intracellular analysis of γ-aminobutyric-acid-associated ion movements in rat sympathetic neurones, J. Physiol. 365, 41 (1985).

J. A. Balschi, V. P. Cirillo and C. S. Springer, Jr., Direct high-resolution nuclear magnetic resonance studies of cation transport in vivo. Na^+ transport in yeast cells, Biophys. J. 38, 323 (1982).

H. Barbier, Développments récents et perspectives des recherches sur l'accumulation du sucre chez la betterave, Sucrerie Française 54 (1981).

J. N. Barrett and K. Graubard, Fluorescent staining of cat motoneurons in vivo with beveled micropipettes, Brain Res. 18, 565 (1970).

R. G. Bates, A. G. Dickson, M. Gratzl, A. Hrabeczy-Pall, E. Lindner and E. Pungor, Determination of mean activity coefficients with ion-selective electrodes, Anal. Chem. 55, 1275 (1983).

G. Baum and M. Lynn, Polymer membrane electrodes: Part II. A potassium ion-selective membrane electrode, Anal. Chim. Acta 65, 393 (1973).

H. Baumgärtl, T. Shigemitsu and D. W. Lübbers, Mikronadelelektroden zur Messung von Ionenaktivitäten in biologischen Geweben, Naturwissenschaften 63, 40 (1976).

C. M. Baumgarten, An improved liquid ion exchanger for chloride ion-selective microelectrodes, Am. J. Physiol. 241, C 258 (1981).

F. Behm, D. Ammann, W. Simon, K. Brunfeldt and J. Halstrøm, Cyclic octa- and decapeptides as ionophores for magnesium, Helv. Chim. Acta 68, 110 (1985).

J.-P. Behr, J.-M. Lehn, A.-C. Dock and D. Moras, Crystal structure of a polyfunctional macrocyclic K⁺ complex provides a solid-state model of a K⁺ channel, Nature **295**, 526 (1982).

G. Bellomo, S. A. Jewell, H. Thor and S. Orrenius, Regulation of intracellular calcium compartmentation: studies with isolated hepatocytes and t-butyl hydroperoxide, Proc. Natl. Acad. Sci. USA **79**, 6842 (1982).

M. R. Bendall and W. P. Aue, Experimental verification of depth pulses applied with surface coils, J. Magn. Res. **54**, 149 (1983).

J. Bendl and E. Pretsch, Conformation analysis of small molecules with PCILO methods, J. Comp. Chem. **3**, 580 (1982).

C. Benninger, J. Kadis and D. A. Prince, Extracellular calcium and potassium changes in hippocampal slices, Brain Res. **187**, 165 (1980).

H. J. C. Berendsen, Water structure, in Theoretical and Experimental Biophysics, A. Cole, ed., Vol. 1, Marcel Dekker, Inc., New York, 1967, p. 1.

M. J. Berridge, Preliminary measurements of intracellular calcium in an insect salivary gland using a calcium-sensitive microelectrode, Cell Calcium **1**, 217 (1980).

D. M. Bers and D. Ellis, Changes of intracellular calcium and sodium activities in sheep heart Purkinje fibres measured with ion-selective micro-electrodes, J. Physiol. **310**, 73 P (1981).

D. M. Bers, A simple method for the accurate determination of free [Ca] in Ca-EGTA solutions, Am. J. Physiol. **242**, C404 (1982).

D. M. Bers and D. Ellis, Intracellular calcium and sodium activity in sheep heart Purkinje fibres: effect of external sodium and intracellular pH, Pflügers Arch. **393**, 171 (1982).

D. M. Bers, Ca influx during single cardiac beats monitored by Ca-microelectrodes: staircase, caffeine and cobalt, Fed. Proc. **42**, 572 (1983).

H. I. Bicher and S. Ohki, Intracellular pH electrode experiments on the giant squid axon. Biochim. Biophys. Acta **255**, 900 (1972).

M. Bickel and G. Cimasoni, The pH of human crevicular fluid measured by a new microanalytical technique, J. Periodontal Res. **20**, 35 (1985).

M. S. Biggs and D. Robson, Advances in the development of extraction resistant flexible PVC compounds, in Polymers in Medicine, E. Chiellini and P. Giusti, eds., Plenum Press, New York, London, 1982, p. 375.

L. C. F. Blackman and R. Harrop, Hydrophilation of glass surfaces. I. Investigation of possible promoters of filmwise condensation, J. Appl. Chem. **18**, 37 (1968).

M. R. Blatt and C. L. Slayman, KCl leakage from microelectrodes and its impact on the membrane parameters of a nonexcitable cell, J. Membr. Biol. **72**, 223 (1983).

J. R. Blinks, F. G. Prendergast and D. G. Allen, Photoproteins as biological calcium indicators, Pharmac. Rev. **28**, 1 (1976).

J. R. Blinks, D. G. Allen, F. G. Prendergast and G. C. Harrer, Photoproteins as models of drug receptors, Life Sci. **22**, 1237 (1978).

J. R. Blinks, W. G. Wier, P. Hess and F. G. Prendergast, Measurement of Ca^{2+} concentrations in living cells, Prog. Biophys. Molec. Biol. **40**, 1 (1982).

T. V. P. Bliss and T. Lømo, Plasticity in a monosynaptic cortical pathway, J. Physiol. **207**, 61 P (1970).

T. V. P. Bliss and T. Lømo, Longlasting potentiation of synaptic transmission in the dentate area of the anaestetized rabbit following stimulation of the perforant path, J. Physiol. **232**, 331 (1973).

T. V. P. Bliss and A. C. Dolphin, What is the mechanism of long-term potentiation in the hippocampus?, Trends Neurosci. **1982**, 289.

R. Bloch, A. Shatkay and H. A. Saroff, Fabrication and evaluation of membranes as specific electrodes for calcium ions, Biophys. J. **7**, 865 (1967).

J. P. C. Boerrigter and A. Lehmenkühler, Microprocessor aided correction of the Ca^{2+} error in the signal of a Na⁺-selective microelectrode, Pflügers Arch. Suppl. to **400**, R56 (1984).

L. Boksányi, O. Liardon and E. Kováts, Chemically modified silicon dioxide surfaces. Reaction of n-alkyldimethylsilanols and n-oxaalkyldimethylsilanols with the hydrated surface of silicon dioxide – the question of the limiting surface concentration, Adv. Coll. Int. Sci. **6**, 95 (1976).

P. J. Bore, L. Chan, D. G. Gadian, G. K. Radda, B. D. Ross, P. Styles and D. J. Taylor, Noninvasive pH_i measurements of human tissue using ^{31}P-NMR, 1982, in Nuccitelli and Deamer 82, p. 527.

W. F. Boron and A. Roos, Comparison of microelectrode, DMO, and methylamine methods for measuring intracellular pH, Am. J. Physiol. **231**, 799 (1976).

S. K. Bosher and R. L. Warren, Very low calcium content of cochlear endolymph, an extracellular fluid, Nature **273**, 377 (1978).

S. K. Bosher, The role of ion-sensitive microelectrodes in the interpretation of endolymphatic changes in the mammalian inner ear, 1981, in Zeuthen 81 a, p. 129.

H. Bostock and P. Grafe, On the conducting block of demyelinated rat spinal roots during long lasting trains of impulses, Pflügers Arch. Suppl. to **402**, R31 (1984).

M. R. Boyett and G. Hart, Factors affecting intracellular sodium activity (a^i_{Na}) during repetitive stimulation in sheep Purkinje fibres, J. Physiol. **357**, 49 P (1984).

D. P. Brezinski, Influence of colloidal charge on response of pH and reference electrodes: the suspension effect, Talanta **30**, 347 (1983).

R. A. Briano, Jr., A reproducible technique for breaking glass micropipettes over a wide range of tip diameters, J. Neurosci. Methods **9**, 31 (1983).

C. A. Briggs, T. H. Brown and D. A. McAfee, Neurophysiology and pharmacology of long-term potentiation in the rat sympathetic ganglion, J. Physiol. **359**, 503 (1985).

F. J. Brinley, Jr., Ion fluxes in the central nervous system, Int. Rev. Neurobiol. **5**, 183 (1963).

F. J. Brinley and A. Scarpa, Ionized magnesium concentration in axoplasm of dialyzed squid axons, FEBS Lett. **50**, 82 (1975).

F. J. Brinley, T. Tiffert and A. Scarpa, The concentration of ionized magnesium in barnacle muscle fibers, J. Physiol. **266**, 545 (1977).

F. J. Brinley, Jr., Calcium buffering in squid axons, Annu. Rev. Biophys. Bioeng. **7**, 363 (1978).

H. M. Brown and J. D. Owen, Micro ion-selective electrodes for intracellular ions, Ion-Selective Electrode Rev. **1**, 145 (1979).

H. M. Brown, J. P. Pemberton and J. D. Owen, A calcium-sensitive microelectrode suitable for intracellular measurement of calcium(II) activity, Anal. Chim. Acta **85**, 261 (1976).

K. T. Brown and D. G. Flaming, Beveling of fine micropipette electrodes by a rapid precision method, Science **185**, 693 (1974).

K. T. Brown and D. G. Flaming, Instrumentation and technique for beveling fine micropipette electrodes, Brain Res. **86**, 172 (1975).

K. T. Brown and D. G. Flaming, New microelectrode techniques for intracellular work in small cells, Neuroscience **2**, 813 (1977).

K. T. Brown and D. G. Flaming, Technique for precision beveling of relatively large micropipettes, J. Neurosci. Meth. **1**, 25 (1979).

R. P. Buck, G. V. Hendrix and J. H. Boles, Theory and responses of neutral carrier membrane for oil soluble anions, IUPAC International Symposium on Selective Ion-Sensitive Electrodes, Cardiff, 1973.

R. P. Buck, J. H. Boles, R. D. Porter and J. A. Margolis, Glass electrode responses interpreted by the solid state homogeneous- and heterogeneous-site membrane potential theory, Anal. Chem. **46**, 255 (1974).

R. P. Buck, Electroanalytical chemistry of membranes, Crit. Rev. Anal. Chem. **5**, 323 (1975).

R. P. Buck, Ion selective electrodes, Anal. Chem. **48**, 23 R (1976).

R. P. Buck, Electrochemistry of ion-selective electrodes, Sensors and Actuators **1**, 197 (1981).

R. Büchi, E. Pretsch and W. Simon, ^{13}C-Kernresonanzspektroskopische Untersuchungen an ionenselektiven Flüssigmembranen, Helv. Chim. Acta **59**, 2327 (1976a).

R. Büchi, E. Pretsch, W. E. Morf and W. Simon, ^{13}C-Kernresonanzspektroskopische und elektromotorische Untersuchungen der Wechselwirkung von neutralen Carriern mit Ionen in Membranen, Helv. Chim. Acta **59**, 2407 (1976b).

C. P. Bührle and U. Sonnhof, Ion fluxes across the membrane of motoneurons during the action of glutamate, 1981, in Syková et al. 81, p. 187.

G. Burckhardt and H. Murer, A cyanine dye as indicator of membrane electrical potential differences in brush border membrane vesicles. Studies with K^+ gradients and Na^+/amino acid cotransport, in Kidney and Body Fluids, L. Takács, ed., Adv. Physiol. Sci. **11**, 409 (1981).

J. Bureš and J. Křivánek, Ionic movements in the brain studied with the aid of washing the cortical surface with an epidural cannula, Physiol. Bohemoslov. **9**, 488 (1960).

W. Burgermeister and R. Winkler-Oswatitsch, Complex Formation of monovalent cations with biofunctional ligands, Topics in Current Chemistry **69**, 91 (1977).

U. Burkert and N. L. Allinger, Molecular Mechanics, ACS Monograph 177, Am. Chem. Soc., Washington, 1982.

P. A. Buser and A. Rougeul-Buser, eds., Cerebral Correlates of Conscious Experience, North-Holland, Publishing Company, Amsterdam, New York, Oxford, 1978.

J. N. Butler, The thermodynamic activity of calcium ion in sodium chloride-calcium chloride electrolytes, Biophys. J. **8**, 1426 (1968).

L. Byerly and W. J. Moody, Intracellular calcium ions and calcium currents in perfused neurones of the snail, *Lymnaea stagnalis*, J. Physiol. **352**, 637 (1984).

D. K. Cabbiness and D. W. Margerum, Macrocyclic effect on the stability of copper(II) tetramine complexes, J. Am. Chem. Soc. **91**, 6540 (1969).

P. C. Caldwell, An investigation of the intracellular pH of crab muscle fibres by means of microglass and micro-tungsten electrodes, J. Physiol. **128**, 169 (1954).

K. Cammann, Working with Ion-Selective Electrodes, Springer Verlag, Berlin, Heidelberg, New York, 1979.

A. K. Campbell, T. J. Lea and C. C. Ashley, Coelenterate photoproteins, 1979, in Ashley and Campbell 79, p. 13.

A. K. Campbell, Intracellular Calcium Its Universal Role as Regulator, John Wiley & Sons, Chichester, Brisbane, Toronto, Singapore, 1983.

F. Capra, The Tao of the Physics, Berkeley, 1975.

E. Carafoli, Mitochondrial uptake of calcium ions and the regulation of cell function, Biochem. Soc. Symp. **39**, 89 (1974).

E. Carafoli and M. Crompton, The regulation of intracellular calcium, Curr. Topics Membr. Transp. **10**, 151 (1978).

E. Carafoli, P. Caroni, M. Chiesi and K. Famulski, Ca^{2+} as a metabolic regulator: mechanisms for the control of its intracellular activity, in Metabolic Compartmentation, H. Sies, ed., Academic Press, London, 1982, p. 521.

E. Carafoli, The regulation of the cellular functions of Ca^{2+}, in Disorders of Mineral Metabolism, F. Bronner and J. W. Coburn, eds., Vol. II, Academic Press, Inc., New York, 1982, p. 1.

N. R. Carlson, Physiology of Behavior, Allyn and Bacon, Inc., Boston, London, Sydney, Toronto, 1977.

P. Caroni, P. Gazzotti, P. Vuilleumier, W. Simon and E. Carafoli, Ca^{2+} transport mediated by a synthetic neutral Ca^{2+}-ionophore in biological membranes, Biochim. Biophys. Acta **470**, 437 (1977).

N. W. Carter, F. C. Rector, R. T. Champion and D. W. Seldin, Measurement of intracellular pH of skeletal muscle with pH-sensitive glass microelectrodes, J. Clin. Invest. **46**, 920 (1967).

D. C. Chang, H. E. Rorschach, B. L. Nichols and C. F. Hazlewood, Implication of diffusion coefficient measurements for the structure of cellular water, 1973, in Hazlewood 73, p. 434.

J. J. Chang, A new technique for beveling the tips of glass capillary micropipettes and microelectrodes, Comp. Biochem. Physiol. **52 A**, 567 (1975).

R. A. Chapman, A. Coray and J. A. S. McGuigan, Na/Ca exchange in mammalian ventricular muscle. A study with Na^+-sensitive micro-electrodes, J. Physiol. **343**, 253 (1983).

R. A. Chapman, J. A. S. McGuigan, G. C. Rodrigo and R. J. Yates, Measurement of intracellular activity of Na ions in frog cardiac muscle at rest and during Na-withdrawal contractures, J. Physiol. **346**, 70 P (1984).

V. K.-H. Chen, A simple piezoelectric drive for glass microelectrodes, J. Phys. E.: Sci. Instr. **11**, 1978.

C. Cheng, Philosophical Aspects of the Mind-Body Problem, The University Press of Hawaii, Honolulu, 1975.

C. W. Chiu, L. H. Lee, C. Y. Wang and G. T. Bryan, Mutagenicity of some commercially available nitro compounds for *Salmonella typhimurium*, Mutation Res. **58**, 11 (1978).

J. Chmielowiec and W. Simon, Alkaline earth cation-complexing chromatography with a neutral ligand chemically bonded to silica gel, Chromatographia **11**, 99 (1978).

T. K. Chowdhury, Fabrication of extremely fine glass micropipette electrodes, J. Sci. Instr. **2**, 1087 (1969).

G. R. J. Christoffersen and E. S. Johansen, Microdesign for a calcium-sensitive electrode, Anal. Chim. Acta **81**, 191 (1976).

M. F. Cicirelli, K. R. Robinson and L. D. Smith, Internal pH of *Xenopus* oocytes: a study of the mechanism and role of pH changes during meiotic maturation, Devel. Biol. **100**, 133 (1983).

M. M. Civan, Intracellular activities of sodium and potassium, Am. J. Physiol. **234**, F261 (1978).

M. M. Civan, Epithelial Ions and Transport, Application of Biophysical Techniques, John Wiley & Sons, New York, Chichester, Brisbane, Toronto, Singapore, 1983.

M. M. Civan, K. Peterson-Yantorno, D. R. DiBona, D. F. Wilson and M. Erecinska, Bioenergetics of Na^+ transport across frog skin: chemical and electrical measurements, Am. J. Physiol. **245**, F691 (1983).

M. M. Civan, Intracellular calcium activity in split frog skin epithelium, Biophys.J. **45**, 140a (1984).

E. Clementi, Lecture notes in chemistry, Vol.2., Springer Verlag, Berlin, Heidelberg, New York, 1976.

E. Clementi, F. Cavallone and R. Scordamaglia, Analytical potentials from "ab initio" computations for the interaction between biomolecules, 1. Water and amino acids, J.Am.Chem.Soc. **99**, 5531 (1977).

C. J. Cohen and H. A. Fozzard, Intracellular K and Na activities in papillary muscle during inotropic interventions, Biophys.J. **25**, 144a (1979).

C. J. Cohen, H. A. Fozzard and S. S. Sheu, Increase in intracellular sodium ion activity during stimulation in mammalian cardiac muscle, Circ. Res. **50**, 651 (1982a).

R. D. Cohen and R. A. Iles, Intracellular pH: measurement, control, and metabolic interrelationships, CRC Critical Rev.Clin.Lab.Sci. **6**, 101 (1975).

R. D. Cohen, R. M. Henderson, R. A. Iles, J. P. Monson and J. A. Smith, The techniques and uses of intracellular pH measurements, Ciba Found. Symp. **87**, 20 (1982b).

M. Cohn and T. R. Hughes, Jr., Nuclear magnetic resonance spectra of adenosine di- and triphosphate. II. Effect of complexing with divalent metal ions, J. Biol. Chem. **237**, 176 (1962).

R. L. Coleman and C. C. Young, Evidence for formation of bicarbonate complexes with Na^+ and K^+ under physiological conditions, Clin.Chem. **27**, 1938 (1981).

J. A. Coles and M. Tsacopoulos, A method of making fine double-barrelled potassium-sensitive micro-electrodes for intracellular recording, J. Physiol. **270**, 12 P (1977).

J. A. Coles, M. Tsacopoulos, P. Rabineau and A. R. Gardner-Medwin, Movement of potassium into glial cells in the retina of the drone, *Apis mellifera,* during photostimulation, 1981, in Syková et al. 81, p. 345.

J. A. Coles and R. K. Orkand, Sodium activity in drone photoreceptors, J. Physiol. **332**, 16 P (1982).

J. A. Coles and R. K. Orkand, Modification of potassium movement through the retina of the drone (*Apis mellifera,* ♂) by glial uptake, J. Physiol. **340**, 157 (1983).

J. A. Coles, J. L. Munoz and F. Deyhimi, Surface and volume resistivity of pyrex glass used for liquid membrane ion-sensitive microelectrodes, 1985, Kessler et al. 85, p.67.

J. A. Coles and R. K. Orkand, Changes in sodium activity during light stimulation in photoreceptors, glia and extracellular space in drone retina, J. Physiol. **362**, 415 (1985).

J. S. Coombs, J. C. Eccles and P. Fatt, The specific ionic conductances and the ionic movements across the motoneuronal membrane that produce the inhibitory postsynaptic potential, J. Physiol. **130**, 326 (1955).

R. L. Coon, N. C. J. Lai and J. P. Kampine, Evaluation of a dual-function pH and pCO_2 in vivo sensor, J.Appl. Physiol. **40**, 625 (1976).

J. A. Connor and P. Hockberger, A novel membrane sodium current induced by injection of cyclic nucleotides into gastropod neurones, J.Physiol. **354**, 139 (1984).

F. W. Cope, Nuclear magnetic resonance evidence for complexing of sodium ions in muscle, Biochem. **54**, 225 (1965).

D. H. Copp, Endocrine control of calcium homeostasis, J. Endocrinol. **43**, 137 (1969).

A. Coray, C. H. Fry, P. Hess, J. A. S. McGuigan and R. Weingart, Resting calcium in sheep cardiac tissue and in frog skeletal muscle measured with ion-selective micro-electrodes, J. Physiol., **305**, 60 P (1980).

A. Coray and J. A. S. McGuigan, Measurement of intracellular ionic calcium concentration in guinea-pig papillary muscle, 1981, in Syková et al. 81, p. 299.

A. Coray and J. A. S. McGuigan, pH_i at rest and during Na withdrawal contractures in ferret ventricle, J. Physiol. **336**, 66 P (1982).

G. Corongiu, E. Clementi, E. Pretsch and W. Simon, Ab initio calculations of the interaction of ions with neutral ligands. I. Pair potentials for Na^+/ ether, Na^+/thioether and Na^+/amide systems, J. Chem. Phys. **70**, 1266 (1979).

G. Corongiu, E. Clementi, E. Pretsch and W. Simon, Ab initio calculations of the interaction of ions with neutral ligands. Pair potentials for Li^+/ ether-, Li^+/thioether-, and Li^+/amide-systems, J. Chem. Phys. **72**, 3096 (1980).

A. K. Covington and N. Kumar, Use of the ionophore antibiotic A23187 in liquid ion-exchange ionselective electrodes, Anal.Chim.Acta **85**, 175 (1976).

A. K. Covington, R. G. Bates and R. A. Durst, Definition of pH scales, standard reference values, measurement of pH and related terminology, Pure Appl.Chem. **55**, 1467 (1983).

A. Craggs, G.J. Moody and J.D.R. Thomas, PVC matrix membrane ion-selective electrodes, J.Chem.Educ. **51**, 541 (1970).

A. Craggs, G.J. Moody and J.D.R. Thomas, Evaluation of calcium ion-selective electrodes based on di(n-alkylphenyl) phosphate sensors and their calibration with ion buffers, Analyst **104**, 412 (1979).

T.A. Cross, C. Pahl, R. Oberhänsli, W.P. Aue, U. Keller and J. Seelig, Ketogenesis in the living rat followed by ^{13}C NMR spectroscopy, Biochem. **23**, 6398 (1984).

M. Dagostino and C.O. Lee, Neutral carrier Na^+- and Ca^{2+}-selective microelectrodes for intracellular application, Biophys.J. **40**, 199 (1982).

G. Dahl and G. Isenberg, Decoupling of heart muscle cells: correlation with increased cytoplasmic calcium activity and with changes of nexus ultrastructure, J. Membr. Biol. **53**, 63 (1980).

J. Daut, The passive electrical properties of guinea-pig ventricular muscle as examined with a voltage-clamp technique, J. Physiol. **330**, 221 (1982a).

J. Daut, The role of intracellular sodium ions in the regulation of cardiac contractility, J. Mol. Cell. Card. **14**, 189 (1982b).

J. Daut and R. Rüdel, The electrogenic sodium pump in guinea-pig ventricular muscle: inhibition of pump current by cardiac glycosides, J. Physiol. **330**, 243 (1982).

J.M. Davidson and R.J. Davidson, eds., The Psychobiology of Consciousness, Plenum Press, New York, London, 1980.

J. Dawson and M.B.A. Djamgoz, Intracellular potassium activities of the muscle cells of a Lepidopteran larva, J.Physiol. **351**, 36 P (1984).

A. de Hemptinne, A double-barrel pH micro-electrode for intracellular use. J. Physiol. **295**, 5 P (1979).

A. de Hemptinne, R. Marrannes and B. Vanheel, Double-barreled intracellular pH electrode: construction and illustration of some results, 1982, in Nuccitelli and Deamer 82, p. 7.

J.W. Deitmer and W.R. Schlue, Intracellular Na^+ and Ca^{2+} in leech retzius neurones during inhibition of the Na^+-K^+ pump, Pflügers Arch. **397**, 195 (1983).

J.W. Deitmer and W.R. Schlue, Na-dependent changes of intracellular calcium in leech sensory neurones, J. Physiol. **357**, 53 P (1984).

S.W. de Laat, W. Wouters, M.M. Marques da Silva Pimenta Guarda and M.A. da Silva Guarda, Intracellular ionic compartmentation, electrical membrane properties, and cell membrane permeability before and during first cleavage in the *Ambystoma* egg, Exp. Cell Res. **91**, 15 (1975).

M. Delpiano and H. Acker, Intracellular ion activity (K^+, Ca^{2+} and Cl^-) and membrane potential of frog muscle in vitro, 1981, in Lübbers et al. 81, p. 206.

M.A. Delpiano and H. Acker, Simultaneous response of the extracellular Ca^{2+} and K^+ activity during hypoxia and hypercapnia and their possible interdependence in the superfused cat carotid body, Pflügers Arch. Suppl. to **402**, R35 (1984).

F. Deyhimi and J.A. Coles, Rapid silylation of a glass surface: choice of reagent and effect of experimental parameters on hydrophobicity, Helv. Chim. Acta **65**, 1752 (1982).

H. Diebler, M. Eigen, G. Ilgenfritz, G. Maass and R. Winkler, Kinetics and mechanism of reactions of main group metal ions with biological carriers, Pure Appl. Chem. **20**, 93 (1969).

B. Dietrich, J.-M. Lehn and J.P. Sauvage, Les cryptates, Tetrahedron Letters **34**, 2885 (1969).

B. Dietrich, J.-M. Lehn, J.P. Sauvage and J. Blanzat, Cryptates – X. Synthèses et propriétés physiques de systèmes diaza-polyoxa-macro-bicycliques, Tetrahedron **29**, 1629 (1973).

B. Dietrich, J. Guilhem, J.-M. Lehn, C. Pascard and E. Sonveaux, 11. Molecular recognition in anion coordination chemistry, Helv. Chim. Acta **67**, 91 (1984).

I. Dietzel, U. Heinemann, G. Hofmeier and H.D. Lux, Transient changes in the size of the extracellular space in the sensorimotor cortex of cats in relation to stimulus-induced changes in potassium concentration, Exp. Brain Res. **40**, 432 (1980).

I. Dietzel, U. Heinemann, G. Hofmeier and H.D. Lux, Changes in the extracellular volume in the cerebral cortex of cats in relation to stimulus induced epileptiform afterdischarges, in Physiology and Pharmacology of Epileptogenic Phenomena, M.R. Klee et al., eds., Raven Press, New York, 1982a, p. 5.

I. Dietzel, U. Heinemann, G. Hofmeier and H.D. Lux, Stimulus-induced changes in extracellular Na^+ and Cl^- concentration in relation to changes in the size of the extracellular space, Exp. Brain. Res. **46**, 73 (1982b).

R. Dingledine and G. Somjen, Calcium dependance of synaptic transmission in the hippocampal slice, Brain Res. **207**, 218 (1981).

R. DiPolo, H. Rojas, J. Vergara, R. Lopez and C. Caputo, Measurements of intracellular ionized calcium in squid giant axons using calcium-selective electrodes, Biochim. Biophys. Acta **728**, 311 (1983).

M. Dobler, Ionophores and Their Structures, J. Wiley & Sons, New York, Chichester, Brisbane, Toronto, 1981.

A. Dörge, R. Rick, K. Gehring and K. Thurau, Preparation of freeze-dried cryosections for quantitative X-ray microanalysis of electrolytes in biological soft tissues, Pflügers Arch. **373**, 85 (1978).

P. J. Donaldson and J. P. Leader, Intracellular ionic activities in the EDL muscle of the mouse, Pflügers Arch. **400**, 166 (1984).

K. Dresdner, R. P. Kline and J. Kupersmith, Extracellular calcium ion depletion in frog ventricle, Biophys. J. **37**, 239 a (1982).

K. P. Dresdner and R. P. Kline, Extracellular calcium depletion in frog ventricular myocardium, J. Physiol. **358**, 57 P (1985).

S. Dütsch, H.-B. Jenny, K. Schlatter, P. Périsset, G. Wolff, J.-T. Clerc, E. Pretsch and W. Simon, Microprocessor-controlled ex vivo monitoring of Na^+ and K^+ concentrations in undiluted urine with ion selective electrodes, Anal. Chem. **57**, 578 (1985).

E. Dufau, H. Acker and D. Sylvester, Triple-barrelled ion-sensitive microelectrode for simultaneous measurements of two extracellular ion activities, Med. Progr. Technol. **9**, 33 (1982).

M. E. Duffey and G. Bebernitz, Intracellular chloride and hydrogen activities in rabbit colon, Fed. Proc. **42**, 1353 (1983).

M. E. Duffey, Intracellular pH and bicarbonate activities in rabbit colon, Am. J. Physiol. **246**, C558 (1984).

J. C. Eccles, An instruction-selection hypothesis of cerebral learning, 1978, in Buser and Rougeul-Buser 78, p. 155.

J. C. Eccles, The Human Psyche, Springer International, Berlin, Heidelberg, 1980.

J. C. Eccles, The modular operation of the cerebral neocortex considered as the material basis of mental events, Neurosci. **6**, 1839 (1981).

J. C. Eccles, Calcium in long-term potentiation as a model for memory, Neurosci. **10**, 1071 (1983).

A. Edelman, S. Curci, I. Samarzija and E. Frömter, Determination of intracellular K^+ activity in rat kidney proximal tubular cells, Pflügers Arch. **378**, 37 (1978).

H. T. Edzes and H. J. C. Berendsen, The physical state of diffusible ions in cells, Ann. Rev. Biophys. Bioeng. **4**, 265 (1975).

J. Ehrenfeld, F. Garcia-Romeu and B. J. Harvey, Electrogenic active proton pump in *Rana esculenta* skin and its role in sodium ion transport, J. Physiol. **359**, 331 (1985).

B. E. Ehrlich and J. M. Diamond, Lithium, membranes, and manic-depressive illness, J. Membr. Biol. **52**, 187 (1980).

G. Eisenman, Glass Electrodes For Hydrogen and Other Cations, Principles and Practice, M. Dekker, Inc., New York, 1967.

G. Eisenman, The ion exchange characteristics of the hydrated surface of Na^+ selective glass electrodes, 1969, in Lavallée et al. 69, p. 32.

D. Ellis, J. W. Deitmer and D. M. Beers, Intracellular pH, Na^+ and Ca^{2+} activity measurements in mammalian heart muscle, 1981, in Lübbers et al. 81, p. 148.

D. Ellis and K. T. MacLeod, Sodium-dependent control of intracellular pH in Purkinje fibres of sheep heart, J. Physiol. **359**, 81 (1985).

E. Eriksson, The significance of pH, ion activities, and membrane potentials in colloidal systems, Science **113**, 418 (1951).

D. Erne, D. Ammann and W. Simon, Liquid membrane pH electrode based on a synthetic proton carrier, Chimia **33**, 88 (1979a).

D. Erne, W. E. Morf, S. Arvanitis, Z. Cimerman, D. Ammann and W. Simon, Durch elektrisch geladene Ionophore induzierter Ionentransport in Modellmembranen mit Selektivität für Magnesium und Calcium, Helv. Chim. Acta **62**, 994 (1979b).

D. Erne, N. Stojanac, D. Ammann, E. Pretsch and W. Simon, Lipophilic amides of EDTA, NTA and iminodiacetic acid as ionophores for alkaline earth metal cations, Helv. Chim. Acta **63**, 2264 (1980a).

D. Erne, N. Stojanac, D. Ammann, P. Hofstetter, E. Pretsch and W. Simon, Lipophilic di- and tri-amides as ionophores for alkaline earth metal cations, Helv. Chim. Acta **63**, 2271 (1980b).

D. Erne, Ionophore mit Selektivität für Erdalkali- und Wasserstoffionen und deren Einsatz in

Flüssigmembranelektroden, Thesis, ETH Zürich, No. 6889, ADAG Administration & Druck AG, Zürich, 1981.

D. Erne, K. V. Schenker, D. Ammann, E. Pretsch and W. Simon, Applicability of a carrier based liquid membrane pH electrode to measurements in acidic solutions, Chimia **35**, 178 (1981).

D. Erne, D. Ammann, A. F. Zhukov, F. Behm, E. Pretsch and W. Simon, Lipophilic diamides as ionophores for alkali and alkaline earth metal cations, Helv. Chim. Acta **65**, 538 (1982).

R. J. Feldmann, The design of computing systems for molecular modeling, Ann. Rev. Biophys. Bioeng. **5**, 477 (1976).

U. Fiedler and J. Růžička, Selectrode – the universal ion-selective electrode. Part VII. A valinomycin-based potassium electrode with nonporous polymer membrane and solid-state inner reference system. Anal. Chim. Acta **67**, 179 (1973).

A. S. Finkel and S. Redman, A shielded microelectrode suitable for single-electrode voltage clamping of neurones in the CNS, J. Neurosci. Meth. **9**, 23 (1983).

D. G. Flaming and K. T. Brown, Micropipette puller design: form of the heating filament and effects of filament width on tip length and diameter, J. Neurosci.Meth. **6**, 91 (1982).

P. Flatman and V. L. Lew, Use of ionophore A 23187 to measure and to control free and bound cytoplasmic Mg in intact red cells, Nature **267**, 360 (1977).

B. Fleet, T. H. Ryan and M. J. D. Brand, Investigation of the factors affecting the response time of a calcium selective liquid membrane electrode, Anal. Chem. **46**, 12 (1974).

N. Fogh-Andersen, T. F. Christiansen, L. Komarmy and O. Siggaard-Andersen, Measurement of free calcium ion in capillary blood and serum, Clin. Chem. **24**, 1545 (1978).

W. Forth, D. Henschler and W. Rummel, eds., Allgemeine und spezielle Pharmakologie und Toxikologie, Wissenschaftsverlag, Mannheim, Wien, Zürich, 1977, p. 396.

K. R. Foster, J. M. Bidinger and D. O. Carpenter, The electrical resistivity of cytoplasm, Biophys. J. **16**, 991 (1976).

E. Frömter and B. Gebler, Electrical properties of amphibian urinary bladder epithelia. III. The cell membrane resistances and the effect of amiloride, Pflügers Arch. **371**, 99 (1977).

E. Frömter, M. Simon and B. Gebler, A double-channel ion-selective microelectrode with the possibility of fluid ejection for localization of the electrode tip in the tissue, 1981, in Lübbers et al. 81, p. 35.

M. Fromm and S. G. Schultz, Some properties of KCl-filled microelectrodes: correlation of potassium "leakage" with tip resistance, J. Membr. Biol. **62**, 239 (1981).

M. Fromm, P. Weskamp and U. Hegel, Versatile piezoelectric driver for cell puncture, Pflügers Arch. **384**, 69 (1980).

D. M. Fry, A scanning electron microscope method for the examination of glass microelectrode tips either before or after use, Experientia **31**, 695 (1975).

M. Fujimoto and T. Kubota, Physicochemical properties of a liquid ion exchanger microelectrode and its application to biological fluids, Jap. J. Physiol. **26**, 631 (1976).

M. Fujimoto and M. Honda, A triple-barreled microelectrode for simultaneous measurements of intracellular Na^+ and K^+ activities and membrane potential in biological cells, Jap. J. Phys. **30**, 859 (1980).

M. Fujimoto, K. Naito and T. Kubota, Electrochemical profile for ion transport across the membrane of proximal tubular cells, Membr. Biochem. **3**, 67 (1980).

A. B. Fulton, How crowded is the cytoplasm?, Cell **30**, 345 (1982).

R. J. J. Funck, W. E. Morf, P. Schulthess, D. Ammann and W. Simon, Bicarbonate-sensitive liquid membrane electrodes based on neutral carriers for hydrogen ions, Anal.Chem. **54**, 423 (1982).

K. W. Fung and K. H. Wong, Potassium- and caesium-selective PVC membrane electrodes based on bis-crown ethers, J. Electroanal. Chem. **111**, 359 (1980).

D. G. Gadian, G. K. Radda, R. E. Richards and P. J. Seeley, ^{31}P NMR in living tissue: the road from a promising to an important tool in biology, 1979, in Shulman 79, p. 463.

D. G. Gadian, G. K. Radda, M. J. Dawson and D. R. Wilkie, pH$_i$ measurements of cardiac and skeletal muscle using ^{31}P-NMR, 1982, in Nuccitelli and Deamer 82, p. 61.

D. G. Gadian, Whole organ metabolism studied by NMR, Ann. Rev. Biophys. Bioeng. **12**, 69 (1983).

V. P. Y. Gadzekpo, J. M. Hungerford, A. M. Kadry, Y. A. Ibrahim and G. D. Christian, Lipophilic lithium ion carrier in a lithium ion selective electrode, Anal. Chem. **57**, 493 (1985).

J. Gajowski, B. Rieckemann and F. Umland, Kaliumselektive Membranelektroden auf der Basis des Kryptanden [2B2B2], Fresenius Z. Anal. Chem. **309**, 343 (1981).

M. Galvan, G. tenBruggencate and R. Senekowitsch, The effects of neuronal stimulation and ouabain upon extracellular K^+ and Ca^{2+} levels in rat isolated sympathetic ganglia, Brain Res. **160**, 544 (1979).

M. Galvan, P. Grafe and G. tenBruggencate, Convulsive actions of 4-amino-pyridine on neurones and extracellular K^+ and Ca^{2+} activities in guinea-pig olfactory cortex slices, in: Physiology and Pharmacology of Epileptogenic Phenomena, M. Klee, ed., 1980.

M. Galvan, A. Dörge, F. Beck and R. Rick, Intracellular electrolyte concentrations in rat sympathetic neurones measured with an electron microprobe, Pflügers Arch. **400**, 274 (1984).

S. M. Gamiño and F. J. Alvarez-Leefmans, Intracellular free magnesium in *Helix aspersa* neurones, Soc. Neurosci. Abstr. **9**, 513 (1983).

S. M. Gamiño and F. J. Alvarez-Leefmans, Microelectrodes containing the neutral ligand ETH-1117 can be used for measuring cytoplasmic free magnesium, Biophys. J. **45**, 87a (1984).

J. F. Garcia-Diaz and W. McD. Armstrong, The steady-state relationship between sodium and chloride transmembrane electrochemical potential differences in *Necturus* gallbladder, J. Membr. Biol. **55**, 213 (1980).

A. R. Gardner-Medwin and C. Nicholson, Measurements of extracellular potassium and calcium concentration during passage of current across the surface of the brain, J. Physiol. **275**, 668 (1977).

P. B. Garlick, G. K. Radda and P. J. Seeley, Studies of acidosis in the ischaemic heart by phosphorus nuclear magnetic resonance, Biochem. J. **184**, 547 (1979).

M. S. Gazzaniga and J. E. LeDoux, The Integrated Mind, Plenum Press, New York, London, 1978.

M. S. Gazzaniga, The Bisected Brain, Meredith Corporation, New York, 1970.

L. A. Geddes, Electrodes and the Measurement of Bioelectric Events, Wiley-Interscience, New York, London, Sydney, Toronto, 1972.

J. Geibel, H. Völkl and F. Lang, A microelectrode for continuous recording of volume fluxes in isolated perfused tubule segments, Pflügers Arch. **400**, 388 (1984).

C. D. Geisler, E. N. Lightfoot, F. P. Schmidt and F. Sy, Diffusion effects of liquid-filled micropipettes: a pseudobinary analysis of electrolyte leakage, IEEE Trans. Biomed. Eng. **19**, 372 (1972).

G. Giebisch, T. Kubota and M. G. O'Regan, Ion-activity measurements in renal tubular epithelium, 1981, in Zeuthen 81a, p. 47.

J. C. Gilkey, L. F. Jaffe, E. B. Ridgway and G. T. Reynolds, A free calcium wave traverses the activating egg of the medaka, *Oryzias latipes,* J. Cell. Biol. **76**, 448 (1978).

R. J. Gillies, J. R. Alger, J. A. den Hollander and R. G. Shulman, Intracellular pH measured by NMR: methods and results, 1982, in Nuccitelli and Deamer 82, p. 79.

F. Giraldez, The sodium pump in *Necturus* gallbladder epithelium, 1985, in Kessler et al. 85, p. 138.

D. Giulian and E. G. Diacumakos, The electrophysiological mapping of compartments within a mammalian cell, J. Cell. Biol. **72**, 86 (1977).

H. G. Glitsch and H. Pusch, On the electrogenic fraction of K-activated Na transport in sheep Purkinje fibres, Pflügers Arch. Suppl. to **392**, R1 (1982).

H. G. Glitsch, H. Pusch, T. Schumacher and F. Verdonck, An identification of the K activated Na pump current in sheep Purkinje fibres, Pflügers Arch. **394**, 256 (1982).

H. G. Glitsch and H. Pusch, Activation of the Na pump in sheep Purkinje fibres by external potassium at various temperatures, Pflügers Arch. Suppl. to **402**, R23 (1984a).

H. G. Glitsch and H. Pusch, On the temperature dependence of the Na pump in sheep Purkinje fibres, Pflügers Arch. **402**, 109 (1984b).

H. G. Glitsch and H. Pusch, On the temperature dependence of active Na transport in sheep Purkinje fibres, Pflügers Arch. Suppl. to **400**, R5 (1984c).

H. G. Glitsch, H. Pusch and T. Schumacher, Inhibition of the Na pump in sheep Purkinje fibres by external sodium, Pflügers Arch. Suppl. to **402**, R22 (1984).

H. G. Glitsch, H. Pusch and T. Schumacher, Temperature dependence of the cardiac Na^+-K^+ pump as studied by Na-sensitive microelectrodes, 1985, in Kessler et al. 85, p. 282.

G. G. Globus, G. Maxwell and I. Savodnik, eds., Consciousness and The Brain, Plenum Press, New York, London, 1976.

R. A. Gorkin and E. Richelson, Lithium ion accumulation by cultured glioma cells, Brain Res. **171**, 365 (1979).

A. L. F. Gorman, S. Levy, E. Nasi and D. Tillotson, Intracellular calcium measured with calcium-sensitive micro-electrodes and arsenazoIII in voltage-clamped *Aplysia* neurones, J. Physiol. **353**, 127 (1984).

T. Gotow, M. Ohba and T. Tomita, Tip potential and resistance of micro-electrodes filled with KCl solution by boiling and nonboiling methods, IEEE Trans. Biomed. Eng. **24**, 366 (1977).

J. L. Gould, Magnetic field sensitivity in animals, Ann. Rev. Physiol. **46**, 585 (1984).

P. Grafe, J. Rimpel, M. M. Reddy and G. tenBruggencate, Changes of intracellular sodium and potassium ion concentrations in frog spinal motoneurons induced by repetitive synaptic stimulation, Neurosci. **7**, 3213 (1982a).

P. Grafe, J. Rimpel, M. M. Reddy and G. tenBruggencate, Lithium distribution across the membrane of motoneurons in the isolated frog spinal cord, Pflügers Arch. **393**, 297 (1982b).

P. Grafe, K. Ballanyi and G. tenBruggencate, Changes of intracellular free ion concentrations, evoked by carbachol or GABA, in rat sympathetic neurons, 1985, in Kessler et al. 85, p. 184.

R. S. Greenwood, W. E. Dodson and S. Goldring, The effect of local anesthetics on the potassium ion-selective electrode, Brain Res. **165**, 171 (1979).

R. Greger, F. Lang and S. Silbernagel, eds., Renal Transport of Organic Substances, Springer Verlag, Berlin, Heidelberg, New York, 1981.

R. Greger, H. Oberleithner, E. Schlatter, A. C. Cassola and C. Weidtke, Chloride activity in cells of isolated perfused cortical thick ascending limbs of rabbit kidney, Pflügers Arch. **399**, 29 (1983).

R. Greger and E. Schlatter, Properties of the lumen membrane of the cortical thick ascending limb of Henle's loop of rabbit kidney, Pflügers Arch. **396**, 315 (1983).

R. Greger and E. Schlatter, Mechanism of NaCl secretion in rectal gland tubules of spiny dogfish (*Squalus acanthias*). II. Effect of inhibitors, Pflügers Arch. **402**, 364 (1984).

R. Greger, C. Weidtke, E. Schlatter, M. Wittner and B. Gebler, Potassium activity in cells of isolated perfused cortical thick ascending limbs of rabbit kidney, Pflügers Arch. **401**, 52 (1984a).

R. Greger, E. Schlatter, F. Wang and J. N. Forrest, Jr., Mechanism of NaCl secretion in rectal gland tubules of spiny dogfish (*Squalus acanthias*). III. Effects of stimulation of secretion by cyclic AMP, Pflügers Arch. **402**, 376 (1984b).

R. Greger and E. Schlatter, Electrolyte activities in Cl^--transporting epithelia: cortical thick ascending limb of rabbit nephron and rectal gland tubules of the spiny dogfish, *Squalus acanthias*, 1985, in Kessler et al. 85, p. 301.

E. Grell, I. Oberbäumer, H. Ruf and H. P. Zingsheim, Elementary steps and dynamic aspects of carrier-mediated cation transport through membranes: the streptogramin antibiotics (group B), 1977, in Biochemistry of Membrane Transport, G. Semenza and E. Carafoli, eds., Springer Verlag, Berlin, Heidelberg, New York, 1977, p. 147.

N. Gresh, P. Claverie and A. Pullman, Intermolecular interactions: reproduction of the results of ab initio supermolecule computations by an additive procedure, Int. J. Quant. Chem.: Quant. Chem. Symp. **13**, 243 (1979).

A. Griffin and J. A. Sechzer, Mandatory versus voluntary regulation of biomedical research, 1983, in Sechzer 83, p. 187.

T. H. Grove, J. J. H. Ackerman, G. K. Radda and P. J. Bore, Analysis of rat heart in vivo by phosphorus nuclear magnetic resonance, Proc. Natl. Acad. Sci. USA **77**, 299 (1980).

M. Güggi, U. Fiedler, E. Pretsch and W. Simon, A lithium ion-selective electrode based on a neutral carrier, Anal. Letters **8**, 857 (1975).

M. Güggi, M. Oehme, E. Pretsch and W. Simon, Neutraler Ionophor für Flüssigmembranelektroden mit hoher Selektivität für Natrium- gegenüber Kalium-Ionen, Helv. Chim. Acta **59**, 2417 (1976).

M. Güggi, E. Pretsch and W. Simon, A barium ion-selective electrode based on the neutral carrier N,N,N',N'-tetraphenyl-3,6,9-trioxaundecane diamide, Anal. Chim. Acta **91**, 107 (1977).

G. G. Guilbault, R. A. Durst, M. S. Frant, H. Freiser, E. H. Hansen, T. S. Light, E. Pungor, G. Rechnitz, N. M. Rice, T. J. Rohm, W. Simon and J. D. R. Thomas, Recommendations for nomenclature of ion-selective electrodes, Pure Appl. Chem. **48**, 127 (1976).

B. L. Gupta and T. A. Hall, Quantitative electron probe X-ray microanalysis of electrolyte elements within epithelial tissue compartments, Fed. Proc. **38**, 144 (1979).

B. L. Gupta, T. A. Hall, S. H. P. Maddrell and R. B. Moreton, Distribution of ions in a fluid-transporting epithelium determined by electron-probe X-ray microanalysis, Nature **264**, 284 (1976).

R. K. Gupta and P. Gupta, Direct observation of resolved resonances from intra- and extracellular sodium-23-ions in NMR studies of intact cells and tissues using dysprosium(III) tripoly phosphate as paramagnetic shift reagent, J. Magn. Res. **47**, 344 (1982).

R. K. Gupta and R. D. Moore, ^{31}P NMR studies of intracellular free Mg^{2+} in intact frog skeletal muscle, J. Biol. Chem. **255**, 3987 (1980).

R. K. Gupta and W. D. Yushok, Noninvasive ^{31}P NMR probes of free Mg^{2+}, Mg ATP, and Mg ADP in intact Ehrlich ascites tumor cells, Proc. Natl.Acad.Sci. **77**, 2487 (1980).

H. L. Haas and J. G. R. Jefferys, Low-calcium field burst discharges of CA1 pyramidal neurones in rat hippocampal slices, J. Physiol. **354**, 185 (1984).

A. Hämmerli, J. Janata and H. M. Brown, Ion-selective electrode for intracellular potassium measurements, Anal. Chem. **52**, 1179 (1980).

M. L. Hair and W. Hertl, Reactions of chlorosilanes with silica surfaces, J. Phys. Chem. **73**, 2372 (1969).

W. J. Hamer and Y.-C. Wu, Osmotic coefficients and mean activity coefficients of uni-univalent electrolytes in water at 25 °C, J. Phys. Chem. Ref. Data **1**, 1047 (1972).

O. P. Hamill, A. Marty, E. Neher, B. Sakmann and F. J. Sigworth, Improved patch-clamp techniques for high-resolution current recording from cells and cell-free membrane patches, Pflügers Arch. **391**, 85 (1981).

Handbook of Chemistry and Physics, 56th ed., Chemical Rubber Publ. Co., Cleveland, Ohio, 1975-76, p. D-153.

C. Hansch and A. Leo, Substituent Constants For Correlation Analysis in Chemistry and Biology, J. Wiley & Sons, New York, Chichester, Brisbane, Toronto, 1979.

A. J. Hansen, Extracellular ion concentration in cerebral ischemia, 1981, in Zeuthen 81 a, p. 239.

A. J. Hansen and. T. Zeuthen, Extracellular ion concentrations during spreading depression and ischemia in the rat brain cortex, Acta Physiol. Scand. **113**, 437 (1981).

H. H. Harary and J. E. Brown, Spatially nonuniform changes in intracellular calcium ion concentrations, Science **224**, 292 (1984).

M. C. Harman and P. A. Poole-Wilson, A liquid ion-exchange intracellular pH microelectrode, J. Physiol. **315**, 1 P (1981).

K. M. Harris and T. J. Teyler, Developmental onset of long-term potentiation in area CA1 of the rat hippocampus, J. Physiol. **346**, 27 (1984).

R. J. Harris and L. Symon, A double ion sensitive micro-electrode for extracellular cerebral cortical measurements, J. Physiol. **312**, 3 P (1981).

R. J. Harris, L. Symon, N. M. Branston and M. Bayhan, Changes in extracellular calcium activity in cerebral ischaemia, J. Cerebral Blood Flow Metab. **1**, 203 (1981).

R. J. Harris, M. Bayhan, N. M. Branston, A. Watson and L. Symon, Modulation of the pathophysiology of primate focal cerebral ischaemia by indomethacin, Stroke **13**, 17 (1982).

R. K. Harris and B. E. Mann, NMR and the Periodic Table, Academic Press, London, New York, San Francisco, 1978.

K. Hartman, S. Luterotti, H. F. Osswald, M. Oehme, P. C. Meier, D. Ammann and W. Simon, Chloride-selective liquid-membrane electrodes based on lipophilic methyl-tri-N-alkyl-ammonium compounds and their applicability to blood serum measurements, Mikrochim. Acta **1978 II**, 235.

B. J. Harvey and R. P. Kernan, Sodium-selective micro-electrode study of apical permeability in frog skin: effects of sodium amiloride and ouabain, J. Physiol. **356**, 359 (1984a).

B. J. Harvey and R. P. Kernan, Intracellular ion activities in frog skin in relation to external sodium and effects of amiloride and/or ouabain, J. Physiol. **349**, 501 (1984b).

J. W. Hastings, G. Mitchell, P. H. Mattingly, J. R. Blinks and M. VanLeeuwen, Response of aequorin bioluminescence to rapid changes in calcium concentration, Nature **222**, 1047 (1969).

C. F. Hazlewood, ed., Physicochemical state of ions and water in living tissues and model systems, Ann. New York Acad. Sci. **204**, 1 (1973).

N. C. Hebert, Properties of microelectrode glasses, 1969, in Lavallée et al. 69, p. 25.

U. Heinemann, H. D. Lux and M. J. Gutnick, Extracellular free calcium and potassium during paroxysmal activity in the cerebral cortex of the cat, Exp. Brain Res. **27**, 237 (1977).

U. Heinemann, A. Konnerth and H. D. Lux, Changes in extracellular free Ca^{2+} and K^+ activity in epileptogenic alumina cream focus in the cerebral cortex of cats, Neurosci. Letters, Suppl. 1, 63 (1978a).

U. Heinemann, H. D. Lux and M. J. Gutnick, Changes in extracellular free calcium and potassium activity in the somatosensory cortex of cats, in Abnormal Neuronal Discharges, N. Chalazonitis and M. Boisson, eds., Raven Press, New York, 1978b, p. 329.

U. Heinemann and A. Konnerth, Changes in extracellular free Ca^{2+} during epileptic activity in chronic alumina cream foci in cats, in Advances in Epileptology, R. Canger, F. Angeleri and J. K. Penry, eds., Raven Press, New York, 1980, p. 371.

U. Heinemann and R. Pumain, Extracellular calcium activity changes in cat sensorimotor cortex induced by iontophoretic application of aminoacids, Exp. Brain Res. **40**, 247 (1980).

U. Heinemann and R. Pumain, Changes in extracellular free Ca^{2+} in the sensorimotor cortex of cats during electrical stimulation and iontophoretic application of amino-acids, 1981 a, in Syková et al. 81, p. 235.

U. Heinemann and R. Pumain, Effects of tetrodotoxin on changes in extra-cellular free calcium induced by repetitive electrical stimulation and iontophoretic application of excitatory amino acids in the sensorimotor cortex of cats, Neurosci. Letters **21**, 87 (1981 b).

U. Heinemann, A. Konnerth and H. D. Lux, Stimulation induced changes in extracellular free calcium in normal cortex and chronic alumina cream foci of cats, Brain Res. 213, 246 (1981).

U. Heinemann, A. Konnerth, J. Louvel, H. D. Lux and R. Pumain, Changes in extracellular free Ca^{2+} in normal and epileptic sensorimotor cortex of cats, in Physiology and Pharmacology of Epileptogenic Phenomena, M. R. Klee et al., eds., Raven Press, New York, 1982, p. 29.

U. Heinemann and J. D. C. Lambert, NMDA and quisqualate induced neuronal depolarisations and changes in extracellular Na^+, Mg^{2+} and Ca^{2+} concentrations in area CA1 of in vitro hippocampal slices, Pflügers Arch. Suppl. to **402**, R30 (1984).

J. M. Heiple and D. L. Taylor, An optical technique for measurement of intracellular pH in single living cells, 1982, in Nuccitelli and Deamer 82, p. 21.

P. Henderson, Zur Thermodynamik der Flüssigkeitsketten, Z. Phys. Chem. **59**, 118 (1907).

W. Hertl, Mechanism of gaseous siloxane reaction with silica, J. Phys. Chem. **72**, 1248 (1968 a).

W. Hertl, Mechanism of gaseous siloxane reaction with silica, J. Phys. Chem. **72**, 3993 (1968 b).

W. Hertl and M. L. Hair, Reaction of hexamethyldisilazane with silica, J. Phys. Chem. **75**, 2181 (1971).

P. Hess and R. Weingart, Intracellular free calcium modified by pH_i in sheep cardiac Purkinje fibres, J. Physiol. **307**, 60 P (1980).

P. Hess and R. Weingart, Free magnesium in cardiac and skeletal muscle measured with ion-selective micro-electrodes, J. Physiol. **318**, 14 P (1981).

P. Hess, P. Metzger and R. Weingart, Free magnesium in sheep, ferret and frog striated muscle at rest measured with ion-selective micro-electrodes, J. Physiol. **333**, 173 (1982).

D. Heuser, J. Astrup, N. A. Lassen, B. Nilsson, K. Norberg and B. K. Siesjö, Are H^+ and K^+ factors for the adjustment of cerebral blood flow to changes in functional state: a microelectrode study, Acta Neurologica Scandinavica, Supplementum 64, **56**, 216 (1977).

D. Heuser, The significance of cortical extracellular H^+, K^+ and Ca^{2+} activities for regulation of local cerebral blood flow unter conditions of enhanced neuronal activity, in Cerebral Vascular Smooth Muscle and Its Control, Ciba Foundation Symp. 56, Elsevier, Excerpta Medica, North-Holland, Amsterdam, Oxford, New York, 1978, p. 339.

D. Heuser, Local ionic control of cerebral microvessels, 1981, in Zeuthen 81 a, p. 85.

C. B. Heyer and H. D. Lux, Unusual properties of the Ca-K system responsible for prolonged action potentials in neurons from the snail *Helix pomatia*, in Abnormal Neuronal Discharges, N. Chalazonitis and M. Boisson, eds., Raven Press, New York, 1978, p. 311.

J. T. Higgins, Jr., B. Gebler and E. Frömter, Electrical properties of amphibian urinary bladder epithelia. II. The cell potential profile in *Necturus maculosus*, Pflügers Arch. **371**, 87 (1977).

J. L. Hill, L. S. Gettes, M. R. Lynch and N. C. Hebert, Flexible valinomycin electrodes for on-line determination of intravascular and myocardial K^+, Am. J. Physiol. **235**, H455 (1978).

J. A. M. Hinke, Glass microelectrodes for measuring intracellular activities of sodium and potassium, Nature **184**, 1257 (1959).

J. A. M. Hinke, Cation-selective microelectrodes for intracellular use, 1967, in Eisenman 67, p. 464.

P. Hník, E. Syková, N. Křiž and F. Vyskočil, Determination of ion activity changes in excitable tissues with ion-selective microelectrodes, 1980, in Koryta 80, p. 129.

A. L. Hodgkin and R. D. Keynes, The mobility and diffusion coefficient of potassium in giant axons from sepia, J. Physiol. **119**, 513 (1953).

G. Hofmeier and H. D. Lux, Time courses of intracellular free calcium and related electrical effects after injection of $CaCl_2$, Pflügers Arch., Suppl. to Vol. **373**, R47 (1978).

G. Hofmeier and H. D. Lux, Intracellular applications of Ca^{2+}-selective microelectrodes in voltage-clamped snail neurons, 1981 a, in Lübbers et al. 81, p. 127.

G. Hofmeier and H. D. Lux, The time courses of intracellular free calcium and related electrical effects after injection of $CaCl_2$ into neurons of the snail, *Helix pomatia*, Pflügers Arch. **391**, 242 (1981 b).

G. Hofmeier and H.D. Lux, the depolarizing action of calcium injected into snail neurons – a mechanism contributing to epileptogenesis, in Physiology and Pharmacology of Epileptogenic Phenomena, M.R. Klee et al., eds., Raven Press, New York, 1982, p.299.

F. Hofmeister, Zur Lehre von der Wirkung der Salze, Archiv. Exp. Pathol. Pharmakol. **24**, 247 (1888).

P.M. Hofstetter, Korrelation des elektrochemischen Verhaltens eines Thioamids als Ionencarrier mit seiner Austauschgeschwindigkeit in Übergangs- und B-Metallkomplexen, Thesis, ETH Zürich No.7128, ADAG Administration und Druck AG Zürich, 1982.

P. Hofstetter, E. Pretsch and W. Simon, NMR-Spektroskopische Untersuchungen der kinetischen Limitierung der Kationenselektivität eines cadmiumselektiven Ionophors, Helv. Chim. Acta **66**, 2103 (1983).

P. Horowitz and W. Hill, The Art of Electronics, Cambridge University Press, Cambridge, London, New York, New Rochelle, Melbourne, Sydney, 1980.

S.B. Horowitz and P.L. Paine, Reference phase analysis of free and bound intracellular solutes. II. Isothermal and isotopic studies of cytoplasmic sodium, potassium, and water, Biophys. J. **25**, 45 (1979).

S.B. Horowitz, P.L. Paine, L. Tluczek and J.K. Reynhout, Reference phase analysis of free and bound intracellular solutes. I. Sodium and potassium in amphibian oocytes, Biophys. J. **25**, 33 (1979).

G. Horvai, T.A. Nieman and E. Pungor, Low resistance liquid membrane ion-selective electrodes, 1985, in Ion-Selective Electrodes, E. Pungor and I. Buzás, eds., Akadémiai Kiadó, Budapest 1985, p.439.

F. Huguenin, Some aspects of non-ionic permeation of NH_3 and CO_2 in mammalian skeletal muscle, 1985, in Kessler et al. 85, p.236.

A. Hulanicki and R. Lewandowski, Some properties of ion-selective electrodes based on poly(vinyl chloride) membranes with liquid-ion-exchanger, Chem. Anal. **19**, 53 (1974).

T. Jacobsen, E.M. Skou and S. Athlung, On the definition of single ion activities, Electrochim. Acta **20**, 523 (1975).

W.E. Jacobus, I.H. Pores, S.K. Lucas, C.H. Kallman, M.L. Weisfeldt and J.T. Flaherty, The role of intracellular pH in the control of normal and ischemic myocardial contractility: A ^{31}P nuclear magnetic resonance and mass spectrometry study, 1982, in Nuccitelli and Deamer 82, p.537.

Z. Janka, I. Szentistványi, A. Juhász and A. Rimanóczy, Steady-state distribution of lithium during cultivation of dissociated brain cells, Experientia **36**, 1071 (1980).

J. Janus, E.-J. Speckmann and A. Lehmenkühler, Relations between extracellular K^+ and Ca^{2+} activities and local field potentials in the spinal cord of the rat during focal and generalized seizure discharges, 1981, in Syková et al. 81, p.181.

J. Janus and A. Lehmenkühler, Changes of extracellular chloride, sodium and calcium activities during stimulus-induced DC potential shifts in the CNS, Pflügers Arch. **389**, R18 (1981 a).

J. Janus and A. Lehmenkühler, A simple procedure for response time reduction of liquid ion-exchanger microelectrodes, Pflügers Arch. **389**, R32 (1981 b).

G.J. Janz and R.P.T. Tomkins, Nonaqueous Electrolytes Handbook, Vol. 1, Academic Press, New York, London, 1972.

A. Jaramillo, S. Lopez, J.B. Justice, Jr., J.D. Salamone and D.B. Neill, Acetylcholine and choline ion-selective microelectrodes, Anal. Chim. Acta **146**, 149 (1983).

H. Jenny, T.R. Nielsen, N.T. Coleman and D.E. Williams, Concerning the measurements of pH, ion activities, and membrane potentials in colloidal systems, Science **112**, 164 (1950).

H.-B. Jenny, D. Ammann, R. Dörig, B. Magyar, R. Asper and W. Simon, Neutral carrier based ion-selective electrode for the determination of Na^+ in urine, Mikrochim. Acta **1980 II**, 125 (1980 a).

H.-B. Jenny, C. Riess, D. Ammann, B. Magyar, R. Asper and W. Simon, Determination of K^+ in diluted and undiluted urine with ion-selective electrodes, Mikrochim. Acta **1980 II**, 309 (1980 b).

M. Joffre, P. Mollard, P. Régondaud, J. Alix, J.P. Poindessault, A. Malassiné and Y.M. Gargouil, Electrophysiological study of single Leydig cells freshly isolated from rat testis. I. Technical approach and recordings of the membrane potential in standard solution, Pflügers Arch. **401**, 239 (1984).

G. Isenberg and G. Dahl, Ultrastructural changes of the gap junction correlated with increased longitudinal resistance (Purkinje fibre), Abstracts of the 49th Meeting of the Deutsche Physiologische Gesellschaft, Springer International, 1978, p. R9.

G. Isenberg, Risk and advantages of using strongly beveled microelectrodes for electrophysiological studies in cardiac Purkinje fibers, Pflügers Arch. **380**, 91 (1979).

G. Isenberg and U. Klockner, Calcium tolerant ventricular myocytes prepared by preincubation in "KB medium", Pflügers Arch. **395**, 6 (1982).

R. Jung, Perception, consciousness and visual attention, 1978, in Buser and Rougeul-Buser 78, p. 15.

IUPAC Commission on Analytical Nomenclature (prepared for publication by G. G. Guilbault), Recommendations for publishing manuscripts on ion-selective electrodes, Ion-Selective Electrode Rev. **1**, 139 (1979).

K. Kafoglis, S. J. Hersey and J. F. White, Microelectrode measurement of K^+ and pH in rabbit gastric glands: effect of histamine, Am. J. Physiol. **246**, G433 (1984).

L. Kaufman, Y. Okada, J. Tripp and H. Weinberg, Evoked neuromagnetic fields, Ann. New York Acad. Sci. **425**, 722 (1984).

M. Kessler, L. C. Clark, Jr., D. W. Lübbers, I. A. Silver and W. Simon, eds., Ion and Enzyme Electrodes in Biology and Medicine, Urban & Schwarzenberg, München, Berlin, Wien, 1976a.

M. Kessler, K. Hájek and W. Simon, Four-barrelled microelectrode for the measurement of potassium-, sodium- and calcium-ion activities, 1976b, in Kessler et al. 76a, p. 136.

M. Kessler, J. Höper, D. Schäfer and R. Strehlau, Measurements of extracellular and of interstitial cation activity (pK, pNa, pCa) with ion-selective electrodes, Biblioth. Anatomica **15**, 237 (1977).

M. Kessler, D. K. Harrison and J. Höper, eds., Ion Measurements in Physiology and Medicine, Springer Verlag, Berlin, Heidelberg, New York, Tokyo, 1985.

R. N. Khuri, Cation and hydrogen microelectrodes in single nephrons, 1969, in Lavallée 69, p. 272.

R. N. Khuri, S. K. Agulian, K. Bogharian, R. Nassar and W. Wise, Intracellular bicarbonate in single cells of *Necturus* kidney proximal tubule, Pflügers Arch. **349**, 295 (1974).

R. N. Khuri, S. M. Agulian, E. L. Boulpae, W. Simon and G. Giebisch, Changes in the intracellular electrochemical potential of Na^+, K^+ and Cl^- in single cells of the proximal tubule of the *Necturus* kidney induced by rapid changes in the extracellular perfusion fluids, Arzneim.-Forsch./Drug Res. **28**, 878 (1978).

R. N. Khuri, Electrochemistry of the nephron, in Membrane Transport in Biology, G. Giebisch, D. C. Tosteson and H. H. Ussing, eds., Vol. 4A, Springer Verlag, Berlin, Heidelberg, New York, 1979, p. 47.

R. N. Khuri and S. K. Agulian, Intracellular electro-chemical studies of single renal tubule cells and muscle fibers, 1981, in Lübbers et al. 81, p. 195.

K. Kimura, T. Maeda, H. Tamura and T. Shono, Potassium-selective PVC membrane electrodes based on bis- and poly(crown ether)s, J. Electroanal. Chem. **95**, 91 (1979a).

K. Kimura, H. Tamura and T. Shono, Caesium-selective PVC membrane electrodes based on bis(crown ether)s, J. Electroanal. Chem. **105**, 335 (1979b).

K. Kimura, H. Tamura and T. Shono, A highly selective ionophore for potassium ions: a lipophilic bis(15-crown-5) derivative, J. Chem. Soc.,Chem. Commun. **1983**, 492.

N. N. L. Kirsch, Die Bedeutung von Komplexbildungs- und Extraktionsgleichgewichten für die Alkali- und Erdalkaliionenselektivität von Flüssigmembranelektroden beruhend auf azyklischen, ungeladenen Liganden, Thesis, ETH Zürich No. 5842, Juris Druck Verlag, Zürich, 1976.

N. N. L. Kirsch and W. Simon, Komplexbildung von Ionophoren vom Typ der Dioxakorksäurediamide mit Alkali- und Erdalkaliionen. Stabilitätskonstanten in Aethanol. Helv. Chim. Acta **59**, 357 (1976).

T. R. Kissel, J. R. Sandifer and N. Zumbulyadis, Sodium binding in human serum, Clin. Chem. **28**, 449 (1982).

S. Kitazawa, K. Kimura, H. Yano and T. Shono, Lipophilic crown-4 derivatives as lithium ionophores, J. Am. Chem. Soc. **106**, 6978 (1984).

S. Kitazawa, K. Kimura, H. Yano and T. Shono, Lithium-selective polymeric membrane electrodes based on dodecylmethyl-14-crown-4, Analyst **110**, 295 (1985).

H. H. Kornhuber, A reconsideration of the brain-mind problem, 1978, in Buser and Rougeul-Buser 78, p. 319.

G. Kortüm, Lehrbuch der Elektrochemie, Verlag Chemie, Weinheim, 1972.

J. Koryta, Medical and Biological Applications of Electrochemical Devices, John Wiley & Sons, Chichester, New York, Brisbane, Toronto, 1980.

J. Koryta and K. Štulík, Ion-Selective Electrodes, 2nd edition, Cambridge University Press, Cambridge, London, New York, New Rochelle, Melbourne, Sidney, 1983.

P. G. Kostyuk and Z. A. Sorokina. On the mechanism of hydrogen ion distribution between cell protoplasm and the medium, in Membrane Transport Metabolism, A. Kleinzeller and A. Kotyk, eds., Academic, New York, 1961, p. 193.

K. Kotera, N. Satake, M. Honda and M. Fujimoto, The measurement of intracellular sodium activities in the bullfrog by means of double-barreled sodium liquid ion-exchanger microelectrodes, Membr. Biochem. **2**, 323 (1979).

R. P. Kraig and C. Nicholson, Sodium liquid ion exchanger microelectrode used to measure large extracellular sodium transients, Science **194**, 725 (1976).

R. P. Kraig and C. Nicholson, Extracellular ionic variations during spreading depression, Neurosci. **3**, 1045 (1978).

R. P. Kraig, C. R. Ferreira-Filho and C. Nicholson, Alkaline and acid transients in cerebellar microenvironment, J. Neurophysiol. **49**, 831 (1983).

S. Krasne, G. Eisenman and G. Szabo, Freezing and melting of lipid bilayers and the mode of action of nonactin, valinomycin, and gramicidin, Science **174**, 412 (1971).

R. H. Kretsinger, The informational role of calcium in the cytosol, in Advances in Cyclic Nucleotide Research, P. Greengard and G. A. Robison, eds., Vol. 11, Raven Press, New York, 1979, p. 1.

B. R. Kripke and T. E. Ogden, A technique for beveling fine micropipettes, Electroencephalogr. Clin. Neurophysiol. **36**, 323 (1974).

N. Křiž and E. Syková, Sensitivity of K$^+$-selective microelectrodes to pH and some biologically active substances, 1981, in Syková et al. 81, p. 25.

K. Krnjević, J. F. Mitchell and J. C. Szerb, Determination of iontophoretic release of acetylcholine from micropipettes, J. Physiol. **165**, 421 (1963).

K. Krnjević, M. E. Morris and R. J. Reiffenstein, Changes in extracellular Ca^{2+} and K$^+$ activity accompanying hippocampal discharges, Can. J. Physiol. Pharmacol. **58**, 579 (1980).

K. Krnjević and M. E. Morris, Electrical and functional correlates of changes in transmembrane ionic gradients produced by neural activity in the central nervous system, 1981, in Zeuthen 81 a, p. 195.

K. Krnjević, M. E. Morris and J. F. MacDonald, Free Ca^{2+} inside cat motoneurons at rest and during activity, Can. Physiol. **13**, 108 (1982).

R. V. Krstić, Ultrastruktur der Säugetierzelle, Springer Verlag, Berlin, Heidelberg, New York, 1976.

H. S. Kruth, Flow cytometry: rapid biochemical analysis of single cells, Anal. Biochem. **125**, 225 (1982).

G. Küchler, H. Beyer, M. Himmel and B. Merrem, Zur Frage der Übertragungseigenschaften von Glasmikroelektroden bei der intracellulären Membranpotentialmessung, Pflügers Arch. **280**, 210 (1964).

A. Kurkdjian and J. Guern, Vacuolar pH measurement in higher plant cells, Plant. Physiol. **67**, 953 (1981).

A. Kurkdjian, Y. Mathieu and J. Guern, Evidence for an action of 2,4-dichlorophenoxyacetic acid on the vacuolar pH of Acer pseudoplatanus in suspension culture, Plant. Sci. Lett. **27**, 77 (1982).

A. C. Kurkdjian and H. Barbier-Brygoo, A hydrogen ion-selective liquid-membrane microelectrode for measurement of the vacuolar pH of plant cells in suspension culture, Anal. Biochem. **132**, 96 (1983).

A. Kurkdjian, H. Quiquampoix, H. Barbier-Brygoo, M. Péan, P. Manigault and J. Guern, Critical evaluation of methods for estimating the vacuolar pH of plant cells, in Biochemistry and Function of Vacuolar ATPase in Fungi and Plants, B. P. Martin, ed., Springer Verlag, 1985, in press.

K. Kusano, R. Miledi and J. Stinnakre, Cholinergic and catecholaminergic receptors in the Xenopus oocyte membrane, J. Physiol. **328**, 143 (1982).

J. H. Ladenson, Direct potentiometric analysis of sodium and potassium in human plasma: evidence for electrolyte interaction with a nonprotein, protein-associated substance(s), J. Lab. Clin. Med. **90**, 654 (1977).

M. G. Lado, S. S. Sheu and H. A. Fozzard, Changes in intracellular Ca^{2+} activity with stimulation in sheep cardiac Purkinje strands, Am. J. Physiol. **243**, H133 (1982).

M. Läubli, O. Dinten, E. Pretsch and W. Simon, Barium selective electrodes based on neutral carriers and their use in the titration of sulfate, Anal. Chem. **57**, 2756 (1985).

P. J. Laming and M. B. A. Djamgoz, A comparison of the selectivities of micro-electrodes incorporating the Orion and Corning liquid ion exchangers for potassium over sodium, J. Neurosci. Methods **8**, 399 (1983).

F. Lang, W. Wang, H. Oberleithner and S. Neuman, Effect of luminal glucose on cell membrane potential and sodium electrochemical gradient in the absence and presence of ouabain, Pflügers Arch. Suppl. to **394**, R22 (1982).

F. Lang, G. Messner, W. Wang and H. Oberleithner, Interaction of intracellular electrolytes and tubular transport, Klin. Wochenschr. **61**, 1029 (1983).

F. Lang, G. Messner, W. Wang, M. Paulmichl, H. Oberleithner and P. Deetjen, The influence of intracellular sodium activity on the transport of glucose in proximal tubule of frog kidney, Pflügers Arch. **401**, 14 (1984).

F. Lang, G. Messner, W. Wang and H. Oberleithner, The effect of ouabain on intracellular ion activities, membrane resistances, and sodium-coupled transport processes, 1985, in Kessler et al. 85, p. 309.

F. Lanter, D. Erne, D. Ammann and W. Simon, Neutral carrier based ion-selective electrode for intracellular magnesium activity studies, Anal. Chem. **52**, 2400 (1980).

F. Lanter, R. A. Steiner, D. Ammann and W. Simon, Critical evaluation of the applicability of neutral carrier-based calcium selective microelectrodes, Anal. Chim. Acta **135**, 51 (1982).

F. Lanter, Herstellung und Charakterisierung von ionenselektiven Carrier-Flüssigmembranmikroelektroden für intra- und extrazelluläre Aktivitätsbestimmungen von physiologisch relevanten Kationen, Thesis ETH Zürich, No. 7076, Juris Druck + Verlag, Zürich, 1982.

U. V. Lassen, A.-M. T. Nielsen, L. Pape and L. O. Simonsen, The membrane potential of Ehrlich ascites tumor cells. Microelectrode measurements and their critical evaluation, J. Membr. Biol. **6**, 269 (1971).

U. V. Lassen and O. Sten-Knudsen, Direct measurements of membrane potential and membrane resistance of human red cells, J. Physiol. **195**, 681 (1968).

P. Latimer, Light scattering vs. microscopy for measuring average cell size and shape, Biophys. J. **27**, 117 (1979).

M. Lavallée, Intracellular pH of rat atrial muscle fibres measured by glass micropipette electrodes, Circ. Res. **15**, 185 (1964).

M. Lavallée, O. F. Schanne and N. C. Hebert, eds., Glass Microelectrodes, John Wiley & Sons, Inc., New York, London, Sidney, Toronto, 1969.

M. Lavallée and G. Szabo, The effect of glass surface conductivity phenomena on the tip potential of glass micropipette electrodes 1969, in Lavallée et al. 69, p. 95.

O. H. LeBlanc, Jr., J. F. Brown, Jr., J. F. Klebe, L. W. Niedrach, G. M. J. Slusarczuk and W. H. Stoddard, Jr., Polymer membrane sensors for continuous intravascular monitoring of blood pH, J. Appl. Phys. **40**, 644 (1976).

O. H. LeBlanc and W. T. Grubb, Long-lived potassium ion selective polymer membrane electrode, Anal. Chem. **48**, 1658 (1976).

P. W. Ledger and M. L. Tanzer, Monensin – a perturbant of cellular physiology, TIBS **9**, 313 (1984).

C. O. Lee and W. McD. Armstrong, State and distribution of potassium and sodium ions in frog skeletal muscle, J. Membr. Biol. **15**, 331 (1974).

C. O. Lee, A. Taylor and E. Windhager, Cytosolic calcium ion activity in epithelial cells of *Necturus* kidney, Nature **287**, 859 (1980a).

C. O. Lee, D. Y. Uhm and K. Dresdner, Sodium-calcium exchange in rabbit heart muscle cells: direct measurement of sarcoplasmic Ca^{2+} activity, Science **209**, 699 (1980b).

C. O. Lee, Ionic activities in cardiac muscle cells and application of ion-selective microelectrodes, Am. J. Physiol. **241**, H459 (1981).

C. O. Lee and D. Y. Uhm, Characteristics of Ca^{2+}-selective microelectrodes and their application to cardiac muscle cells, 1981, in Syková et al. 81, p. 317.

C. O. Lee and M. Dagostino, Effect of strophanthidin on intracellular Na ion activity and twitch tension of constantly driven canine cardiac Purkinje fibres, Biophys. J. **40**, 185 (1982).

H. C. Lee, J. G. Forte and D. Epel, The use of fluorescent amines for the measurement of pH_i applications in liposomes, gastric microsomes, and sea urchin gametes, 1982, in Nuccitelli and Deamer 82, p. 135.

J.-J. Leguay and J. Guern, Quantitative effects of 2,4-dichlorophenoxyacetic acid on growth of suspension-cultured *Acer pseudoplatanus* cells, Plant Physiol. **56**, 356 (1975).

A. Lehmenkühler, W. Zidek, M. Staschen and H. Caspers, Cortical pH and pCa in relation to DC potential shifts during spreading depression and asphyxiation, 1981, in Syková et al. 81, p. 225.

A. Lehmenkühler, Transient alkaline shift in cortical tissue pH during the onset of spreading depression and of anoxic negative DC-potential shift, Pflügers Arch. Suppl.to **394**, R49 (1982).

A. Lehmenkühler, M.Staschen and H.Caspers, Depth profile of extracellular pH in the brain cortex during seizure activity, Pflügers Arch.Suppl.to **394**, R50 (1982).

A. Lehmkühler, H.Caspers and U.Kersting, Relations between DC potentials, extracellular ion activities, and extracellular volume fraction in the cerebral cortex with changes in pCO_2, 1985, in Kessler et al.85, p.199.

J.-M. Lehn, J.P. Sauvage and B.Dietrich, Cryptates. Cation exchange rates, J.Am.Chem.Soc. **92**, 2916 (1970).

J.-M. Lehn, Design of organic complexing agents. Strategies towards properties, Structure and Bonding **16**, 1 (1973).

J.-M. Lehn, Cryptates: The chemistry of macropolycyclic inclusion complexes, Accounts Chem.Res. **11**, 49 (1978).

A.A. Lev and W.McD. Armstrong, Ionic activities in cells, Curr.Topics Membr.Transp. **6**, 59 (1975).

S. Levy, Intracellular free Ca concentration is not a direct indicator of the receptor sensitivity in *Limulus* ventral eye, Biophys.J. **37**, 85a (1982).

S. Levy, D.Tillotson and A.L.F. Gorman, Intracellular Ca^{2+} gradient associated with Ca^{2+} channel activation measured in a nerve cell body, Biophys.J. **37**, 182a (1982).

S. Levy and A.Fein, Relationship between light sensitivity and intracellular free Ca concentration in *Limulus* ventral photoreceptors, J.Gen. Physiol. **85**, 805 (1985).

L.M. Lewis, T.W. Flechtner, J.Kerkay, K.H. Pearson and S. Nakamoto, Bis(2-ethylhexyl)phthalate concentrations in the serum of hemodialysis patients, Clin.Chem. **24**, 741 (1978).

S.A. Lewis, N.K. Wills and D.C. Eaton, Basolateral membrane potential of a tight epithelium: ionic diffusion and electrogenic pumps, J.Membr.Biol. **41**, 117 (1978).

S.A. Lewis and N.K. Wills, Resistive artifacts in liquid-ion exchanger microelectrode estimates of Na^+ activity in epithelial cells, Biophys.J. **31**, 127 (1980).

S.A. Lewis and N.K. Wills, Applications and interpretations of ion-specific microelectrodes in tight epithelia, 1981, in Zeuthen 81a, p.3.

B. Libet, Subjective and neuronal time factors in conscious sensory experience, studied in man, and their implications for the mind-brain relationship, in The Search for Absolute Values in a Changing World, Vol. II, The International Cultural Foundation Press, New York, 1978a, p.971.

B. Libet, Neuronal vs. subjective timing for a conscious sensory experience, 1978b, in Buser and Rougeul-Buser 78, p.69.

B. Lindemann, Impalement artifacts in microelectrode recordings of epithelial membrane potentials, Biophys.J. **15**, 1161 (1975).

E. Lindner, K.Tóth and E.Pungor, Response time curves of ion-selective electrodes, Anal.Chem **48**, 1071 (1976).

E. Lindner, K.Tóth, E.Pungor, W.E. Morf and W.Simon, Response time studies on neutral carrier ion-selective membrane electrodes, Anal.Chem. **50**, 1627 (1978).

E. Lindner, K.Tóth, E.Pungor, F.Behm, P.Oggenfuss, D.H. Welti, D.Ammann, W.E. Morf, E.Pretsch and W.Simon, Lead-selective neutral carrier based liquid membrane electrode, Anal.Chem. **56**, 1127 (1984).

E. Lindner, K.Tóth and E.Pungor, Problems related to the definition of response time, Pure Appl.Chem., 1985, in press.

G. Ling and R.W. Gerard, The normal membrane potential of frog sartorius fiber, J.Cell.Comp. Physiol. **34**, 383 (1949).

G.N. Ling, A Physical Theory of the Living State: the Association-Induction Hypothesis, Blaisdell Publishing Company, New York, London, 1962.

G.N. Ling and F.W. Cope, Potassium ion: is the bulk of intracellular K^+ adsorbed?, Science **163**, 1335 (1969).

G.N. Ling, Measurements of potassium ion activity in the cytoplasm of living cells, Nature **221**, 386 (1969).

G.N. Ling, C.Miller and M.M. Ochsenfeld, The physical state of solutes and water in living cells according to the association-induction hypothesis, 1973, in Hazlewood 73, p.6.

G.N. Ling and M.M. Ochsenfeld, Mobility of potassium ion in frog muscle cells, both living and dead, Science **181**, 78 (1973).

G. N. Ling, The cellular resting and action potentials: interpretation based on the association-induction hypothesis, Physiol. Chem. Phys. **14**, 47 (1982).

C. Liu, M. Chin, B. L. T. Prosser, N. J. Palleroni, J. W. Westley and P. A. Miller, X-14885 A, a novel divalent cation ionophore produced by a streptomyces culture: discovery, fermentation, biological as well as ionophore properties and taxonomy of the producing culture, J. Antibiotics **36**, 1118 (1983).

W. R. Loewenstein and Y. Kanno, The electrical conductance and potential across the membrane of some cell nuclei, J. Cell. Biol. **16**, 421 (1963).

J. R. Lopez, L. Alamo, C. Caputo, J. Vergara and R. DiPolo, Determination of ionized Mg concentration in skeletal muscle fibers with magnesium selective microelectrodes, Biophys. J. **41**, 179 a (1983 a).

J. R. Lopez, L. Alamo, C. Caputo, R. DiPolo and J. Vergara, Determination of ionic calcium in frog skeletal muscle fibres, Biophys. J. **43**, 1 (1983 b).

J. R. Lopez, L. Alamo and C. Caputo, Constant level of intracellular free magnesium concentration during muscular activity and fatigue, Biophys. J. **45**, 232 a (1984 a).

J. R. Lopez, L. Alamo, C. Caputo, J. Vergara and R. DiPolo, Direct measurement of intracellular free magnesium in frog skeletal muscle using magnesium-selective microelectrodes, Biochim. Biophys. Acta **804**, 1 (1984 b).

J. Louvel and U. Heinemann, Diminution de la concentration extracellulaire des ions calcium lors des crises épileptiques focales induites par l'oenanthotoxine dans le cortex du Chat, C. R. Acad. Sc. Paris, **291**, 997 (1980).

C. R. Lowe, Biosensors, Trends in Biotechnology **2**, 59 (1984).

D. W. Lübbers, H. Acker, R. P. Buck, G. Eisenman, M. Kessler and W. Simon, eds., Progress in Enzyme and Ion-Selective Electrodes, Springer Verlag, Berlin, Heidelberg, New York, 1981.

A. Lundin and M. Baltscheffsky, Measurement of photophosphorylation and ATPase using purified firefly luciferase, Meth. Enzymol. **57**, 50 (1978).

W. K. Lutz, H.-K. Wipf and W. Simon, Alkalikationen-Spezifität und Träger-Eigenschaften der Antibiotica Nigericin und Monensin, Helv. Chim. Acta **53**, 1741 (1970).

H. D. Lux and E. Neher, The equilibration time course of $[K^+]_o$ in cat cortex, Exp. Brain Res. **17**, 190 (1973).

H. D. Lux, Fast recording ion specific microelectrodes: their use in pharmacological studies in the CNS, Neuropharmacology **13**, 509 (1974).

H. D. Lux and G. Hofmeier, Kinetics of the calcium dependent potassium current in Helix neurons, Pflüger Arch. Suppl. to **373**, R47 (1978).

H. D. Lux and C. B. Heyer, A new electrogenic calcium-potassium system, The Neurosciences, F. O. Schmitt and F. G. Worden, eds., 4th Study Programme, MIT Press, Cambridge, MA, 1979, p. 601.

H. D. Lux and G. Hofmeier, Effects of calcium currents and intracellular free calcium in *Helix* neurones, 1979, in Ashley and Campbell 79, p. 409.

H. D. Lux, G. Hofmeier and J. B. Aldenhoff, Intracellular free calcium affects electric membrane properties. A study with calcium-selective microelectrodes and with arsenazo III in *Helix* neurons, 1981, in Syková et al. 81, p. 99.

M. Maj-Żurawska, D. Erne, D. Ammann and W. Simon, Lipophilic synthetic monoamides of dicarboxylic acids as ionophores for alkaline earth metal cations, Helv. Chim. Acta **65**, 55 (1982).

B. L. Maloff, S. P. Scordilis and H. Tedeschi, Membrane potential of mitochondria measured with microelectrodes, Science **195**, 898 (1977).

E. Marban, T. J. Rink, R. W. Tsien and R. Y. Tsien, Free calcium in heart muscle at rest and during contraction measured with Ca^{2+}-sensitive microelectrodes, Nature **286**, 845 (1980).

Y. Marcus, Introduction to Liquid State Chemistry, John Wiley & Sons, London, New York, Sidney, Toronto, 1977.

R. Margalit and G. Eisenman, Ionic permeation of lipid bilayer membranes mediated by a neutral, noncyclic Li^+-selective carrier having imide and ether ligands. I. Selectivity among monovalent cations, J. Membr. Biol. **61**, 209 (1981).

P. L. Markovic and J. O. Osburn, Dynamic response of some ion-selective electrodes, Am. Inst. Chem. Eng. J. **19**, 504 (1973).

W. S. Marshall and S. D. Klyce, Cell finder speeds impalaments with microelectrodes, Pflügers Arch. **391**, 258 (1981).

J.-B. Martin, R. Bligny, F. Rebeille, R. Douce, J.-J. Leguay, Y. Mathieu and J. Guern, A ^{31}P nuclear magnetic resonance study of intracellular pH of plant cells cultivated in liquid medium, Plant. Physiol. **70**, 1156 (1982).

T. Maruizumi, H. Miyagi, Y. Takata and T. Kobayashi, Characterization of Na ion-sensitive solvent polymeric membranes based on a neutral carrier, J. Appl. Polym. Sci. **30**, 487 (1985).

F. Math and J. L. Davrainville, Postnatal variations of extracellular free calcium levels in the rat. Influence of undernutrition, Experientia **35**, 1355 (1979).

F. Math and J. L. Davrainville, Electrophysiological study on the postnatal development of mitral cell activity in the rat olfactory bulb, Brain Res. **190**, 243 (1980).

Ph. Matile, Enzyme der Vakuolen aus Wurzelzellen von Maiskeimlingen. Ein Beitrag zur funktionellen Bedeutung der Vakuole bei der intrazellulären Verdauung, Z. Naturforsch. **21b**, 871 (1966).

Y. Matsumura, S. Aoki, K. Kajino and M. Fujimoto, The double-barreled microelectrode for the measurement of intracellular pH, using liquid ion-exchanger, and its biological application, in Advances Physiol. Sci, Vol. 11, L. Takacs, ed., Pergamon Press, Akadémiai Kiadó, 1980 a, p. 387.

Y. Matsumura, K. Kajino and M. Fujimoto, Measurement of intracellular pH of bullfrog skeletal muscle and renal tubular cells with double-barreled antimony microelectrodes, Membr. Biochem. **3**, 99 (1980 b).

F. P. McGlone, I. J. Russell and O. Sand, Measurement of calcium ion concentrations in the lateral line cupulae of *Xenopus laevis*, J. Exp. Biol. **83**, 123 (1979).

F. C. McLean and A. B. Hastings, A biological method for the estimation of calcium ion concentration, J. Biol. Chem. **107**, 337 (1934).

P. C. Meier, D. Ammann, H. F. Osswald and W. Simon, Ion-selective electrodes in clinical chemistry, Med. Progr. Technol. **5**, 1 (1977).

P. C. Meier, D. Ammann, W. E. Morf and W. Simon, Liquid-membrane ion-selective electrodes and their biomedical applications, in Medical and Biomedical Applications of Electrochemical Devices, 1980, in Koryta 80, p. 13.

P. C. Meier, Two-parameter Debye-Hückel approximation for the evaluation of mean activity coefficients of 109 electrolytes, Anal. Chim. Acta **136**, 363 (1982).

P. C. Meier, F. Lanter, D. Ammann, R. A. Steiner and W. Simon, Applicability of available ion-selective liquid-membrane microelectrodes to intracellular ion-activity measurements, Pflügers Arch. **393**, 23 (1982).

P. C. Meier, W. E. Morf, M. Läubli and W. Simon, Evaluation of the optimum composition of neutral-carrier membrane electrodes with incorporated cation-exchanger sites, Anal. Chim. Acta **156**, 1 (1984).

E. Metzger, D. Ammann, U. Schefer, E. Pretsch and W. Simon, Lipophilic neutral carriers for lithium selective liquid membrane electrodes, Chimia **38**, 440 (1984).

A. P. Minton and J. Wilf, Effect of macromolecular crowding upon the structure and function of an enzyme: glyceraldehyde-3-phosphate dehydrogenase, Biochem. **20**, 4821 (1981).

M. S. Mohan and R. G. Bates, Calibration of ion-selective electrodes for use in biological fluids, Clin. Chem. **21**, 864 (1975).

G. J. Moody, R. B. Oke and J. D. R. Thomas, A calcium-sensitive electrode based on a liquid ion exchanger in a poly(vinyl chloride) matrix, Analyst **95**, 910 (1970).

G. J. Moody and J. D. R. Thomas, Selective Ion-Sensitive Electrodes, Merrow, Watford, Hertfordshire, 1971.

C. Moore and B. C. Pressman, Mechanism of action of valinomycin on mitochondria, Biochim. Biophys. Res. Commun. **15**, 562 (1964).

M. Moreau, J. P. Vilain and P. Guerrier, Free calcium changes associated with hormone action in amphibian oocytes, Develop. Biol. **78**, 201 (1980).

W. E. Morf and W. Simon, Berechnung von freien Hydratationsenthalpien und Koordinationszahlen für Kationen aus leicht zugänglichen Parametern, Helv. Chim. Acta **54**, 794 (1971 a).

W. E. Morf and W. Simon, Abschätzung der Alkali- und Erdalkali-Ionenselektivität von elektrisch neutralen Träger-Antibiotica ("Carrier-Antibiotica") und Modellverbindungen, Helv. Chim. Acta **54**, 2683 (1971 b).

W. E. Morf, D. Ammann, E. Pretsch and W. Simon, Carrier antibiotics and model compounds as components of selective ion-sensitive electrodes, Pure Appl. Chem. **36**, 421 (1973).

W. E. Morf, D. Ammann and W. Simon, Elimination of the anion interference in neutral carrier cation-selective membrane electrodes, Chimia **28**, 65 (1974 a).

W. E. Morf, G. Kahr and W. Simon, Reduction of the anion interference in neutral carrier liquid-membrane electrodes responsive to cations, Anal. Lett. **7** (1), 9 (1974 b).

W. E. Morf, E. Lindner and W. Simon, Theoretical treatment of the dynamic response of ion-selective membrane electrodes, Anal. Chem. **47**, 1596 (1975).

W. E. Morf, Calculation of liquid-junction potentials and membrane potentials on the basis of the Planck theory, Anal. Chem. **49**, 810 (1977).

W. E. Morf and W. Simon, Ion-selective electrodes based on neutral carriers, in Ion-selective Electrodes in Analytical Chemistry, H. Freiser, ed., Plenum Press, New York, London, Washington, Boston, 1978.

W. E. Morf, D. Ammann, R. Bissig, E. Pretsch and W. Simon, Cation selectivity of neutral macrocyclic and nonmacrocyclic complexing agents in membranes, in Progress in Macrocyclic Chemistry, R. M. Izatt and J. J. Christensen, eds., Vol. 1, John Wiley & Sons, New York, Chichester, Brisbane, Toronto, 1979, p. 1.

W. E. Morf, the Principles of Ion-Selective Electrodes and of Membrane Transport, Akadémiai Kiadó, Budapest, 1981/ Elsevier, Amsterdam, New York, 1981.

J. G. Morin, Coelenterate bioluminescence, in Coelenterate Biology: Reviews and New Perspectives, L. Muscatine and H. M. Lenhoff, eds., Academic Press, New York, 1974, p. 397.

M. E. Morris, Measurements of ion activity in the CNS: extracellular K^+ and Ca^{2+} in the hippocampus, 1981, in Syková et al. 81, p. 241.

M. E. Morris and K. Krnjević, Slow diffusion of Ca^{2+} in the rat's hippocampus, Can. J. Physiol. Pharmacol. **59**, 1022 (1981).

M. E. Morris, K. Krnjević and N. Ropert, Changes in free Ca^{2+} recorded inside hippocampal pyramidal neurons in response to fimbrial stimulation, Soc. Neurosci. Abs. **9**, 395 (1983).

H. Moser, Intracellular pH regulation in the sensory neurone of stretch receptor of the crayfish (*Astacus fluviatilis*), J. Physiol. **362**, 23 (1985).

V. B. Mountcastle, Some neural mechanisms for directed attention, 1978, in Buser and Rougeul-Buser 78, p. 37.

M. Muhammed, The formation of trilaurylammonium carbonate, Acta Chem. Scand. **26**, 412 (1972).

M. H. Muheim, Fabrication of well defined micropipette tips by hydrofluoric acid etching, Pflügers Arch. **372**, 101 (1977).

J.-L. Munoz, F. Deyhimi and J. A. Coles, Silanization of glass in the making of ion-sensitive microelectrodes, J. Neurosci. Methods **8**, 231 (1983).

E. Murphy, K. Coll, T. L. Rich and J. R. Williamson, Hormonal effects on calcium homeostasis in isolated hepatocytes, J. Biol. Chem. **255**, 6600 (1980).

W. A. C. Mutch and A. J. Hansen, Extracellular pH changes during spreading depression and cerebral ischemia: mechanisms of brain pH regulation, J. Cerebr. Blood Flow Metabol. **4**, 17 (1984).

W. A. C. Mutch and A. J. Hansen, Brain extracellular pH changes during alterations in substrate supply, 1985, in Kessler et al. 85, p. 189.

G. Navon, S. Ogawa, R. G. Shulman and T. Yamane, High-resolution ^{31}P nuclear magnetic resonance studies of metabolism in aerobic *Escherichia coli* cells, Proc. Natl. Acad. Sci. **74**, 888 (1977).

W. L. Nastuk and A. L. Hodgkin, The electrical activity of single muscle fibers, J. Cell. Comp. Physiol. **35**, 39 (1950).

E. Neher, Elektronische Meßtechnik in der Physiologie, Springer Verlag, Berlin, Heidelberg, New York, 1974.

D. J. Nelson, J. Ehrenfeld and B. Lindemann, Volume changes and potential artifacts of epithelial cells of frog skin following impalement with microelectrodes filled with 3 M KCl, J. Membr. Biol., Special Issue **40**, 91 (1978).

C. Nicholson, R. Steinberg, H. Stöckle and G. tenBruggencate, Calcium decrease associated with aminopyridine-induced potassium increase in cat cerebellum, Neurosci. Letters **3**, 315 (1976).

C. Nicholson, G. tenBruggencate, R. Steinberg and H. Stöckle, Calcium modulation in brain extracellular microenvironment demonstrated with ion-selective micropipette, Proc. Natl. Acad. Sci. **74**, 1287 (1977).

C. Nicholson, R. P. Kraig, G. tenBruggencate, H. Stöckle and R. Steinberg, Potassium, calcium, chloride and sodium changes in extracellular space during spreading depression in cerebellum, Arzneim.-Forsch./Drug Res. **28**, 874 (1978a).

C. Nicholson, G. tenBruggencate, H. Stöckle and R. Steinberg, Calcium and potassium changes in extracellular microenvironment of cat cerebellar cortex, J. Neurophysiol. **41**, 1026 (1978b).

C. Nicholson, Brain cell microenvironment as a communication channel, in The Neurosciences, F. O. Schmitt and F. G. Worden, eds., MIT Press, Cambridge Mass., 1979.

C. Nicholson and J. M. Phillips, Diffusion of anions and cations in the extracellular microenvironment of the brain, J. Physiol. **296**, 66 P (1979).

C. Nicholson, J. M. Phillips and A. R. Gardner-Medwin, Diffusion from an iontophoretic point source in the brain: role of tortuosity and volume fraction, Brain Res. **169**, 580 (1979).

C. Nicholson, Measurement of extracellular ions in the brain, Trends in Neurosciences **3**, 216 (1980a).

C. Nicholson, Modulation of extracellular calcium and its functional implications, Fed. Proc. **39**, 1519 (1980b).

C. Nicholson and R. P. Kraig, The behavior of extracellular ions during spreading depression, 1981, in Zeuthen 81 a, p. 217.

C. Nicholson, J. M. Phillips, C. Tobias and R. P. Kraig, Extracellular potassium, calcium and volume profiles during spreading depression, 1981, in Syková et al. 81, p. 211.

C. Nicholson, R. P. Kraig, C. R. Ferreira-Filho and P. Thompson, Hydrogen ion variations and their interpretation in the microenvironment of the vertebrate brain, 1985, in Kessler et al. 85, p. 229.

B. P. Nicolsky, Theory of the glass electrode, Zh. Fis. Khim. **10**, 495 (1937a).

B. P. Nicolsky, Theory of the glass electrode I., Acta physicochimica U. R. S. S. **7**, 597 (1937b).

J. R. Nilsson and J. R. Coleman, Calcium-rich, refractile granules in *Tetrahymena pyriformis* and their possible role in the intracellular ion-regulation, J. Cell. Sci. **24**, 311 (1977).

R. Nuccitelli and D. W. Deamer, eds., Intracellular pH: Its Measurement, Regulation and Utilization in Cellular Functions, Kroc Foundation Series, Vol. 15, A. R. Liss, Inc., New York, 1982.

H. Oberleithner, T. Kubota and G. Giebisch, Potassium (K^+) transport and intracellular K^+ activity in distal tubules of *Amphiuma*, Fed. Proc. **39**, 4266 (1980).

H. Oberleithner, F. Lang, G. Giebisch and P. Deetjen, The effect of furosemide (F) on intracellular Na^+-activity (Na_i^+) in early distal tubule of *Amphiuma*, Pflügers Arch. Suppl. to **39**, 66 (1981).

H. Oberleithner, F. Lang, W. Wang and G. Giebisch, Effects of inhibition of chloride transport on intracellular sodium activity in distal amphibian nephron, Pflügers Arch. **394**, 55 (1982a).

H. Oberleithner, G. Giebisch, F. Lang and W. Wang, Cellular mechanism of the furosemide sensitive transport system in the kidney, Klin. Wochenschr. **60**, 1173 (1982b).

H. Oberleithner, F. Lang and W. Wenhui, Potassium dependence of the sodium chloride cotransport mechanism in the distal amphibian nephron, Pflügers Arch. Suppl. to **392**, R14 (1982c).

H. Oberleithner, F. Lang, R. Greger, W. Wang and G. Giebisch, Effect of luminal potassium on cellular sodium activity in the early distal tubule of *Amphiuma* kidney, Pflügers Arch. **396**, 34 (1983a).

H. Oberleithner, F. Lang, W. Wang, G. Messner and P. Deetjen, Evidence for an amiloride sensitive Na^+ pathway in the amphibian diluting segment induced by K^+ adaptation, Pflügers Arch. **399**, 166 (1983b).

H. Oberleithner, F. Lang, G. Messner and W. Wang, Mechanism of hydrogen ion transport in the diluting segment of frog kidney, Pflügers Arch. **402**, 272 (1984a).

H. Oberleithner, F. Lang, G. Messner and W. Wang, Evidence for Na^+-dependent H^+ secretion in frog diluting segment, Pflügers Arch. Suppl. to **400**, R20 (1984b).

J. O'Doherty, J. F. Garcia-Diaz and W. McD. Armstrong, Sodium-selective liquid ion-exchanger microelectrodes for intracellular measurements, Science **203**, 1349 (1979).

J. O'Doherty, S. J. Youmans and W. McD. Armstrong, Calcium regulation during stimulus-secretion coupling: continuous measurement of intracellular calcium activities, Science **209**, 510 (1980).

J. O'Doherty and R. Stark, Transmembrane and transepithelial movement of calcium during stimulus-secretion coupling, Am. J. Physiol. **241**, G150 (1981a).

J. O'Doherty and R. J. Stark, Measurement of intracellular calcium activities, 1981b, in Syková et al. 81, p. 91.

J. O'Doherty and R. J. Stark, Stimulation of pancreatic acinar secretion: increases in cytosolic calcium and sodium, Am. J. Physiol. **242**, G513 (1982).

M. Oehme, M. Kessler and W. Simon, Neutral carrier Ca^{2+}-microelectrode, Chimia **30**, 204 (1976).

M. Oehme and W. Simon, Microelectrode for potassium ions based on a neutral carrier and comparison of its characteristics with a cation exchanger sensor, Anal. Chim. Acta **86**, 21, (1976).

M. Oehme, Beitrag zur Entwicklung ionenselektiver Mini- und Mikroelektroden und zu deren Meßtechnik, Thesis, ETH Zürich, Nr. 5953, Juris Druck + Verlag, Zürich, 1977.

U. Oesch and W. Simon, Kinetische Betrachtung der Verteilung von elektrisch neutralen Ionophoren zwischen einer Flüssigmembran und einer wässerigen Phase, Helv. Chim. Acta **62**, 754 (1979).

U. Oesch, D. Ammann, E. Pretsch and W. Simon, Ionophore extrem hoher Lipophilie als selektive Komponenten für Flüssigmembranelektroden, Helv. Chim. Acta **62**, 2073 (1979).

U. Oesch and W. Simon, Life time of neutral carrier based ion-selective liquid-membrane electrodes, Anal. Chem. **52**, 692 (1980).

U. Oesch, S. Caras and J. Janata, Field effect transistors sensitive to sodium and ammonium ions, Anal. Chem. **53**, 1983 (1981).

U. Oesch, O. Dinten, D. Ammann and W. Simon, Life time of neutral carrier based membranes in aqueous systems and blood serum, 1985 a, Kessler et al. 85, p. 42.

U. Oesch, O. Dinten and W. Simon, 1985 b, in preparation.

U. Oesch, P. Anker, D. Ammann and W. Simon, Membrane technological optimization of ion-selective electrodes based on solvent polymeric membranes for clinical applications, 1985 c, in Ion-Selective Electrodes, E. Pungor and I. Buzás, eds., 1985, p. 81.

U. Oesch and W. Simon, 1985, in preparation.

Y. Ogawa, H. Harafugi and N. Kurebayashi, Comparison of the characteristics of four metallochromic dyes as potential calcium indicators for biological experiments, J. Biochem. **87**, 1293 (1980).

T. E. Ogden, M. C. Citron and R. Pierantoni, The jet stream microbeveler: an inexpensive way to bevel ultrafine glass micropipettes, Science, **201**, 469 (1978).

S. Ohkuma and B. Poole, Fluorescence probe measurement of the intralysosomal pH in living cells and the perturbation of pH by various agents, Proc. Natl. Acad. Sci. USA **75**, 3327 (1978).

S. T. Ohnishi and S. Ebashi, Spectrophotometrical measurement of instantaneous calcium binding of the relaxing factor of muscle, J. Biochem. **54**, 506 (1963).

S. T. Ohnishi, Characterization of the murexide method: Dual-wavelength spectrophotometry of cations under physiological conditions, Anal. Biochem. **85**, 165 (1978).

Y. Okada and A. Inouye, Tip potential and fixed charges on the glass wall of microelectrode, Experientia **31**, 545 (1975).

Y. Okada and A. Inouye, Studies on the origin of the tip potential of glass microelectrode, Biophys. Struct. Mechan. **2**, 31 (1976).

U. Olsher, The lipophilic macrocyclic polyether 2,3,9,10-dibenzo-1,4,8,11- tetraoxacyclotetradeca-2,9-diene (Dibenzo-14-crown-4): a selective ionophore for lithium ions, J. Am. Chem. Soc. **104**, 4006 (1982).

F. Oosawa, Polyelectrolytes, Marcel Dekker, Inc., New York, 1971.

R. K. Orkand, I. Dietzel and J. A. Coles, Light-induced changes in extracellular volume in the retina of the drone, *Apis mellifera*, Neurosci. Letters **45**, 273 (1984).

F. W. Orme, Liquid ion-exchanger microelectrodes, 1969, in Lavallée et al. 69, p. 376.

H. F. Osswald, R. E. Dohner, T. Meier, P. C. Meier and W. Simon, Flow-through system of high stability for the measurement of ion activities in clinical chemistry, Chimia **31**, 50 (1977).

N. Otake and M. Mitani, Ionophorous properties of antibiotic-6016, a novel magnesium selective ionophore, Agric. Biol. Chem. **43**, 1543 (1979).

Y. A. Ovchinnikov, V. T. Ivanov and A. M. Shkrob, Membrane-Active Complexones, B. B. A. Library, Vol. 12, Elsevier, Amsterdam, 1974.

J. D. Owen, H. M. Brown and J. P. Pemberton, Ca^{2+} in the *Aplysia* giant cell and the *Balanus eburneus* muscle fibre, Biophys. J. **16**, 34 a (1976).

K. R. Page, L. S. Kelday and D. J. F. Bowling, The diffusion of KCl from micro-electrodes, J. Exp. Bot. **32**, 55 (1981).

H. Pallmann, Die Wasserstoffaktivität in Dispersionen und kolloiddispersen Systemen, Kolloidchem. Beihefte **30**, 334 (1930).

L. G. Palmer and M. M. Civan, Distribution of Na^+, K^+ and Cl^- between nucleus and cytoplasm in *Chironomus* salivary gland cells, J. Membr. Biol. **33**, 41 (1977).

L. G. Palmer, T. J. Century and M. M. Civan, Activity coefficients of intracellular Na^+ and K^+ during development of frog oocytes, J. Membr. Biol. **40**, 25 (1978).

W. Patnode and D. F. Wilcock, Methylpolysiloxanes, J. Am. Chem. Soc. **68**, 358 (1946).

M. Paulmichl, G. Gstraunthaler and F. Lang, Electrical properties of Madin Darby canine kidney (MDCK) cells, Pflügers Arch. Suppl. to **402**, R9 (1984).

H. Pauly, Über den physikalisch-chemischen Zustand des Wassers und der Elektrolyte in der lebenden Zelle, Biophysik **10**, 7 (1973).

C.J. Pedersen, Cyclic polyethers and their complexes with metal salts, J.Am.Chem.Soc. **89**, 2495 (1967).

W. Penfield, The Mystery of the Mind, Princeton University Press, Princeton, 1975.

J.P. Peri and A.L. Hensley, Jr., The surface structure of silica gel, J.Phys.Chem. **72**, 2926 (1968).

H. Petsche, I.B. Müller-Paschinger, H.Pockberger, O.Prohaska, P.Rappelsberger and R.Vollmer, Depth profiles of electrocortical activities and cortical architectonics, in Architectonics of the Cerebral Cortex, M.A.B. Brazier and H.Petsche, eds., Raven Press, New York, 1978, p.257.

W. Pfeffer, Osmotische Untersuchungen. Studien zur Zellmechanik, Verlag W.Engelmann, Leipzig, 1877.

H. Pfister and H.Pauly, Chemical potential of KCl and its ion constituents in concentrated protein salt solutions, J.Polymer Sci. Part C **39**, 179 (1972).

J.M. Phillips and C.Nicholson, Anion permeability in spreading depression investigated with ion-sensitive microelectrodes, Brain Res. **173**, 567 (1979).

A. Picard and M.Dorée, Hormone-induced parthenogenetic activation of mature starfish oocytes, Exp.Cell.Res. **145**, 315 (1983a).

A. Picard and M.Dorée, Intracellular microinjection of alkaline buffers reversibly inhibits the initial phase of hormone action in meiosis reinitiation of starfish oocytes, Develop.Biol. **97**, 184 (1983b).

A. Picard and M.Dorée, Is calcium the second messenger of 1-methyladenine in meiosis reinitiation of starfish oocytes, Exp.Cell.Res. **145**, 325 (1983c).

J. Pick, K.Tóth, M.Vašák, E.Pungor and W.Simon, Development of a silicone-rubber potassium membrane electrode, Anal.Chim.Acta **64**, 477 (1973).

D. Pinkel, Flow cytometry and sorting, Anal.Chem. **54**, 503 A (1982).

L.A.R. Pioda, V.Stankova and W.Simon, Highly selective potassium ion responsive liquid-membrane electrode, Anal.Lett. **2**, 665 (1969).

P.I. Polimeni and E.Page, Magnesium in heart muscle, Circ.Res. **33**, 367 (1973).

P.A. Poole-Wilson and S.R. Seabrooke, Relationship betwen intracellular pH and contractility in guinea-pig papillary muscles, J.Physiol. **365**, 63 P (1985).

K.R. Popper and J.C. Eccles, The Self and Its Brain, Springer International, Berlin, Heidelberg, New York, 1977.

K.R. Popper, Objective Knowledge. An Evolutionary Approach, Oxford University Press, Oxford, 1979.

T. Pozzan, T.J. Rink and R.Y. Tsien, Intracellular free Ca^{2+} in intact lymphocytes, J.Physiol. **318**, 12 P (1981).

J.M. Pratt, Inorganic Chemistry of Vitamin B_{12}, Academic Press, London, 1972.

F.G. Prendergast, The use of photoproteins in the detection and quantitation of Ca^{2+} in biological systems, Trends Anal.Chem. **1**, 378 (1982).

B.C. Pressman, E.J. Harris, W.S. Jagger and J.H. Johnson, Antibiotic-mediated transport of alkali ions across lipid barriers, Proc. Natl. Acad. Sci. **58**, 1949 (1967).

E. Pretsch, D.Ammann and W.Simon, Design of ion carriers and their application in ion selective electrodes, Research Development **25**, 20 (1974).

E. Pretsch, R.Büchi, D.Ammann and W.Simon, Lipophilic complexing agents designed for use in ion-selective liquid membrane electrodes, in Analytical Chemistry, Essays in Memory of Anders Ringbom, E.Wänninen, ed., Pergamon Press, Oxford, New York, 1977, p.321.

E. Pretsch, D.Ammann, H.F. Osswald, M.Güggi and W.Simon, Ionophore vom Typ der 3-Oxapentandiamide, Helv.Chim.Acta **63**, 191 (1980).

E. Pretsch, J.Bendl, P.Portmann and M.Welti, Application of quantum chemical calculations in the design of ion carriers, in Steric Effects in Biomolecules, G.Naray-Szabo, ed., Akadémiai Kiadó, Budapest, 1982, p.82.

E. Pretsch, D.Wegmann, D.Ammann, A.Bezegh, O.Dinten, M.W. Läubli, W.E. Morf, U.Oesch, K.Sugahara, H.Weiss and W.Simon, Effects of lipophilic charged sites on the electromotive behaviour of liquid membrane electrodes, 1985, in Kessler et al.85, p.11.

K.H. Pribram, The mind/brain issue as a scientific problem, in The Search for Absolute Values in a Changing World, Vol.II, The International Cultural Foundation Press, New York, 1978, p.979.

D.A. Prince, C.Benninger and J.Kadis, Evoked ionic alterations in brain slices, 1981, in Syková et al.81, p.247.

O. Prohaska, Dünnschichtelektroden für die Hirnforschung, Elektronikschau **1**, 26 (1983).

M. Prudhomme and G.Jeminet, Semi-synthesis of A23187 (calcimycin) analogs, Experientia **39**, 256 (1983).

L.R. Pucacco and N.W. Carter, A glass-membrane pH microelectrode, Anal.Biochem. **73**, 501 (1976).

L.R. Pucacco and N.W. Carter, A submicrometer glass-membrane pH micro-electrode, Anal.Biochem. **89**, 151 (1978).

R. Pumain, I. Kurcewicz and J.Louvel, Fast extracellular calcium transients: involvement in epileptic processes, Science **222**, 177 (1983).

E. Pungor and Y.Umezawa, Response time in electrochemical cells containing ion-selective electrodes, Anal.Chem. **55**, 1432 (1983).

R.D. Purves, The physics of iontophoretic pipettes, J. Neurosci. Meth. **1**, 165 (1979).

R.D. Purves, Microelectrode Methods for Intracellular Recording and Ionophoresis, Biological Techniques Series, Vol.6, Academic Press, London, New York, Toronto, Sidney, San Francisco, 1981.

G.K. Radda, D.G. Gadian and B.D. Ross, Energy metabolism and cellular pH in normal and pathological conditions. A new look through ^{31}phosphorus nuclear magnetic resonance, Ciba Foundation Symposium **87**, 36 (1982).

M.D. Reboiras, H. Pfister and H. Pauly, Activity coefficients of salts in highly concentrated protein solutions. I.Alkali chlorides in isoionic bovine serum albumin solutions. Biophys.Chem. **9**, 37 (1978).

R.Rangarajan and G.A. Rechnitz, Dynamic response of ion-selective membrane electrodes, Anal.Chem. **47**, 324 (1975).

S.J. Rehfeld, J.Barkeley and H.F. Loken, Effect of pH and NaCl on measurements of ionized calcium in matrices of serum and human albumin with a new calcium-selective electrode, Clin.Chem. **30**, 304 (1984).

W. Rehwald, J.Geibel, E.Gstrein and H.Oberleithner, A microelectrode for continuous monitoring of glucose concentration in isolated perfused tubule segments, Pflügers Arch. **400**, 398 (1984).

L. Reuss and S.A. Weinman, Intracellular ionic activities and transmembrane electrochemical potential differences in gallbladder epithelium, J.Membr. Biol. **49**, 345 (1979).

L. Reuss and L.Costantin, Cl^-/HCO_3^- at the apical membrane of *Necturus* gallbladder, J.Gen. Physiol. **83**, 801 (1984).

G.T. Reynolds, Localisation of free ionized calcium in cells by means of image intensification, 1979, in Ashley and Campbell 79, p.227.

R. Rick, M. Horster, A. Dörge and K.Thurau, Determination of electrolytes in small biological fluid samples using energy dispersive X-ray micro-analysis, Pflügers Arch. **369**, 95 (1977).

E.B. Ridgway and C.C. Ashley, Calcium transients in single muscle fibres, Biochim. Biophys. Res.Commun. **39**, 229 (1967).

J. Riemer, C.-J. Mayer and G. Ulbrecht, Determination of membrane potential in smooth muscle cells using microelectrodes with reduced tip potential, Pflügers Arch. **349**, 267 (1974).

T.J. Rink and R.Y. Tsien, Calcium-selective micro-electrodes with bevelled, submicron tips containing poly(vinylchloride)-gelled neutral-ligand sensor, J.Physiol. **308**, 5 P (1980).

T.J. Rink, R.Y. Tsien and A.E. Warner, Free calcium in *Xenopus* embryos measured with ion-selective microelectrodes, Nature **283**, 658 (1980).

T.J. Rink, R.Y. Tsien and T. Pozzan, Cytoplasmic pH and free Mg^{2+} in lymphocytes, J.Cell.Biol. **95**, 189 (1982).

W.G. Robertson and R.W. Marshall, Calcium measurements in serum and plasma – total and ionized, CRC Crit. Rev.Clin. Lab. Sci. **11**, 271 (1979).

G.R. Robinson and B.I.H. Scott, A new method of estimating micropipette tip diameter, Experientia **29**, 1039 (1973).

R.A. Robinson and R.M. Stokes, Electrolyte Solutions, Butterworths, London, 1968.

A. Roos and W.F. Boron, Intracellular pH, Physiol. Rev. **61**, 296 (1981).

A. Roos and D.W. Keifer, Estimation of intracellular pH from distribution of weak electrolytes, 1982, in Nuccitelli and Deamer 82, p.55.

B. Rose and W.R. Loewenstein, Permeability of cell junction depends on local cytoplasmic calcium activity, Nature **254**, 250 (1975a).

B. Rose and W.R. Loewenstein, Calcium ion distribution in cytoplasm visualized by aequorin: diffusion in cytosol restricted by energized sequestering, Science **190**, 1204 (1975b).

H. Rottenberg, T. Grunwald and M. Avron, Determination of pH in chloroplasts. 1. Distribution of [^{14}C] methylamine. Eur. J. Biochem. **25**, 54 (1972).

A. Rougeul-Buser, J.-J. Bouyer and P. Buser, Transitional states of awareness and short-term fluctuations of selective attention: neurophysiological correlates and hypotheses, 1978, in Buser and Rougeul-Buser 78, p. 215.

A. N. Rowan, Of Mice, Models, and Men, A Critical Evaluation of Animal Research, State University of New York Press, Albany, 1984.

H. Ruprecht, P. Oggenfuss, U. Oesch, D. Ammann, W. E. Morf and W. Simon, Ion transport in artificial membranes induced by neutral ionophores, Inorg. Chim. Acta **79**, 67 (1983).

J. Růžička, E. H. Hansen and J. C. Tjell, Selectrode – the universal ion-selective electrode. Part VI. The calcium(II) selectrode employing a new ion exchanger in a nonporous membrane and a solid-state reference system, Anal. Chim. Acta **67**, 155 (1973).

O. Ryba and J. Petránek, Interference of permeable anions in potassium-sensitive membrane electrodes based on valinomycin and dimethyl-dibenzo-30-crown-10, J. Electroanal. Chem. **67**, 321 (1976).

M. Sankar and R. G. Bates, Buffers for the physiological pH range: Thermodynamic constants of 3-(N-morpholino)propane sulfonic acid from 5 to 50°C, Anal. Chem. **50**, 1922 (1978).

A. Scarpa, Indicators of free magnesium, Biochemistry **13**, 2789 (1974).

A. Scarpa, Measurement of calcium ion concentrations with metallochromic indicators, 1979, in Ashley and Campbell 79, p. 85.

A. Scarpa and F. J. Brinley, In situ measurements of free cytosolic magnesium ions, Fed. Proc. **40**, 2646 (1981).

A. Scarpa, Cell ion measurement with metallochromic indicators, Techniques in Cellular Physiology **P127**, 1 (1982).

O. F. Schäfer, The properties of poly(vinylisobutyl ether) as a matrix for ion-selective electrodes, Anal. Chim. Acta **87**, 495 (1976).

J. G. Schindler, G. Stork, H.-J. Strüh and W. Schäl, Siloxanverbindungen als aktive Komponenten in ionenselektiven Elektrodenmembranen, Fresenius Z. Anal. Chem. **290**, 45 (1978).

J. G. Schindler and M. M. Schindler, Bioelektrochemische Membranelektroden, Walter de Gruyter, Berlin, New York, 1983.

W. R. Schlue and J. W. Deitmer, Evidence for Na$^+$-Ca^{2+} exchange across the neuronal membrane in the leech CNS, Pflügers Arch. Suppl. to **394**, R48 (1982).

W. R. Schlue and W. Wuttke, Potassium activity in leech neurophile glial cells changes with external potassium concentration, Brain Res. **270**, 368 (1983).

W. R. Schlue and R. C. Thomas, A dual mechanism for intracellular pH regulation by leech neurones, J. Physiol. **364**, 327 (1985).

W. R. Schlue, W. Wuttke and J. W. Deitmer, Ion activity measurements in extracellular spaces, nerve and glial cells in the central nervous system of the leech, 1985, in Kessler et al. 85, p. 166.

K. Schmid and G. Böhmer, Positionierung von Mikroelektroden: Vorwärts im μm-Bereich, Elektronik **2**, 48 (1985).

J. K. Schneider, P. Hofstetter, E. Pretsch, D. Ammann and W. Simon, N,N,N′,N′-Tetrabutyl-3,6-dioxaoctan-dithioamid, Ionophor mit Selektivität für Cd^{2+}, Helv. Chim. Acta **63**, 217 (1980).

R. Scholer and W. Simon, Membranelektrode zur selektiven, potentiometrischen Erfassung organischer Kationen, Helv. Chim. Acta **55**, 171 (1972).

H. Scholze, Glas. Natur, Struktur und Eigenschaften, 2. Auflage, Springer Verlag, Berlin, Heidelberg, New York, 1977.

R. A. Schümperli, H. Oetliker and R. Weingart, Effect of 50% external sodium in solutions of normal and twice normal tonicity on internal sodium activity in frog skeletal muscle, Pflügers Arch. **393**, 51 (1982).

P. Schulthess, Y. Shijo, H. V. Pham, E. Pretsch, D. Ammann and W. Simon, A hydrogen ion-selective liquid-membrane electrode based on tri-n-dodecylamine as neutral carrier, Anal. Chim. Acta **131**, 111 (1981).

P. Schulthess, D. Ammann, W. Simon, C. Caderas, R. Stepánek and B. Kräutler, A lipophilic derivative of vitamin B$_{12}$ as a selective carrier for anions, Helv. Chim. Acta **67**, 1026 (1984).

P. Schulthess, D. Ammann, B. Kräutler, C. Caderas, R. Stepánek and W. Simon, Nitrite selective liquid membrane electrode, Anal. Chem. **57**, 1397 (1985).

P. Schuster, W. Jakubetz and W. Marius, Molecular models for the solvation of small ions and polar molecules, Topics in Current Chemistry, **60**, 1 (1975).

G. Schwarzenbach, Der Chelateffekt, Helv. Chim. Acta **35**, 2433 (1952).

R. Scordamaglia, F. Cavallone and E. Clementi, Analytical potentials from "ab initio" computations for the interaction between biomolecules. 2. Water with the four bases of DNA, J. Am. Chem. Soc. **99**, 5545 (1977).

J. A. Sechzer, ed., The role of animals in biomedical research, Ann. New York Acad. Sci., Vol. 406, 1983.

J. Šenkyr, D. Ammann, P. C. Meier, W. E. Morf, E. Pretsch and W. Simon, Uranyl ion selective electrode based on a new synthetic neutral carrier, Anal. Chem. **51**, 786 (1979).

H. Seto, A. Jyo and N. Ishibashi, Anomalous response of liquid membrane electrode based on neutral carrier, Chem. Lett. **1975**, 483.

I. Shabunova and F. Vyskočil, Postdenervation changes of intracellular potassium and sodium measured by ion selective microelectrodes in rat soleus and extensor digitorum longus muscle fibres, Pflügers Arch. **394**, 161 (1982).

A. E. Shamoo, ed., Carriers and channels in biological systems, Ann. New York Acad. Sci., Vol. 264, 1975.

A Shanzer, D. Samuel and R. Korenstein, Lipophilic lithium ion carriers, J. Am. Chem. Soc. **105**, 3815 (1983).

S. S. Sheu and H. A. Fozzard, Transmembrane Na^+ and Ca^{2+} electrochemical gradients in cardiac muscle and their relationship to force development, J. Gen. Physiol. **80**, 325 (1982).

H. Shimazaki, Role of Müller cells and potassium activity in the vertebrate retina with respect to b-wave generation, Thesis, University of Georgia, Athens, Georgia, 1983.

O. Shimomura, F. H. Johnson and Y. Saiga, Extraction, purification, and properties of aequorin, a bioluminescent protein from the luminous hydromedusan, *Aequorea*, J. Cell. Comp. Physiol. **59**, 223 (1962).

O. Shimomura and F. H. Johnson, Partial purification and properties of the *Chaetopterus* luminescence system, in Bioluminescence in Progress, F. H. Johnson and Y. Haneda, eds., Princeton University Press, Princeton, 1966, p. 495.

O. Shimomura and F. H. Johnson, Structure of the light-emitting moiety of aequorin, Biochem. **11**, 1602 (1972).

O. Shimomura and F. H. Johnson, Chemistry of the calcium-sensitive photoprotein aequorin, 1979, in Ashley and Campbell 79, p. 73.

T. Shono, M. Okahara, I. Ikeda, K. Kimura and H. Tamura, Sodium-selective PVC membrane electrodes based on bis(12-crown-4)s, J. Electroanal. Chem. **132**, 99 (1982).

M. Shporer and M. M. Civan, The state of water and alkali cations within the intracellular fluids: the contribution of NMR spectroscopy, in Current Topics in Membranes and Transport, F. Bronner and A. Kleinzeller, eds., Vol. 9, Academic Press, New York, San Francisco, London, 1977, p. 1.

R. G. Shulman, ed., Biological Applications of Magnetic Resonance, Academic Press, New York, San Francisco, London, 1979.

H. Sies, On metabolic compartmentation: introductory remarks, in Metabolic Compartmentation, in H. Sies, ed., Academic Press, London, 1982, p. 1.

O. Siggaard-Andersen, N. Fogh-Andersen and J. Thode, Elimination of the erythrocyte effect on the liquid junction potential in potentiometric measurements on whole blood using mixed salt bridge solutions, Scand. J. Clin. Lab. Invest. Suppl. 165, **43**, 43 (1983).

O. Siggaard-Andersen, J. Thode and N. Fogh-Andersen, What is "ionized calcium", Scand. J. Clin. Lab. Invest. Suppl. 165, **43**, 11 (1983).

I. A. Silver, Measurement of pH and ionic composition of pericellular sites, Phil. Trans. R. Soc. Lond. B. **271**, 261 (1975).

I. A. Silver, Multi-parameter electrodes, 1976, in Kessler et al. 76, p. 119.

W. Simon, W. E. Morf and P. C. Meier, Specifity for alkali and alkaline earth cations of synthetic and natural organic complexing agents in membranes, in Structure and Bonding, Vol. 16, J. D. Dunitz, P. Hemmerich, J. A. Ibers, C. K. Jorgensen, J. B. Neilands, D. Reinen, R. J. P. Williams, eds., Springer Verlag, Berlin, Heidelberg, New York, 1973, p. 113.

W. Simon, D. Ammann, H. F. Osswald, P. C. Meier and R. E. Dohner, Applicability of ion-selective electrodes in automated systems for clinical analysis, in Advances in Automated Analysis. Technicon International Congress 1976, E. C. Barton et al., eds., Vol. 1, Mediad Inc. Tarrytown, New York, 1977 a, p. 59.

W. Simon, W. E. Morf and D. Ammann, Calcium ionophores, in Calcium-Binding Proteins and

Calcium Function, R.H. Wasserman, R.A. Corradino, E.Carafoli, R.H. Kretsinger, D.H. MacLennan and F.L. Siegel, eds., North-Holland, New York, Amsterdam, Oxford, 1977b, p.50.

W. Simon, D.Ammann, M.Oehme and W.E. Morf, Calcium-selective electrodes, Annals New York Acad.Sci. **307**, 52 (1978a).

W. Simon, D.Ammann, H.Osswald, M.Oehme and P.C. Meier, Limitations of liquid membrane electrodes, Arzneimittel-Forschung/Drug Research **28**, 707 (1978b).

W. Simon, D.Ammann, R.A. Dörig, D.Erne, R.J.J. Funck, H.-B. Jenny, E.Pretsch and R.A. Steiner, Flüssigmembranelektroden mit neutralen Ionophoren, in Fortschritte in der elektrochemischen Analytik – Theorie und Anwendungen, Karl-Marx-Universität, ed., Karl-Marx-Universität, Leipzig, 1980, p.95.

W. Simon, D.Ammann, P.Anker, U.Oesch and D.M. Band, Ion-selective electrodes and their clinical application in the continuous ion monitoring, Annals New York Acad.Sci. **428**, 279 (1984a).

W. Simon, E.Pretsch, W.E. Morf, D.Ammann, U.Oesch and O.Dinten, Design and application of neutral carrier-based ion-selective electrodes, Analyst **109**, 207 (1984b).

R.D. Slocum and S.J. Roux, Cellular and subcellular localization of calcium in gravistimulated oat coleoptiles and its possible significance in the establishment of tropic curvature, Planta **157**, 481 (1983).

F.M. Snell, Some electrical properties of fine-tipped pipette micro-electrodes, 1969, in Lavallée et al.69, p.111.

J.H. Sokol, C.O. Lee and F.J. Lupo, Measurement of the free calcium ion concentration in sheep cardiac Purkinje fibers with neutral carrier Ca^{2+}- selective microelectrodes, Biophys.J. **25**, 143a (1979).

R.L. Solsky, Ion-selective electrodes in biomedical analysis, CRC Crit.Rev.Anal.Chem. **14**, 1 (1982).

G.G. Somjen, Stimulus-evoked and seizure-related responses of extracellular calcium activity in spinal cord compared to those in cerebral cortex, J. Neurophysiol. **44**, 617 (1980).

G.G. Somjen, The why and how of measuring the activity of ions in extracellular fluid of spinal cord and cerebral cortex, 1981, in Zeuthen 81a, p.175.

G. Somjen, R.Dingledine, B.Connors and B.Allen, Extracellular potassium and calcium activities in the mammalian spinal cord, and the effect of changing ion levels on mammalian neural tissues, 1981b, in Syková et al.81, p.159.

U. Sonnhof, R.Förderer, W.Schneider and H.Kettenmann, Cell puncturing with a step motor driven manipulator with simultaneous measurement of displacement, Pflügers Arch. **392**, 295 (1982).

R.M. Spanswick and A.G. Miller, Measurement of the cytoplasmic pH in *Nitella translucens,* Plant Physiol. **59**, 664 (1977).

E.-J. Speckmann, C.E. Elger and A.Lehmenkühler, Penicillin activity in brain tissue: a method for continuous measurement, Electroencephalogr.Clin. Neurophysiol. **56**, 664 (1983).

R.W. Sperry, The great cerebral commissure, Sci.Amer. **210**, 42 (1964).

R.W. Sperry, Science and Moral Priority, Columbia University Press, New York, 1983.

S.F. Spicker and H.T. Engelhardt, Jr., eds., Philosophical Dimensions of the Neuro-Medical Sciences, Reidel Publishing Company, Dordrecht, Holland, 1975.

R.J. Stark and J.O'Doherty, Intracellular Na^+ and K^+ activities during insulin stimulation of rat soleus muscle, Am.J.Physiol. **242**, E193 (1982).

Z. Štefanac and W.Simon, In-Vitro-Verhalten von Makrotetroliden in Membranen als Grundlage für hochselektive kationenspezifische Elektrodensysteme, Chimia **20**, 436 (1966).

Z. Štefanac and W.Simon, Ion specific electrochemical behavior of macrotetrolides in membranes, Microchem.J. **12**, 125 (1967).

R. Steinberg and G.tenBruggencate, Dependence of extracellular Ca^{2+} upon active transport mechanism in cerebellar cortex, Pflügers Arch. **373**, R68 (1978).

R.A. Steiner, M.Oehme, D.Ammann and W.Simon, Neutral carrier sodium ion-selective microelectrode for intracellular studies, Anal.Chem. **51**, 351 (1979).

R.A. Steiner, Entwicklung von kationenselektiven Carrier-Flüssigmembranmikroelektroden für Aktivitätsbestimmungen im Intra- und Extrazellulärraum, Thesis, ETH Zürich, No.7099, ADAG Administration & Druck AG, Zürich, 1982.

J. Stinnakre, Defection and measurement of intracellular calcium. A comparison of techniques. Trends Neurosci. **4**, 46 (1981).

H. Stöckle and G. tenBruggencate, Climbing fiber-mediated rhythmic modulations of potassium and calcium in cat cerebellar cortex, Exp. Neurol. **61**, 226 (1978).

H. Stöckle and G. tenBruggencate, Fluctuation of extracellular potassium and calcium in the cerebellar cortex related to climbing fiber activity, Neurosci. **5**, 893 (1980).

H. Streb and I. Schulz, Regulation of cytosolic free Ca^{2+} concentration in acinar cells of rat pancreas, Am. J. Physiol. **245**, G347 (1983).

K. Suzuki and E. Frömter, The potential and resistance profile of *Necturus* gallbladder cells, Pflügers Arch. **371**, 109 (1977).

K. Suzuki, V. Rohliček and E. Frömter, A quasi-totally shielded, low-capacitance glass-microelectrode with suitable amplifiers for high-frequency intracellular potential and impedance measurements, Pflügers Arch. **378**, 141 (1978).

E. Syková, P. Hník and L. Vyklický, eds., Ion-Selective Microelectrodes and Their Use in Excitable Tissues, Plenum Press, New York, London, 1981.

G. Szabo, G. Eisenman, R. Laprade, S. M. Ciani and S. Krasne, Experimentally observed effects of carriers on the electrical properties of bilayer membranes-equilibrium domain, 1973, in Membranes, Vol. 2, G. Eisenman, ed., Marcel Dekker, Inc., New York, 1973, p. 179.

J. Szentagothái, The local neuronal apparatus of the cerebral cortex, 1978, in Buser and Rougeul-Buser 78, p. 131.

I. Tabushi, Y. Kuroda and K. Yokota, A,B,D,F-Tetrasubstituted β-cyclodextrin as artificial channel compound, Tetrahedron Letters **23**, 4601 (1982).

H. Tamura, K. Kimura and T. Shono, Thallium(I)-selective PVC membrane electrodes based on bis(crown ether)s, J. Electroanal. Chem. **115**, 115 (1980).

H. Tamura, K. Kimura and T. Shono, Coated wire sodium- and potassium-selective electrodes based on bis(crown ether) compounds, Anal. Chem. **54**, 1224 (1982).

C. Tanford, S. A. Swanson and W. S. Shore, Hydrogen ion equilibria of bovine serum albumin, J. Am. Chem. Soc. **77**, 6414 (1955).

I. Tasaki and I. Singer, Some problems involved in electric measurements of biological systems, Ann. New York Acad. Sci. **148**, 36 (1968).

M. Tauchi and R. Kikuchi, A simple method for beveling micropipettes for intracellular recording and current injection, Pflügers Arch. **368**, 153 (1977).

C. V. Taylor and D. M. Whitaker, Potentiometric determination in the protoplasm and cell-sap of *Nitella*. Protoplasma **3**, 1 (1927).

P. S. Taylor and R. C. Thomas, The effect of leakage on micro-electrode measurements of intracellular sodium activity in crab muscle fibres, J. Physiol. **352**, 539 (1984).

G. tenBruggencate, C. Nicholson and R. Steinberg, Rhythmic modulation of extracellular Ca^{2+}- and K^+-levels in the cerebellar cortex related to climbing fiber activity, Pflügers Arch. **368**, R37 (1977).

G. tenBruggencate and R. Steinberg, Effects of ouabain and adenosine on extracellular Ca^{2+} and K^+, as measured with ion selective microelectrodes in cerebellar cortex, Naunyn-Schmiedeberg's Archives of Pharmacology, **302**, R55 (1978a).

G. tenBruggencate, R. Steinberg, H. Stöckle and C. Nicholson, Modulation of extracellular Ca^{2+}- and K^+-levels in the mammalian cerebellar cortex, in Iontophoresis and Transmitter Mechanisms in the Mammalian Central Nervous System, R. W. Ryall and J. S. Kelly, eds., Elsevier/North-Holland, Amsterdam, 1978b, p. 412.

G. tenBruggencate, A. Ullrich, M. Galvan, H. Förstl and P. Baierl, Effects of lithium application upon extracellular potassium structures of the peripheral and central nervous system of rats, 1981, in Lübbers et al. 81, p. 135.

S. M. Theg and W. Junge, The effect of low concentrations of uncouplers on the detectability of proton deposition in thylakoids. Evidence for subcompartmentation and preexisting pH differences in the dark, Biochim. Biophys. Acta **723**, 294 (1983).

A. P. Thoma, A. Viviani-Nauer, S. Arvanitis, W. E. Morf and W. Simon, Mechanism of neutral carrier mediated ion transport through ion-selective bulk membranes, Anal. Chem. **49**, 1567 (1977).

J. A. Thomas, P. C. Kolbeck and T. A. Langworthy, Spectrophotometric determination of cytoplasmic and mitochondrial pH transitions using trapped pH indicators, 1982, in Nuccitelli and Deamer 82, p. 105.

M. V. Thomas, Techniques in Calcium Research, Academic Press, London, 1982.

R. C. Thomas, Membrane current and intracellular sodium changes in a snail neurone during extrusion of injected sodium, J. Physiol. **201**, 495 (1969).

R.C. Thomas, New design for sodium-sensitive glass microelectrode, J. Physiol. **210**, 82 P (1970).

R.C. Thomas, Intracellular pH of snail neurones measured with new pH-sensitive glass microelectrode, J. Physiol. **238**, 159 (1974).

R.C. Thomas, W. Simon and M. Oehme, Lithium accumulation by snail neurones measured by a new Li^+-sensitive microelectrode, Nature **258**, 754 (1975).

R.C. Thomas, Construction and properties of recessed-tip micro-electrodes for sodium and chloride ions and pH, 1976, in Kessler et al. 76, p. 141.

R.C. Thomas, The role of bicarbonate, chloride and sodium ions in the regulation of intracellular pH in snail neurones, J. Physiol. **273**, 317 (1977).

R.C. Thomas, Ion-Sensitive Intracellular Microelectrodes. How to make and use them, Academic Press, London, New York, San Francisco, 1978.

R.C. Thomas and C.J. Cohen, A liquid ion-exchanger alternative to KCl for filling intracellular reference microelectrodes, Pflügers Arch. **390**, 96 (1981).

R.C. Thomas, Experimental displacement of intracellular pH and the mechanism of its subsequent recovery, J. Physiol. **354**, 3 P (1984).

H.T. Tien, Bilayer Lipid Membranes (BLM), Marcel Dekker, Inc., New York, 1974.

U. Tietze and Ch. Schenk, Halbleiter-Schaltungstechnik, Springer Verlag, Berlin, Heidelberg, New York, 1980.

T. Tomita, Single and coaxial microelectrodes in the study of the retina, 1969, in Lavallée et al. 69, p. 124.

J.L. Toner, D.S. Daniel and S.M. Geer, Hemispherands and sodium-selective compositions and electrodes containing same, United States Patent 4,476,007, 1984.

J.L. Treasure, D.W. Ploth and T. Treasure, Continuous measurement of potassium concentration in blood during hemodialysis with an ion-specific electrode, in Kessler et al. 85, p. 297.

A.S. Troshin, Problems of Cell Permeability, Pergamon Press, Oxford, London, Edinburgh, New York, Toronto, Paris, Braunschweig, 1966.

G. Trube, D. Pelzer and H.-M. Piper, The importance of membrane integrity for the measurement of electrical properties of isolated rat heart myocytes, Pflügers Arch. Suppl. to **394**, R12 (1982).

M. Tsacopoulos and A. Lehmenkühler, A double-barrelled Pt-microelectrode for simultaneous measurement of PO_2 and bioelectrical activity in excitable tissues, Experientia **33**, 1337 (1977).

M. Tsacopoulos, R.K. Orkand, J.A. Coles, S. Levy and S. Poitry, Oxygen uptake occurs faster than sodium pumping in bee retina after a light flash, Nature **301**, 604 (1983).

R.Y. Tsien, New calcium indicators and buffers with high selectivity against magnesium and proteins: Design, synthesis and properties of prototype structures, Biochemistry **19**, 2396 (1980).

R.Y. Tsien and T.J. Rink, Neutral carrier ion-selective microelectrodes for measurement of intracellular free calcium, Biochim. Biophys. Acta **599**, 623 (1980).

R.Y. Tsien, A non-disruptive technique for loading calcium buffers and indicators into cells, Nature **290**, 527 (1981).

R.Y. Tsien and T.J. Rink, Ca^{2+}-selective electrodes: a novel PVC-gelled neutral carrier mixture compared with other currently available sensors, J. Neurosci. Meth. **4**, 73 (1981).

J.T. Tupper and H. Tedeschi, Microelectrode studies on the membrane properties of isolated mitochondria, Proc. Natl. Acad. Sci. **63**, 370 (1969).

I. Uemasu and Y. Umezawa, Comparison of definitions of response times for copper(II) ion selective electrodes, Anal. Chem. **54**, 1198 (1982).

I. Uemasu and Y. Umezawa, Single ion activity at high ionic strengths with ion-selective electrodes and the Debye-Hückel equation, Anal. Chem. **55**, 386 (1983).

K. Ugurbil, R.G. Shulman and T.R. Brown, High-resolution ^{31}P and ^{13}C nuclear magnetic resonance studies of *Escherichia coli* cells in vivo, 1979, in Shulman 79, p. 537.

E. Ujec, O. Keller, V. Pavlík and J. Machek, The electrical parameters of a coaxial, low-resistance, potassium-selective microelectrode, Physiol. Bohemoslov. **27**, 570 (1978).

E. Ujec, O. Keller, J. Machek and V. Pavlík, Low impedance coaxial K^+ selective microelectrodes, Pflügers Arch. **382**, 189 (1979).

E. Ujec, E.E.O. Keller, N. Kříž, V. Pavlík and J. Machek, Low-impedance, coaxial, ion-selective, double-barrel microelectrodes and their use in biological measurements, Bioelectrochem. Bioenergetics **7**, 363 (1980).

E. Ujec, O. Keller, N. Kříž, V. Pavlík and J. Machek, Double-barrel ion-selective [K^+, Ca^{2+}, Cl^-] coaxial microelectrodes (ISCM) for measurements of small and rapid changes in ion activities, 1981, in Syková et al. 81, p. 41.

A. Ullrich, P. Baierl and G. tenBruggencate, Extracellular potassium in rat cerebellar cortex during acute and chronic lithium application, Brain Res. **192**, 287 (1980).

A. Ullrich, R. Steinberg, P. Baierl and G. tenBruggencate, Changes in extracellular potassium and calcium in rat cerebellar cortex related to local inhibition of the sodium pump, Pflügers Arch. **395**, 108 (1982).

M. Vassalle and C. O. Lee, The relationship among intracellular sodium activity, calcium, and strophanthidin inotropy in canine cardiac Purkinje fibers, J. Gen. Physiol. **83**, 287 (1984).

C. A. Vega and R. G. Bates, Standards for pH measurements in isotonic saline media of ionic strength $I = 0.16$, Anal. Chem. **50**, 1295 (1978).

F. Vögtle, H. Puff, E. Friedrichs and W. M. Müller, Selective inclusion and orientation of chloroform in the molecular cavity of a 30-membered hexalactam host, J. Chem. Soc., Chem. Commun. **1982**, 1398.

F. Vögtle, T. Kleiner, R. Leppkes, M. W. Läubli, D. Ammann and W. Simon, Neutrale Ionophore mit Selektivität für Na^+, Chem. Ber. **116**, 2028 (1983).

H. Völkl, J. Geibel, W. Rehwald and F. Lang, A microelectrode for continuous monitoring of redox activity in isolated perfused tubule segments, Pflügers Arch. **400**, 393 (1984).

L. L. Voronin, Long-term potentiation in the hippocampus, Neurosci. **10**, 1051 (1983).

P. Vuilleumier, P. Gazzotti, E. Carafoli and W. Simon, The translocation of Ca^{2+} across phospholipid bilayers induced by a synthetic neutral Ca^{2+}-ionophore, Biochim. Biophys. Acta **467**, 12 (1977).

F. Vyskočil and N. Kříž, Modifications of single and double-barrel potassium specific microelectrodes for physiological experiments, Pflügers Arch. **337**, 265 (1972).

W. J. Waddell and T. C. Butler, Calculation of intracellular pH from the distribution of 5,5-dimethyl-2,4-oxazolidinedione (DMO). Application to skeletal muscle of the dog. J. Clin. Invest. **38**, 720 (1959).

W. J. Wadman and U. Heinemann, Laminar profiles of $[K^+]_o$ and $]Ca^{2+}]_o$ in region CA1 of the hippocampus of kindled rats, 1985, in Kessler et al. 85, p. 221.

H. Wakasugi, T. Kimura, W. Haase, A. Kribben, R. Kaufmann and I. Schulz, Calcium uptake into acini from rat pancreas: evidence for intracellular ATP-dependent calcium sequestration, J. Membr. Biol. **65**, 205 (1982).

J. Walden, E.-J. Speckmann and A. Lehmenkühler, Continuous measurement of pentylenetetrazol (PTZ) by a liquid ion exchanger microelectrode, Pflügers Arch. Suppl. to **392**, R43 (1982).

J. Walden, A. Lehmenkühler, E.-J. Speckmann and O. W. Witte, Continuous measurement of pentylenetetrazol concentration by a liquid ion exchanger microelectrode, J. Neurosci. Methods **11**, 187 (1984).

J. L. Walker, Jr., Ion specific liquid ion exchanger microelectrodes, Anal. Chem. **43**, 89 A (1971).

J. H. Wang and E. Copeland, Equilibrium potentials of membrane electrodes, Proc. Natl. Acad. Sci. **70**, 1909 (1973).

W. Wang, H. Oberleithner, F. Lang and G. Messner, cAMP hyperpolarizes the cell membrane potential and decreases intracellular Na^+ activity in frog proximal tubule, Pflügers Arch. Suppl. to **394**, R27 (1982).

W. Wang, H. Oberleithner and F. Lang, The effect of cAMP on the cell membrane potential and intracellular ion activities in proximal tubule of *Rana esculenta,* Pflügers Arch. **396**, 192 (1983).

W. Wang, G. Messner, H. Oberleithner, F. Lang and P. Deetjen, The effect of ouabain on intracellular activities of K^+, Na^+, Cl^-, H^+ and Ca^{2+} in proximal tubules of frog kidneys, Pflügers Arch. **401**, 6 (1984).

J. A. Wasserstrom, D. J. Schwartz and H. A. Fozzard, Catecholamine effects on intracellular sodium activity and tension in dog heart, Am. J. Physiol. **243**, H670 (1982).

B. S. Weakly, A Beginner's Handbook in Biological Transmission Electron Microscopy, Churchill Livingstone, Edinburgh, London, Melbourne, New York, 1981.

D. J. Webb and R. Nuccitelli, Intracellular pH changes accompanying the activation of development in frog eggs: comparison of pH microelectrodes and ^{31}P-NMR measurements, 1982, in Nuccitelli and Deamer 82, p. 293.

E. Weber and F. Vögtle, Progress in crown ether chemistry (Part IV B), Kontakte **1**, 24 (1981).

D. Wegmann, H. Weiss, D. Ammann, W. E. Morf, E. Pretsch, K. Sugahara and W. Simon, Anionselective liquid membrane electrodes based on lipophilic quaternary ammonium compounds, Mikrochim. Acta **1984 III**, 1.

R. Weiler, Triangular tubing: new, fast-filling microelectrodes for the intracellular use of HRP, Naturwissenschaften **69**, 285 (1982).

R. Weingart and P. Hess, Interaction between pH_i and pCa_i in cardiac tissue, 1981, in Syková et al. 81, p. 307.

R. Weingart and G. Niemeyer, Assessment of preretinal pH by ionselective microelectrodes, Experientia **39**, 642 (1983).

R. Weingart and P. Hess, Free calcium in sheep cardiac tissue and frog skeletal muscle measured with Ca^{2+}-selective microelectrodes, Pflügers Arch. **402**, 1 (1984).

S. A. Weinman and L. Reuss, Na^+-H^+ exchange and Na^+ entry across the apical membrane of *Necturus* gallbladder, J. Gen. Physiol. **83**, 57 (1984).

M. Welti, E. Pretsch, E. Clementi and W. Simon, Interaction of Ca^{2+} and Mg^{2+} with ionophores studied by using a pair-potential model based on ab initio calculations, Helv. Chim. Acta **65**, 1996 (1982).

R. J. Werrlein, Cells and tissue culture systems, 1981, in Zeuthen 81 a, p. 257.

J. W. Westley, Ch. M. Lin, J. F. Blount, L. H. Sello, N. Troupe and P. A. Miller, Isolation and characterization of a novel polyether antibiotic of the pyrrolether class, antibiotic X-14885 A, J. Antibiotics **36**, 1275 (1983).

W. J. Whalen, The pO_2 in isolated muscle measured with an intracellular electrode, 1969, in Lavallée et al. 69, p. 396.

H. G. Wieser and M. G. Yasargil, Selective amygdalohippocampectomy as a surgical treatment of mesiobasal limbic epilepsy, Surg. Neurol. **17**, 445 (1982).

H. G. Wieser, Electroclinical Features of the Psychomotor Seizure, Fischer, Stuttgart, New York; Butterworths, London, 1983.

R. J. P. Williams, The biochemistry of sodium, potassium, magnesium, and calcium, Quart. Rev. Chem. Soc. **24**, 331 (1970).

N. K. Wills and S. A. Lewis, Intracellular Na^+ activity as a junction of Na^+ transport rate across a tight epithelium, Biophys. J. **30**, 181 (1980).

N. K. Wills, Apical membrane potassium and chloride permeabilities in surface cells of rabbit descending colon epithelium, J. Physiol. **358**, 433 (1985).

K. Winnefeld and H. Schröter, Zur Problematik von Chloridaktivitätsmessungen in Protein-(Hämoglobin)Lösungen, Z. Med. Lab. Diagn. **22**, 350 (1981).

S. Wischnitzer, Introduction to Electron Microscopy, Pergamon Press, New York, 1981.

E. M. Wright, Electrophysiology of plasma membrane vesicles, Am. J. Physiol. **246**, F363 (1984).

B. W. Wright, M. L. Lee, S. W. Graham, L. V. Phillips and D. M. Hercules, Glass surface analytical studies in the preparation of open tubular columns for gas chromatography, J. Chromatogr. **199**, 355 (1980).

S. T. Wu, G. M. Pieper, J. M. Salhany and R. S. Eliot, Measurement of free magnesium in perfused and ischemic arrested heart muscle. A quantitative phosphorus-31 nuclear magnetic resonance and multiequilibria analysis, Biochemistry **20**, 7399 (1981).

H.-R. Wuhrmann, W. E. Morf and W. Simon, Modellberechnung der EMK und der Ionenselektivität von Membranelektroden-Meßketten, Helv. Chim. Acta **56**, 1011 (1973).

P. Wuhrmann, H. Ineichen, U. Riesen-Willi and M. Lezzi, Change in nuclear potassium electrochemical activity and puffing of potassium-sensitive salivary chromosome regions during *Chironomus* development, Proc. Natl. Acad. Sci. **76**, 806 (1979).

P. Wuhrmann and M. Lezzi, K^+-and Na^+-activity measurements in larval salivary glands of *Chironomus tentans,* Europ. J. Cell. Biol. **22**, 473 (1980).

U. Wuthier, H. V. Pham, R. Zünd, D. Welti, R. J. J. Funck, A. Bezegh, D. Ammann, E. Pretsch and W. Simon, Tin organic compounds as neutral carriers for anion selective electrodes, Anal. Chem. **56**, 535 (1984).

W. Wuttke and W. R. Schlue, Potassium activity in leech glial cells measured with ion-sensitive microelectrodes, Pflügers Arch. Suppl. to **394**, R47 (1982).

H. Yamaguchi and N. L. Stephens, Determination of intracellular pH of airway smooth muscle using recessed tip pH microelectrode, in Proc. Int. Congr. Physiol. Sci., Paris, 1977, Vol. 13, p. 824.

M. Yamauchi, A. Jyo and N. Ishibashi, Potassium ion-selective membrane electrodes based on naphtho-15-crown-5, Anal. Chim. Acta **136**, 399 (1982).

H. J. C. Yeh, F. J. Brinley, Jr. and E. D. Becker, Nuclear magnetic resonance studies on intracellular sodium in human erythrocytes and frog muscle, Biophys. J. **13**, 56 (1973).

S. Yoshikami, J. S. George and W. A. Hagins, Light-induced calcium fluxes from outer segment layer of vertebrate retinas, Nature **286**, 395 (1980a).

S. Yoshikami, J. George and W. A. Hagins, Light causes large fast Ca^{2+} efflux from outer segments of live retinal rods, Fed. Proc. **39**, 2066 (1980b).

K. Yoshitomi and E. Frömter, Cell pH of rat renal proximal tubule in vivo and the conductive nature of peritubular HCO_3^- (OH^-) exit, Pflügers Arch. **402**, 300 (1984a).

K. Yoshitomi and E. Frömter, The intracellular pH of rat kidney proximal tubular cells in vivo, Pflügers Arch. Suppl. to **400**, R20 (1984b).

T. Zeuthen and C. Monge, Intra- and extracellular gradients of electrical potential and ion activities of the epithelial cells of the rabbit ileum in vivo recorded by microelectrodes, Phil. Trans. R. Soc. Lond. B. **71**, 277 (1975).

T. Zeuthen, Gradients of chemical and electrical potentials in the gallbladder, J. Physiol. **256**, 32 P (1976a).

T. Zeuthen, A double-barrelled Na^+-sensitive microelectrode, J. Physiol. **254**, 8 P (1976b).

T. Zeuthen, Intracellular gradients of electrical potential in the epithelial cells of the *Necturus* gallbladder, J. Membr. Biol. **33**, 281 (1977).

T. Zeuthen, Intracellular gradients of ion activities in the epithelial cells of the *Necturus* gallbladder recorded with ion-selective microelectrodes, J. Membr. Biol. **39**, 185 (1978).

T. Zeuthen, How to make and use double-barrelled ion-selective microelectrodes, in Current Topics in Membranes and Transport, E. Boulpaep and G. Giebisch, eds., Vol. 13, Academic Press, New York, 1980.

T. Zeuthen, ed., The Application of Ion-Selective Microelectrodes, Research Monographs in Cell and Tissue Physiology, Vol. 4, Elsevier/North-Holland Biomedical Press, Amsterdam, New York, Oxford, 1981a.

T. Zeuthen, Ion transport in leaky epithelia studied with ion-selective microelectrodes, 1981b, in Zeuthen 81a, p. 27.

T. Zeuthen, Relations between intracellular ion activities and extracellular osmolarity in *Necturus* gallbladder epithelium, J. Membr. Biol. **66**, 109 (1982).

A. F. Zhukov, D. Erne, D. Ammann, M. Güggi, E. Pretsch and W. Simon, Improved lithium ion-selective electrode based on a lipophilic diamide as neutral carrier, Anal. Chim. Acta **131**, 117 (1981).

12 Subject Index

Progress in Clinical Biochemistry and Medicine

Editors: **E. Beaulieu, D. T. Forman, L. Jaenicke, J. A. Kellen, Y. Nagai, G. F. Springer, L. Träger, J. L. Wittliff**

Volume 1

Essential and Non-Essential Metals – Metabolites with Antibiotic Activity – Pharmacology of Benzodiazepines – Interferon Gamma Research

1984. 42 figures. VII, 203 pages. ISBN 3-540-13605-3

Contents: *M. Costa, A. J. Kraker, S. R. Patierno:* Toxicity and Carcinogenicity of Essential and Non-Essential Metals. – *R. G. Werner:* Secondary Metabolites with Antibiotic Activity from the Primary Metabolism of Aromatic Amino Acids. – *U. Klotz:* Clinical Pharmacology of Benzodiazepines. – *H. Kirchner:* Interferon Gamma.

Volume 2

Oncogenes and Human Cancer – Blood Groups in Cancer Copper and Inflammation – Human Insulin

1985. 25 figures. VII, 163 pages. ISBN 3-540-15567-8

Contents: *T. L. J. Boehm:* Oncogenes and the Genetic Dissection of Human Cancer: Implications for Basic Research and Clinical Medicine. – *W. J. Kuhns, F. J. Primus:* Alterations of Blood Groups and Blood Group Precursors in Cancer. – *U. Deuschle, U. Weser:* Copper and Inflammation. – *R. Obermeier, M. Zoltobrocki:* Human Insulin – Chemistry, Biological Characteristics and Clinical Use.

Volume 3

Metabolic Control in Diabetes Mellitus – Beta Adrenoceptor Blocking Drugs – NMR Analysis of Cancer Cells – Immunoassay in the Clinical Laboratory – Cyclosporine

1986. 68 figures. VIII, 192 pages. ISBN 3-540-16249-6

Contents: *W. Berger, R. Flückiger:* Monitoring of Metabolic Control in Diabetes Mellitus: Methodological and Clinical Aspects. – *H. G. Köppe:* Recent Chemical Developments in the Field of Beta Adrenoceptor Blocking Drugs. – *C. E. Mountford, K. T. Holmes, I. C. P. Smith:* NMR Analysis of Cancer Cells. – *E. L. Nickoloff:* The Role of Immunoassay in the Clinical Laboratory. – *R. M. Wenger, T. G. Payne, M. H. Schreier:* Cyclosporine: Chemistry, Structure-Activity Relationships and Mode of Action.

Springer-Verlag
Berlin Heidelberg
New York Tokyo

Design and Synthesis of Organic Molecules Based on Molecular Recognition

Proceedings of the XVIIIth Solvay Conference
on Chemistry
Brussels, November 28–December 01, 1983

Editor: **G. van Binst**
Chairmen of the conference: **E. Katchalski-Katzir, V. Prelog**

1986. Approx. 95 figures. Approx. 410 pages.
ISBN 3-540-16123-6

Contents: Introduction. – Molecular Recognition in Biochemical Processes. Thermodynamics, Kinetics and Stereochemistry in Molecular Recognition. – Synthetic Models of **"Hosts"**. – Design of new **"Guests"**. – Concluding Remarks. – List of Authors.

One can learn a considerable amount from Nature, which uses molecular recognition to achieve selectivity in a degree so far unattainable to mere mortals. To analyze the structural features applied by Nature, to accomplish high molecular recognition, and to simulate these features by synthesis have recently become favorite occupations of chemists. This monograph contains the chairman's introduction, in which he summarizes the main points at issue, together with the contributions of the renowned scientists who participated.

Springer-Verlag
Berlin Heidelberg
New York Tokyo

Springer

UNIVERSITY OF DELAWARE LIBRARY

Please return this book as soon as you have finished
with it. In order to avoid a fine it must be returned by
the latest date stamped below.

APR 1 4 1987
APR 2 6 1987
MAY 2 6 1987
JUN 1 5 1987

JUL 1 5 1987

JUL 3   1987
APR 8   1990

DEC 2   1990

JAN 2 6 1995

FEB 0 1 2005
        OCT 1 5 2014